ISNM
International Series of
Numerical Mathematics
Vol. 111

Edited by
K.-H. Hoffmann, München
H. D. Mittelmann, Tempe
J. Todd, Pasadena

Optimal Control

Calculus of Variations, Optimal Control Theory and Numerical Methods

Edited by

R. Bulirsch
A. Miele
J. Stoer
K.H. Well

Springer Basel AG

Editors

Prof. Dr. R. Bulirsch
Mathematisches Institut
TH München
Postfach 20 24 20
D-80290 München 2
Germany

Prof. Dr. J. Stoer
Inst. f. Angewandte
Mathematik u. Statistik
Am Hubland
D-97074 Würzburg
Germany

Dr. A. Miele
Dept. of Mechanical Engineering
and Materials Science
Post Office Box 1892
Houston, Texas 77251-1892
USA

Prof. Dr. K. Well
Inst. f. Flugmechanik
u. Flugregelung
Universität Stuttgart
Forststr. 86
D-70176 Stuttgart
Germany

Library of Congress Cataloging-in-Publication Data
Optimal control : calculus of variations, optimal control theory, and
  numerical methods / edited by R. Bulirsch ... [et al.].
    p. cm. – (International series of numerical mathematics :
vol. 111)

  1. Calculus of variations--Congresses.   2. Control theory-
Congesses.   3. Mathematical optimization–Congresses.
I. Bulirsch, Roland.   II. Series: International series of numerical
mathematics ; v. 111.
QA315.068   1993
003'.5–dc20

Deutsche Bibliothek Cataloging-in-Publication Data
Optimal Control : calculus of variations, optimal control theory
and numerical methods / ed. by R. Bulirsch ... – Basel ; Boston
; Berlin : Birkhäuser, 1993
  (International series of numerical mathematics ; Vol. 111)

  ISBN 978-3-0348-7541-7   ISBN 978-3-0348-7539-4 (eBook)
  DOI 10.1007/978-3-0348-7539-4

NE: Bulirsch, Roland [Hrsg.] ; GT

© 1993 Springer Basel AG
Originally published by Birkhäuser Verlag, P.O. Box 133, CH-4010 Basel, Switzerland in 1993
Camera-ready copy prepared by the editors
Printed on acid-free paper produced from chlorine-free pulp
Cover design: Heinz Hiltbrunner, Basel

9 8 7 6 5 4 3 2 1

Contents

Analysis and Synthesis of Nonlinear Systems

Applications to Mechanical and Aerospace Systems

Foreword

The conference on Optimal Control - Variationsrechnung und Optimalsteuerungen was held at the Mathematisches Forschungsinstitut of the University of Freiburg during the period May 26 to June 1, 1991. It was the second conference with this title in Oberwolfach, the first one took place in 1986. There were 52 participants, 8 from the United States of America, 6 from Russia and Byelorussia, several from France, Poland, and Austria, and many from Germany - the first time from a unified Germany.

In comparison to the material covered during the first meeting, this manuscript contains new theoretical and practical advances for analyzing and synthesizing optimal controls of dynamical systems governed by partial and ordinary differential equations. New necessary and sufficient conditions for optimality are given. Recent advances in numerical methods are due to new techniques for solving large sized nonlinear programs with sparse Hessians and due to a combination of direct and indirect methods for solving the multi-point boundary-value problem associated with the optimization problem. The construction of feedback controls for nonlinear systems is the third major contribution of this book. Recent advances in the theory of problems with uncertainty, on decomposition methods of nonlinear systems, and on new techniques for constructing feedback controls for state- and control constrained linear quadratic systems are presented. The book has solutions for many complex practical optimal control problems.

The authors gratefully acknowlege the assistance of the reviewers and the help of Mr. Tilmann Raible, who combined the various LaTeX-files into their present form.

Roland Bulirsch
Mathematisches Institut
TU München
Postfach 202420
8000 München 2
Germany

Angelo Miele
Department of Mechanical
Engineering and Material Sciences
P.O.Box 1892
Houston, Texas 77251-1892
USA

Josef Stoer
Angewandte Mathematik und
Statistik
Universität Würzburg
Am Hubland
8700 Würzburg
Germany

Klaus H. Well
Institut fr Flugmechanik
und Flugregelung
Universität Stuttgart
Forststraße 86
7000 Stuttgart 1
Germany

Participants

M. D. ARDEMA, University of Santa Clara, Santa Clara, California, USA

H. BENKER, Technische Hochschule Leuna-Merseburg, Merseburg, Germany

J. T. BETTS, Boeing Computer Services, Seattle, Washington, USA

L. BITTNER , Universität Greifswald, Greifswald, Germany

H. G. BOCK, Universität Heidelberg, Heidelberg, Germany

V. G. BOLTYANSKI, Scientific Institute of Systems, Moscow, Russia

H. BOURDACHE-SIGUERDIDJANE, Ecole Superieure d´ Electricite,Gif-sur-Yvette,
France

M. BROKATE, Universität Kaiserslautern, Kaiserslautern, Germany

R. BULIRSCH, Technische Universität München, Munich, Germany

A. CALISE, Georgia Institute of Technology, Atlanta, Georgia, USA

R. CALLIES, Technische Universität München, Munich, Germany

F. L. CHERNOUSKO, USSR Academy of Science, Moscow, Russia

K. CHUDEJ, Technische Universität München, Munich, Germany

E. M. CLIFF, Virginia Polytechnic Institute and State University, Blacksburg, Virginia,
USA

V. V. DIKUSAR, USSR Academy of Sciences, Moscow, Russia

B. DITTMAR, Pädagogische Hochschule Halle-Kothen, Halle, Germany

G. FEICHTINGER, Technische Universität Wien, Vienna, Austria

W. GRIMM, Universität Stuttgart, Stuttgart, Germany

HANDSCHUG, Technische Hochschule Leuna-Merseburg, Merseburg, Germany

K.-H. HOFFMANN, Technische Universität München, Munich, Germany

C. JÄNSCH, DLR Oberpfaffenhofen, Weßling, Germany

W. KAMPOWSKY, Ernst-Moritz-Arndt-Universität, Greifswald, Germany

F. M. KIRILLOVA, Institute of Mathematics, Minsk, Byelorussia

R. KLÖTZLER, Karl-Marx-Universität, Leipzig, Germany

V. B. KOLMANOVSKIJ, Electronic Engineering (MIEM), Moscow, Russia

D. KRAFT, Fachhochschule München, Munich, Germany

B. KUGELMANN, Technische Universität München, Munich, Germany

G. LEITMANN, University of California, Berkeley, California, USA

K. LESCH, Technische Universität München, Munich, Germany

H. MAURER, Numerische u. Instrumentelle Mathematik, Münster, Germany

K. D. MEASE, University of Princeton, Princeton, USA

A. A. MELIKYAN, USSR Academy of Science, Moscow, Russia

A. MIELE, Rice University, Houston, Texas, USA

L. MIKULSKI, Czarnowiejska 103/16,PL-30-049, Cracow, Poland

H. J. OBERLE, Universität Hamburg, Hamburg, Germany

H. J. PESCH, Technische Universität München, Munich, Germany

S. PICKENHAIN, Fachbereich Mathematik, Leipzig, Germany

Preface

The contributions presented at the meeting can be grouped into *Optimality Conditions
and Algorithms, Numerical Methods, Analysis and Synthesis of Nonlinear Systems,
Applications to Mechanical and Aeropsace Systems.*

The first chapter *Optimality Conditions and Algorithms* begins with a generalization of the maximum principle. The importance of the maximum principle consists not
only in the results themselves, but also in its proof and the possibilities of generalization. There are many similar optimization problems and each differs one from another
by specific conditions. The tent method is a general tool for finding necessary conditions
in different extremal problems. The next contribution also deals with generalization of
the maximum principle with emphasis on duality. The third article presents two algorithms using the maximum principle for computing lumped parameter optimal control
problems with special considerations towards distributed parameter systems with non-
convex functionals. Article No. 4 considers optimal control problems governed by a
nonlinear initial boundary value problem for a parabolic differential equation of second
order. Here, the control functions appear in the coefficient functions of the differential equation and the Neumann boundary condition. Using a decomposition condition
and a basic lemma of convexification it is possible to develop necessary conditions in
form of a minimum principle. The next article considers a class of nonlinear optimal
control problems for a parabolic partial differential equation with constraints on the
control and state variables. The problem is non-convex due to the nonlinearity of
the differential equation. Therefore, the known first order optimality conditions are
not sufficient for optimality. Second order optimality conditions are derived and it is
shown how these conditions can be applied to show convergence of the optimal controls for certain numerical approximations of the optimal control problem. The sixth
article considers necessary conditions for several control problems governed by integral
equations of the Volterra and Fredholm type, of integral equations with delay, and
integro-differential equations. Both Lagrange functionals and generalized Mayer functionals are considered. These optimality conditions lead in special cases to a sequence
of controls converging to a solution in the sense of the functional. Articles No. 7 and 8
deal with optimal control problems governed by ordinary differential equations. While
No. 7 presents algorithms for solving nonlinear equations via the solution of optimal
control problems, No. 8 is concerned with necessary conditions for minimax and maximin problems of the Chebyshev type. These conditions are applicable to problems
with single and multiple maxima. By using Contesous domain of maneuverabilty it is
shown that when the maxima are isolated single points, the control is generally continuous at the jump point in the minimax problems and discontinuous in the maximax
problems in which the first time derivative of the maximax function contains the control variable. The theory is applied to the problem of maximizing the flight radius

in a closed circuit glide of a hypervelocity vehicle and to a maximax optimal control problem in which the control appears explicitly with the first time derivative of the maximax function.

The chapter *Numerical Methods* contains articles concerned with recent developments in numerical methods for solving optimal control problems governed by ordinary differential and parabolic partial differential equations. The first article presents one of the most effective numerical techniques for the solution of trajectory optimization and optimal control problems. It combines a nonlinear programming algorithm with a discretization of the trajectory dynamics. The resulting mathematical programming problem is characterized by matrices, which are large and sparse. Constraints on the path of the trajectory are treated as algebraic inequalities to be satisfied by the solution of the nonlinear program. An approach is described which exploits the sparse structure based on quadratic programming which uses a Schur-Complement method in conjunction with a symmetric indefinite multifrontal sparse linear algebra package. The second article contains an algorithm that discretizes the control and the state variables by a similar transcription method, and solves the nonlinear program by sequential quadratic programming. The solution of the transcribed problem is utilized to obtain estimates for the adjoint variables of the original optimal control problem. These estimates are used to solve the boundary value problem associated with the optimality conditions via an indirect multiple shooting method. Examples for optimal robot trajectories and the ascent of the lower stage of a two-stage-to-orbit launch vehicle are given. The third article reviews some of the recent results on the convergence of reduced SQP methods and compares their behaviour with full SQP methods. As an example a paraboic control problem is solved.

The third chapter *Analysis and Synthesis of Nonlinear Systems* starts with an article concerned with constructing suboptimal feedback control in such a way that the dynamic system reaches the terminal state in a finite amount of time. This control is obtained through decomposition of the system into subsystems with one degree of freedom each and applying the methodology of differential game theory. The obtained feedback control is robust with respect to small disturbances and parameter variations. An application to robot control is discussed. The second article synthesizes a feedback strategy for a nonlinear system under uncertainty and gives an example. The third article analyses stability conditions by presenting algebraic stability conditions which are equivalent to the known characteristic linear equation of eigenvalues. The theory is illustrated by the determination of analytical solutions of the trajectories of an optimal feedback controlled spacecraft angular momentum. From this result the stability conditions are deduced. The fourth article contains a summary of the contributions of the authors towards developing feedback controls for nonlinear systems. Optimal controls under uncertainty and restrictions on controls and state are synthesized and possible applications for the construction of estimators, identificators, and controllers are indicated. The fifth paper analyses minimum time admissible controls for an optimal control problem from arbitrary initial states. Switching curves of the optimal control are constructed and the dependence on system parameters is investigated numerically. The last article of the chapter synthesizes feedback controls for a system whose output function is given under state constraints , in the presence of measurement errors, and with bounds on the controls. The approach is based on the theory of analytical centers.

The fourth chapter *Applications to Mechanical and Aerospace Systems* contains two applications to mechanical systems, the remaining ones are concerned with the computation of trajectories for Aerospace vehicles. The first paper models a computer disc drive as a singularly perturbed linear dynamic system. A time optimal control law is developed from the analysis of two reduced order systems. Attention is focused on the dependence of the control switching times on the small parameters of the system. Simulation results show that the control law gives improved results relative to existing methods, except when movements on the disc are very short. The second article is concerned with determining the optimal shape of a thin-walled elastic beam. The goal is to minimze the volume or to maximize the first and second eigenvalue. The control is the width of the profile. The problem is formulated as an optimal control problem and solved with the indirect multiple shooting method. The third article addresses the computational procedure to obtain estimates for the Lagrange multiplier functions as starting values for an indirect shooting method. An application is given. The next article describes optimal periodic trajectories for maximizing range per fuel consumed by an aircraft. Applying the optimality conditions, it is shown that a singular arc may exist, and that the chattering control appears for a more realistic fuel consumption model. It is shown that significant reductions in fuel consumption are possible for this type of aircraft. Convergence problems of the numerical algorithm are discussed. The fifth paper analyses numerically the influence of the dynamic pressure constraint on the optimal control of a high performance aircraft. Different types of controls for various combinations of active control- and state constraints are given, and the optimal switching structure for different boundary conditions is given. The sixth paper presents numerical solutions for the ascent of a two-stage-to-orbit hypersonic vehicle. A realistic mathematical model of such a vehicle is constructed using nonlinear least square algorithms for data approximation. Solutions are computed with a multiple shooting algorithm. Estimates of the adjoint variables are generated via a direct collocation method. Paper No. 7 gives a computational method for determinimg the controllability domain of aerospace vehicles during ascent and reentry. A feedback agorithm is given which computes control corrections while satisfying state- and control constraints as well as interior point constraints. Finally, the last paper applies a multiple shooting method to the concurrent vehicle and trajectory design of an interplanetary space probe.

Optimality Conditions and
Algorithms

International Series of Numerical Mathematics, Vol. 111, ©1993 Birkhäuser Verlag Basel

Tent Method in Optimal Control Theory

Vladimir G. Boltyanski

1   Introduction

The mathematical theory of optimal control (and more generally, the theory of extremal problems) plays an important role in applied mathematics. The central results of the theory (such as the Kuhn-Tucker theorem, the maximum principle, and other criteria of optimality) are important tools of applied mathematics.

At the same time, specific statements of extremal problems arising in different applied situations vary widely. Often, a new statement of an extremal problem falls out of the frame of the standard formulation of the Kuhn-Tucker theorem or the maximum principle. This circumstance leads to new versions of these theorems. Today, the Kuhn-Tucker theorem is not the only result, and this umbrella name covers a group of similar theorems, different from one another because of the specific conditions added on. The same situation occurs with respect to the maximum principle and other criteria.

Fortunately, there are unified, general methods applicable to most extremal problems. From this point of view, acquiring the knowledge of a general method is more important than the listing of criteria for numerous concrete extremal problems. Indeed, a general method helps one to obtain the specific optimization criteriom for a new statement of an extremal problem via a reasoning scheme that is more or less standard.

The *tent method* is such a general method. The first version of the method was worked out by Boltyanski in 1975 [1] after studying the well known Dubovitski-Miljutin paper [2]. This version of the tent method was finite-dimensional. Nevertheless, it allowed one to obtain simple proofs of the Kuhn-Tucker theorem and the maximum principle, as well as the solutions of different concrete extremal problems. Later, the author developed a new version of the tent method (cf. [3,4] where further references are given) with the "curse of finite-dimensionality" taken off. In the modern statement, the tent method is more general than the Dubovitski-Miljutin method, although both methods have much in common. The differences between the methods will be indicated in the sequel.

As Dubovitski and Miljutin have written in their papers, the main necessary criteriom in their theory (called the *Euler equation* in their papers and in this paper as well) was discovered by them after examining the proof of the Farkas lemma (that is well known in mathematical programming) and Boltyanski's first proof of the maximum principle [5]. So, the tent method and the Dubovitski-Miljutin method one are twins.

2   Statements of Extremal Problems

Let B be a Banach space, and let $\Omega_1, \ldots, \Omega_s$ be its subsets, such that their intersection is nonempty,

$$\Omega_1 \cap \ldots \cap \Omega_s \neq \emptyset.$$

Let further f be a scalar function that is defined on an open set $G \subset B$ containing Ω.

2.1 Abstract Extremal Problem. *Given the sets $\Omega_1, \ldots, \Omega_s$ in the Banach space B, find the minimum of the function f on the set $\Omega = \Omega_1 \cap \ldots \cap \Omega_s$.*

This problem generalizes the statement of many concrete extremal problems which may be obtained by specifying the sets Ω_i and the function f.

Example 2.1 Let us consider a mathematical programming problem that consists in looking for the minimum of the function f on the set Ω, which is defined by the system

$$(2.1) \quad g_1(x) = 0, \ldots, g_k(x) = 0, g_{k+1}(x) \leq 0, \ldots, g_s(x) \leq 0 \quad (x \in B),$$

where $g_1, \ldots, g_s$ are scalar functions. This problem is easily reduced to the above abstract extremal problem. Indeed, the equalities $g_i(x) = 0$ define the sets Ω_i ($i = 1, \ldots, k$) and the inequalities $g_j(x) \leq 0$ define the sets Ω_j ($j = k + 1, \ldots, s$). Consequently, the system (2.1) defines the intersection of the sets $\Omega_1, \ldots, \Omega_s$.

Example 2.2 A simple problem of optimal control is an another illustration. We consider the system $\dot{x} = g(x, u)$, where $x = (x_1, \ldots, x_n) \in \mathbb{R}^n$ is the state and u is the control, that belongs to a region $U \subset \mathbb{R}^d$. As control $u(t)$, we may take an arbitrary piecewise continuous function with values in U. We suppose that an initial state x_* and a terminal set M are given. Also, a scalar function $f(x)$ on $\mathbb{R}^n$ is given. The problem is to find a control $u(t)$, $t_* \leq t \leq t_0$, such that the corresponding trajectory $x(t)$ starting at the point x_* (i.e. $x(t_*) = x_*$) reaches at time t_0 a point of the set M and, under these conditions, yields the minimum of the function f at the terminal point,

$$x_0 = x(t_0) \in M, \qquad f(x_0) \rightarrow min.$$

To understand this problem in the frame of the shore ideas, we denote by Ω_1 the domain of controllability of the system $\dot{x} = g(x, u)$, $u \in U$, i.e., the set of all the states reachable by the system under the control $u(t) \in U$ starting from x_*. Further, we now denote by Ω_2 the terminal set M. Thus, it is necessary to find the minimum of the function f on the set $\Omega_1 \cap \Omega_2$.

The following theorem contains a formal solution of the abstract extremal problem considered above. In order to formulate the theorem, we introduce one more set. Let $y_0 = f(x_0)$ be the value of the function f at the point $x_0 \in \Omega$. We denote by $\Omega_0 = \Omega_0(x_0)$ the union of all the points $x \in \Omega$ where $f(x) < y_0$ and the point x_0,

$$\Omega_0 = \{x : f(x) < y_0\} \cup \{x_0\}.$$

Theorem 2.1 *Let x_0 be a point of the set $\Omega = \Omega_1 \cap \ldots \cap \Omega_s$. The function f considered on the set Ω takes its minimal value at the point x_0 if and only if the intersection $\Omega_0 \cap \Omega$ consists of the only point x_0,*

$$(2.2) \quad \Omega_0 \cap \Omega = \Omega_0 \cap \Omega_1 \cap \ldots \cap \Omega_s = \{x_0\}.$$

Proof If there exists a point $x' \in \Omega_0 \cap \Omega$ distinct from x_0, then $f(x') < f(x_0)$ (according to the definition of the set Ω_0), i.e., the value of the function f at the point $x' \in \Omega$ is less than its value at the point x_0. But this is impossible, since x_0 is a minimum point. So, the condition (2.2) is necessary. The sufficiency is verified similarly.

Remark 2.1 The Theorem 2.1 may be applied not only in the case where the minimumpoint is unique, but in the general case as well. Indeed, assume that there exists another minimumpoint x_0' of the function f considered on the set Ω, i.e. $f(x_0) = f(x_0') = y_0$ and $f(x) \geq y_0$ for every $x \in \Omega$. Then, the sets

$$\Omega_0 = \{x : f(x) < y_0\} \cup \{x_0\}, \quad \Omega_0' = \{x : f(x) < y_0\} \cup \{x_0'\}$$

do not coincide. Under these conditions, the following equalities hold:

$$\Omega_0 \cap \Omega = \{x_0\}, \quad \Omega_0' \cap \Omega = \{x_0'\},$$

and the first of them means that x_0 is a minimumpoint, while the second one means that x_0' is a minimumpoint.

Theorem 2.1 reduces the previous abstract extremal problem to the following one.

2.2 Abstract Intersection Problem. *Given the sets $\Omega_0, \Omega_1, \ldots, \Omega_s$ with a common point x_0 in the Banach space B, find the conditions under which the intersection $\Omega_0 \cap \Omega_1 \cap \ldots \cap \Omega_s$ consists only of the point x_0.*

This problem is more convenient than the previous one on account of its symmetry. Also, this problem applies to a wider category of concrete extremal problems.

Example 2.3 Let p scalar functions $f_1, \ldots, f_p$ be given on an open set $G \in B$ that contains the intersection $\Omega = \Omega_1 \cap \ldots \cap \Omega_s$. Let further the function f be defined as the maximum of these functions,

$$(2.3) \quad f(x) = \max(f_1(x), \ldots, f_p(x)), \quad x \in G.$$

We consider the *minimax problem*, i.e., we are interested in finding the minimum of the function (2.3) on the set Ω. Let x_0 be a point of the set Ω, and let $y_0 = f(x_0)$ be the value of the function f at this point. We put

$$\Omega_0 = \{x : f(x) < y_0\} \cup \{x_0\},$$
$$\Omega_0^{(i)} = \{x : f_i(x) < y_0\} \cup \{x_0\}, \qquad i = 1, \ldots, p.$$

It is easily shown by virtue of (2.3) that the relation

$$\Omega_0 = \Omega_0^{(1)} \cap \ldots \cap \Omega_0^{(p)}$$

holds. Consequently, equation (2.2) takes the form

$$(2.4) \quad \Omega_0 \cap \Omega = \Omega_0^{(1)} \cap \ldots \cap \Omega_0^{(p)} \cap \Omega_1 \cap \ldots \cap \Omega_s = \{x_0\}.$$

Thus, the minimum problem is solved by the relation (2.4), that is, this problem is reduced to the abstract intersection problem. By the way, the relation (2.4) is more convenient than (2.2). For example, if the functions $f_1, \ldots, f_p$ are smooth, then working with the sets $\Omega_0^{(i)}$ is more convenient than working with the Ω_0, because in general the function (2.3) is not smooth and may have a complicated structure.

Example 2.4 As in Example 2.3, let p scalar functions $f_1, \ldots, f_p$ be given on the set G. A point $x_0 \in \Omega$ is said to be a *Pareto point* with respect to the functions $f_1, \ldots, f_p$, if there is no point $x \in \Omega$ such that the inequalities

$$f_1(x) \leq f_1(x_0), \ldots, \ f_p(x) \leq f_p(x_0)$$

hold and at least one of them is strong. In other words, x_0 is a Pareto point if it follows from the relations

$$f_1(x) \leq f_1(x_0), \ldots, \ f_p(x) \leq f_p(x_0), \qquad x \in \Omega,$$

that all these inequalities are in fact equalities. Looking for Pareto points is easily reduced to the scheme of the abstract intersection problem. Indeed, let us fix a point $x_0 \in \Omega$, and let us denote by Γ the set of all points $x \in \Omega$ satisfying the system of equalities

$$f_1(x) = f_1(x_0), \ldots, \ f_p(x) = f_p(x_0).$$

Further, let us consider the sets

$$\Omega_{(i)}^{(0)} = (\{x : f_i(x) \leq f_i(x_0)\} \setminus \Gamma) \cup \{x_0\}, \qquad i = 1, \ldots, p.$$

Then, the following proposition is true and can be proved in the same way as Theorem 2.1.

Proposition 2.1. *A point $x \in \Omega$ is a Pareto point with respect to the functions $f_1, \ldots, f_p$ if and only if the intersection*

$$\Omega_{(1)}^{0} \cap \ldots \cap \Omega_{(p)}^{0} \cap \Omega_1 \cap \ldots \cap \Omega_s$$

consists of only the point x_0.

There are many other extremal problems which reduce to the general scheme of the abstract intersection problem. Also, there exist other ways of reducing concrete extremal problems to the abstract extremal problem, and these ways are connected with leaving the state space $\mathbb{R}^n$. For example, let us return to Example 2.2. Let $x_0 \in M$ be a point reached under the admissible control $u(t)$. Then, we can consider the following sets in the space $\mathbb{R}^{n+1} = \mathbb{R}^n \times \mathbb{R}$:

(i) the set Ω_1^* that consists of all the points $(x(t), f(x(t)))$ for all admissible trajectories $x(t)$;

(ii) the set $\Omega_2^* = M \times I\!\!R$;

(iii) the ray Ω_0^* emanating from the point $x_0^* = (x_0, f(x_0))$ and going in the direction of the $(n+1)$th negative semiaxis.

It is easily shown that x_0 is a solution of the above optimization problem if and only if the intersection $\Omega_0^* \cap \Omega_1^* \cap \Omega_2^*$ consists of only the point x_0^*. Thus, the abstract intersection problem is a general model of different extremal problems.

3 o-Mappings

The *tent method* is the tool for solving the previous abstract intersection problem. The idea is to replace each of the sets $\Omega_0, \Omega_1, \ldots, \Omega_s$ by a linear approximation in order to pass from the relation

$$(3.1) \qquad \Omega_0 \cap \Omega_1 \cap \ldots \cap \Omega_s = \{x_0\}$$

to simpler conditions in terms of the linear approximations. We shall suppose that, for every set Ω_i $(i = 0, \ldots, s)$, a convex cone K_i is choosen as a linear approximation of Ω_i near the point x_0. In the sequel, the cone K_i is said to be *a tent* of the set Ω_i atpreviouspreviousprevious the point x_0. But, in order to introduce the general definition of a tent, we shall consider in this section some preliminary notions.

A *plane* in a Banach space B is a set of form $x_0 + L$, where L is a subspace of B. Thus, in general, a plane does not contain the origin $0 \in B$. If the plane $x_0 + L$ is a closed set in B (i.e., if L is a closed subspace), then we shall say that $x_0 + L$ is a *closed plane*.

Let $M \subset B$ be a convex set. The minimal plane that contains M is called the *support plane* of M and is denoted by *aff* M. We shall say that a convex set $M \subset B$ is *standard* if *aff* M is a closed plane and the *relative interior ri* M of the set M is nonempty. We recall that a point $x \in M$ belongs to the relative interior *ri* M if there exists a neighborhood U of x such that $U \cap$ *aff* $M \subset M$. In particulary, if the interior *int* M of the convex set M in the space B is nonempty, then M is a standard convex set, called *convex body* in the space B.

A set $K \subset B$ is said to be a *cone* with apex at the origin if, for every point $x \in K$ and every positive number λ, the point λx belongs to K. If K is a cone with apex at the origin, then the set $x_0 + K$ is said to be a cone with apex at x_0. In the sequel, we usually consider standard convex cones with apex at a point x_0.

Definition 3.1 Let B_1, B_2 be Banach spaces, and let $G_1 \subset B_1$ be an open set. A mapping $f : G_1 \to B_2$ is said to be *smooth* if it has the Fréchet derivative f'_x, $x \in G$, and this derivative is continuous with respect to x.

Definition 3.2 Let B_1, B_2 be Banach spaces, and let $U_1 \subset B_1$ be a neighborhood of the origin $0 \in B_1$. A mapping $U_1 \to B_2$ is said to be an *o-mapping* if it is smooth and its Fréchet derivative at the origin is equal to zero. For simplicity, every o-mapping will be denoted by o.

The following two lemmas may be deduced directly from general theorems of functional analysis (cf., for example, pp. 199 and 207 of the book [6]).

Lemma 3.1 *Let $q_1 : B_1 \to B_2$, $q_2 : B_2 \to B_3$ be continuous linear operators, and let $f_1 = q_1 + o$, $f_2 = q_2 + o$ be local mappings (i.e., they are defined on some neighborhoods of the origins of the spaces B_1, B_2). Then, $f_2 \circ f_1 = q_2 \circ q_1 + o$.*

Lemma 3.2 *Let $f = e + o$, where e is the identity mapping of a Banach space B. Then, there exist neighborhoods V and W of the origin $0 \in B$ such that f maps homeomorphically V on W and $f^{-1} = e + o$.*

Lemma 3.3 *Let $q : B_1 \to B_2$ be a continuous linear operator, and let $f = q + o$. Let further $q(z) \in int\, C_2$, where $C_2 \subset B_2$ be a convex cone with apex 0. Then, there exist a convex cone $C_1 \subset B_1$ with apex 0 and a neighborhood U_1 of the origin $0 \in B_1$ such that $z \in int\, C_1$ and $f(C_1 \cap U_1) \subset C_2$.*

Proof Let ϵ be a positive number such that, if $\|y\| \leq (\|q\| + \|z\| + 1)\epsilon$, then $q(z) + y \in C_2$. We may suppose that $\epsilon < 1$. Let C_1 be the set of all the points $\lambda(z + h)$ where $\|h\| \leq \epsilon$, $\lambda \geq 0$. Then, $C_1 \subset B_1$ is a convex cone and $z \in int\, C_1$. Finally, we put $\varphi = f - q$. Then, φ is an o-mapping, and hence there exists a neighborhood U_1 of the origin $0 \in B_1$ such that $\|\varphi(x)\| \leq \epsilon \|x\|$ for every $x \in U_1$.

Let $x \in C_1 \cap U_1$, $x \neq 0$. Then, $x = \lambda(z + h)$ where $\lambda > 0$, $\|h\| < \epsilon$. Consequently,

$$\left\| \frac{f(x)}{\lambda} - q(z) \right\| = \left\| \frac{q(x) - \varphi(x)}{\lambda} - q(z) \right\| = \left\| q(h) + \frac{\varphi(x)}{\lambda} \right\|$$

$$\leq \|q\|\, \|h\| + \frac{1}{\lambda} \epsilon \|x\| \leq \|q\|\epsilon + \epsilon \left\| \frac{x}{\lambda} \right\| = \epsilon(\|q\| + \|z + h\|)$$

$$\leq \epsilon(\|q\| + \|z\| + 1).$$

In other words, $\frac{f(x)}{\lambda} = q(z) + y$, where $\|y\| \leq \epsilon(\|q\| + \|z\| + 1)$. This means that $\frac{f(x)}{\lambda} \in C_2$, and hence $f(x) \in C_2$. Thus, $f(C_1 \cap U_1) \subset C_2$.

4   Notion of Tent

Now we are going to pass to the definition of tent. Two simple examples supply a preliminary visual understanding of the notion of tent. If Ω is a smooth hypersurface in $I\!\!R^n$, then the tangent hyperplane of Ω at a point $x_0 \in \Omega$ is a tent of Ω at this point. Further, let Ω be a convex body in $I\!\!R^n$, and let x_0 be its boundary point. Then, the *support cone* of Ω at the point x_0 is a tent of Ω at this point. We recall that the support cone of Ω at the point x_0 is the closure of the union of all rays emanating from x_0 and passing through a point of Ω distinct of x_0.

Definition 4.1 Let $\Omega \subset B$ be a set containing a point x_0, and let $K \subset B$ be a standard closed cone with the apex x_0. The cone K is said to be a *tent* of the set Ω at the point x_0 if, for every point $z \in ri\,K$, there exists a convex cone $Q_z \subset K$ with apex x_0 and a mapping $\psi_z : \Sigma_z \to B$, where Σ_z is a neighborhood of the point x_0, such that

(i) $aff\,Q_z = aff\,K$,

(ii) $z \in ri\,Q_z$,

(iii) $\psi_z(x) = x + o(x - x_0)$,

(iv) $\psi_z(Q_z \cap \Sigma_z) \subset \Omega$.

    It follows immediately from this definition that, if K is a tent of the set Ω at the point x_0, and if $K_1 \subset K$ is a closed standard cone with the same apex x_0, then K_1 is a tent of Ω at the point x_0, as well.

    The following lemma describes simpler conditions under which a cone K is a tent of a set Ω. This particular case is convenient for many applications.

Lemma 4.1 *Let Ω be a set containing a point x_0, and let K be a standard closed cone with apex x_0. The following conditions are sufficient for K to be a tent of Ω at the point x_0: there exist a neighborhood Σ of the point x_0 and a mapping $\psi : \Sigma \to B$, such that*

(i) $\psi(x) = x + o(x - x_0)$,

(ii) $\psi(K \cap \Sigma) \subset \Omega$.

    Indeed, for every point $z \in ri\,K$, we may put $Q_z = K$, $\psi_z = \psi$, and conditions (i) – (iv) of Definition 4.1 will be satisfied.

    Now, we prove three theorems containing the simplest cases of tents.

Theorem 4.1 *Let $g = h + o$, where $h : B \to B_1$ is a continuous linear operator. If there exists a subspace $L \subset B$ such that the operator h considered only on L possesses the inverse operator $k : B_1 \to L$, then the kernel K of the operator $h : B \to B_1$ is a tent of the manifold $\Omega = g^{-1}(0)$ at the point 0.*

Proof We put $p = k \circ h$, $q = e - p$, where $e : L \to L$ is the identity. Then, $p(x) = x$ for every $x \in L$. Hence, for every $x \in B$, the relation

$$p(q(x)) = p(x) - p(p(x)) = p(x) - p(x) = 0$$

is true, i.e., $q(x) \in p^{-1}(0) = h^{-1}(0) = K$. We put $f = k \circ g = p + o$, $\psi = f + q$. Then, $\varphi = (p + o) + (e - p) = e + o$. According to Lemma 3.2, there exist neighborhoods V and Σ of the origin such that φ maps homeomorphically V onto Σ and the mapping $\psi = \varphi^{-1}$ has the form $\psi = e + o$. For every point $x \in K \cap \Sigma$, the relation

$$f(\psi(x)) = \varphi(\psi(x)) - q(\psi(x)) = x - q(\psi(x)) \in K$$

is true. Also, $f(\psi(x)) = k(g(\psi(x))) \in L$. Consequently,

$$f(\psi(x)) \in K \cap L = \{0\},$$

i.e., $f(\psi(x)) = 0$. Hence, $\psi(x) \in f^{-1}(0) = g^{-1}(0) = \Omega$. Thus, $\psi(K \cap \Sigma) \subset \Omega$ and, by virtue of Lemma 4.1, K is a tent of the set Ω at the point 0.

Consequence 4.1 *Let L be a closed subspace of a Banach space B. Let further $g = h + o$, where $h : B \to L$ is a continuous linear operator such that $h(x) = x$ on L. Then $K = h^{-1}(0)$ is a tent of the manifold $\Omega = g^{-1}(0)$ at the point 0.*

Theorem 4.2 *Let $f(x)$ be a real function defined on a neighborhood of a point $x_0 \in B$. Let us suppose that*

$$f(x) = f(x_0) + l(x - x_0) + o(x - x_0),$$

where l is a nontrivial continuous linear functional. We put

$$\Omega_1 = \{x : f(x) \le f(x_0)\}, \quad \Omega_0 = \{x : f(x) < f(x_0)\} \cup \{x_0\},$$
$$K_0 = \{x : l(x) \le l(x_0)\}.$$

Then, K_0 is a tent of each of the sets Ω_0, Ω at the point x_0.

Proof It is sufficient to consider the case $x_0 = 0$. Let $z \in ri\, K_0$, i.e., $l(z) < 0$. The set $W_z = \{w : l(w) < -\frac{1}{2}l(z)\} \subset B$ is open and contains the origin. Hence, the set $z + W_z$ is a neighborhood of the point z. We denote by Q_z^* the cone with apex at the origin that is generalized by the set $z + W_z$. If $x \in Q_z^*$ and $x \neq 0$, i.e., $x = \lambda(z + w)$, $\lambda > 0$, $w \in W_z$, then

$$l(x) = l(\lambda(z + w)) = \lambda(l(z) + l(w))$$

$$(4.1) \quad < \lambda(l(z) - \frac{1}{2}l(z)) = \frac{1}{2}\lambda l(z) < 0.$$

Hence, $Q_z^* \subset K_0$.

Now, we put $h = \frac{1}{2}l$, $f_1 = f - h = h + o$, and denote by C the negative real semiaxis. Then, $C \subset \mathbb{R}$ is a cone with apex at the origin and $h(z) = \frac{1}{2}l(z) < 0$, i.e., $h(z) \in int\, C$. According to Lemma 3.3, there exists a cone Q_z^{**} with apex at the origin and a neighborhood Σ_z of the origin in B such that $z \in int\, Q_z^{**}$ and $f_1(Q_z^{**} \cap \Sigma_z) \subset C$, i.e., $f_1(x) \leq 0$ for every $x \in Q_z^{**} \cap \Sigma_z$. In other words,

$$(4.2) \quad f_1(x) = f(x) - \frac{1}{2}l(x) \leq 0$$

for every $x \in Q_z^{**} \cap \Sigma_z$.

Finally, we put $Q_z = Q_z^* \cap Q_z^{**} \subset K_0$. Then, $z \in int\, Q_z$ and $aff\, Q_z = B = aff\, K_0$, i.e., conditions (i) and (ii) of Definition 4.1 are satisfied. If $x \in Q_z \cap \Sigma_z$, $x \neq 0$, then according to relations (4.1) and (4.2),

$$f(x) = \frac{1}{2}l(x) + (f(x) - \frac{1}{2}l(x)) < 0.$$

Hence, $Q_z \cap \Sigma_z \subset \Omega_0$. Consequently, if we denote by ψ_z the identity $e : B \to B$, then conditions (iii) and (iv) of Definition 4.1 will be satisfied, too. Thus, K_0 is a tent of the set Ω_0 at the origin. Since $\Omega_0 \subset \Omega_1$, then K_0 is a tent of the set Ω_1 at the origin as well.

Theorem 4.3 *Let $\Omega \subset B$ be a standard convex set, let $0 \in \Omega$, and let K be a support cone of the set Ω at the point 0, i.e.,*

$$K = cl\left(\bigcup_{\lambda > 0} (\lambda \Omega)\right).$$

Then, K is a tent of the set Ω at the origin.

Proof Let $z \in ri\,K$. Then, the point z belongs to the set $\Omega^* = \bigcup_{\lambda > 0}(\lambda\Omega)$. Indeed, if $z \notin \Omega^*$, then there exists a plane Γ through z that is a hyperplane in the subspace $aff\,\Omega^* = aff\,\Omega$, such that the standard convex set Ω^* is situated in a closed half-space with boundary Γ. Consequently, the set $K = cl\,(\Omega^*)$ is situated in this half-space, too, contradicting the inclusion $z \in ri\,K$.

Since $z \in \Omega^*$, then there exists $\lambda_0 > 0$, such that $z \in \lambda_0\Omega$. Let $\lambda_1 > \lambda_0$. We prove that $z \in ri\,(\lambda_1\Omega)$. Conversely let us admit, that $z \notin ri\,(\lambda_1\Omega)$. Then, there exists a hyperplane Γ through z in $aff\,\Omega$ such that the set $\lambda_1\Omega$ is situated in a closed half-subspace Π in $aff\,\Omega$ with boundary Γ. Let λ', λ'' be positive numbers such that $\lambda' < 1 < \lambda'' < \frac{\lambda_1}{\lambda_0}$. Then, $\lambda'z \in \lambda_1\Omega$, $\lambda''z \in \lambda_1\Omega$, and z is an interior point of the segment $[\lambda'z,\ \lambda''z]$. Since Γ is a support hyperplane in $aff\,\Omega$ of the set $\lambda_1\Omega$ at the point z, then the segment $[\lambda'z,\ \lambda''z]$ is contained in Γ, and therefore $0 \in \Gamma$. Consequently, $\lambda\Pi = \Pi$ for every $\lambda > 0$. It follows by virtue of the inclusion $\lambda_1\Omega \subset \Pi$ that $\lambda\Omega \subset \Pi$ for every $\lambda > 0$, i.e., $\Omega^* \subset \Pi$. Hence, $K \subset \Pi$, i.e., Γ is a support hyperplane in $aff\,\Omega$ of the cone K, contradicting the inclusion $z \in ri\,K$. This contradiction shows that $z \in ri\,(\lambda_1\Omega)$.

We put $\lambda = \frac{1}{\lambda_1}$. Then, $\lambda z \in ri\,\Omega$. Let ϵ be a positive number such that $x \in \Omega$, if $x \in aff\,\Omega$ and $\|x - \lambda z\| < \epsilon$. We may suppose that $\epsilon < \frac{1}{2}\|\lambda z\|$. Let us denote by $\Sigma_z \subset B$ the open ball of the radius ϵ centered at the origin. We put

$$U_z = (aff\,\Omega) \cap (\lambda z + \Sigma_z).$$

Then, $U_z \subset \Omega$. We denote by Q_z the cone with apex 0 that is generated by the set U_z. Then, $z \in ri\,Q_z$ and $aff\,Q_z = aff\,\Omega = aff\,K$, i.e., conditions (i) and (ii) of Definition 4.1 are satisfied. Let us denote by ψ_z the identity mapping of the space B. Then condition (iii) of Definition 4.1 is obviously satisfied. We show that condition (iv) is satisfied as well. Indeed, let x be a nonzero element of $\psi_z(\Sigma_z \cap Q_z)$, i.e., $x \in \Sigma_z \cap Q_z$, $x \neq 0$. Then, $\|x\| < \epsilon$ and $x = \kappa(\lambda z + h)$, where $h \in aff\,\Omega$, $\|h\| < \epsilon$, $\kappa > 0$, and consequently $\lambda z + h = \frac{1}{\kappa}x$. If $\kappa \geq 1$, then the inequalities $\|\frac{1}{\kappa}x\| \leq \|x\| < \epsilon$ take place, and hence

$$\|\lambda z\| = \left\|\frac{1}{\kappa}x - h\right\| \leq \left\|\frac{1}{\kappa}x\right\| + \|h\| < \epsilon + \epsilon = 2\epsilon,$$

contradicting the inequality $\epsilon < \frac{1}{2}\|\lambda z\|$ (cf. the definition of the number ϵ). Consequently, $\kappa < 1$. Finally, since

$$\lambda z + h \in U_z \subset \Omega,$$

then $\kappa(\lambda z + h) \in \Omega$, i.e., $x \in \Omega$. Thus, $\psi_z(\Sigma_z \cap Q_z) = \Sigma_z \cap Q_z \subset \Omega$, i.e., condition (iv) holds.

Consequence 4.2 *Let $\Omega \subset B$ be a standard convex set, let $x_0 \in \Omega$, and let K be the support cone of Ω at the point x_0. Then, K is a tent of the set Ω at the point x_0.*

The following theorem explains the importance of tents for the solution of the abstract intersection problem.

Theorem 4.4 *If a tent K of a set $\Omega \subset B$ at a point $x_0 \in \Omega$ contains a ray l emanating from the point x_0, then the set Ω contains a point distinct from x_0.*

Proof Let $z \in l$, $z \neq x_0$. Since $l \subset K$, then l is a tent of the set Ω at the point x_0. Consequently, according to conditions (iii) and (iv) of Definition 4.1, there exists a mapping $\psi_z : \Sigma_z \to B$, where Σ_z is a neighborhood of the point x_0 such that $\psi_z(x) = x + o(x - x_0)$ and $\psi_z(l \cap \Sigma_z) \subset \Omega$. Let ϵ be a positive number such that $\|o(x - x_0)\| < \frac{1}{2}\|x - x_0\|$, if $\|x - x_0\| < \epsilon$. We choose a point $x \in l$, $x \neq x_0$, with $\|x - x_0\| < \epsilon$ and $x \in \Sigma_z$. Then,

$$\|\psi_z(x) - x_0\| = \|x - x_0 + o(x - x_0)\|$$
$$\geq \|x - x_0\| - \|o(x - x_0)\| \geq \frac{1}{2}\|x - x_0\| > 0.$$

Consequently, $\psi_z(x) \neq x_0$. Since $x \in l \cap \Sigma_z$, then $\psi_z(x) \in \Omega$. Thus, $\psi_z(x)$ is a point of the set Ω distinct from x_0.

Remark 4.1 The above reasoning shows that the set Ω contains a curve Λ for which l is the tangent ray at the point x_0. Indeed, if x runs through a segment $[x_0, x_1] \subset l$, then $\psi_z(x) \in \Omega$. This means that $\psi_z(x)$, $x \in [x_0, x_1]$ is a parametric description of a curve that is contained in Ω. Since $\psi_z(x) = x + o(x - x_0)$, then l is the tangent ray of the curve at the point x_0.

Consequence 4.3 *We have the following necessary condition of the validity of the relation (3.1): any tent of the set $\Omega_0 \cap \Omega_1 \cap \ldots \cap \Omega_s$ consists of only the point x_0.*

It seems intuitively clear that, in a usual situation, the intersection $K_0 \cap K_1 \cap \ldots \cap K_s$ of the tents $K_0, K_1, \ldots, K_s$ of the sets $\Omega_0, \Omega_1, \ldots, \Omega_s$ is a tent of the intersection set. For example, if Ω_0 and Ω_1 are two hypersurfaces in $I\!\!R^n$ with a common point x_0 and if K_0, K_1 are their tangent hyperplanes at the point x_0, then usually (i.e., if K_0 and K_1 are not coincide) the plane $K_0 \cap K_1$ is a tent of the set $\Omega_0 \cap \Omega_1$ at the point x_0. Indeed, $K_0 \cap K_1$ is the $(n-2)$-dimensional tangent plane of the $(n-2)$-dimensional surface $\Omega_0 \cap \Omega_1$.

But this usual situation does not take place in general case. Indeed, if Ω_0 and Ω_1 are two spheres in $I\!\!R^n$ tangent to each other at the point x_0, and if $K_0 = K_1$ is their common tangent hyperplane at x_0, then the intersection $\Omega_0 \cap \Omega_1$ consists only of the point x_0, but $K_0 \cap K_1$ is the hyperplane, and $K_0 \cap K_1$ is not a tent of the set $\Omega_0 \cap \Omega_1$ at the point x_0. This means that $K_0 \cap K_1 \cap \ldots \cap K_s$ is not always a tent of the intersection set $\Omega_0 \cap \Omega_1 \cap \ldots \cap \Omega_s$. In the next two sections, we shall obtain conditions under which this situation takes place.

5 Subspaces in General Position for a Banach Space

In the sequel, we shall give statements without proofs. The reader can find complete proofs in the paper [3] and the book [4], where the theory considers not only Banach spaces, but also linear topological spaces. Nevertheless, we shall consider in the following statements only the case of Banach spaces.

Let L be a closed subspace of a Banach space B, and let a be an element of B. As usual, the number $\min \|a - x\|$, $x \in L$, is said to be the *distance* of the point a from the subspace L. We shall denote this distance by $d(a, L)$.

Definition 5.1 A system $Q_0, Q_1, \ldots, Q_s$ $(s \geq 1)$ of closed subspaces of B is said to be in *general position* if, for every $\epsilon > 0$, there exists a positive number δ such that the inequalities $d(a, Q_i) \leq \delta \|a\|$, $i = 0, 1, \ldots, s$, imply $d(a, Q_0 \cap Q_1 \cap \ldots \cap Q_s) \leq \epsilon \|a\|$.

In particular, let Q_0, Q_1 be two closed subspaces of B, and let $Q_0 \cap Q_1 = \{0\}$. In this case, the subspaces Q_0, Q_1 are in general position if and only if there exists $\delta > 0$ such that, for every unit vectors $a_0 \in Q_0$, $a_1 \in Q_1$, the inequality $\|a_0 - a_1\| > \delta$ holds. This circumstance may be considered as indicative of the existence of an nonzero angle between the subspaces Q_0 and Q_1 which satisfy the condition $Q_0 \cap Q_1 = \{0\}$. In [3] and [4], there are examples illustrating cases when the subspaces are not in general position.

Let us note that, if B is a finite-dimensional space, then each of its subspaces $Q_0, Q_1, \ldots, Q_s$ is in general position.

Definition 5.2 A system $Q_0, Q_1, \ldots, Q_s$ $(s \geq 1)$ of closed subspaces of a Banach space B is said to possess the *property of general intersection* if every two subspaces L_1, L_2 (each representable as intersection of several of the subspaces $Q_0, Q_1, \ldots, Q_s$) are in general position.

Theorem 5.1 *Let $Q_0, Q_1, \ldots, Q_s$ be a system of closed subspaces of a Banach space B. This system possesses the property of general intersection if and only if each subsystem containing at least two subspaces is in general position.*

Theorem 5.2 *Two closed subspaces Q_1, Q_2 of a Banach space B are in general position if and only if there exists a number $\kappa > 0$ such that every element $x \in Q_1 + Q_2$ may be represented in the form $x = x_1 + x_2$, where $x_1 \in Q_1$, $x_2 \in Q_2$, $\|x_1\| \leq \kappa\|x\|$, $\|x_2\| \leq \kappa\|x\|$.*

Theorem 5.3 *Two closed subspaces Q_1, Q_2 of a Banach space B are in general position if and only if the subspace $Q_1 + Q_2$ is closed.*

Theorem 5.4 *Two closed subspaces Q_1, Q_2 of a Banach space B are in general position if and only if, for any relatively open sets $G_1 \subset Q_1$, $G_2 \subset Q_2$, the sum $G_1 + G_2$ is an open set of the subspace $Q_1 + Q_2$.*

Theorem 5.5 *Two closed subspaces Q_1, Q_2 of a Banach space B that have the only common element 0 and satisfy the condition $\mathrm{cl}\,(Q_1 + Q_2) = B$ are in general position if and only if B is represented as the direct sum of the subspaces Q_1, Q_2, i.e., $B = Q_1 \oplus Q_2$. More generally, let Q_1, Q_2 be closed subspaces of a Banach space B which satisfy the condition $\mathrm{cl}\,(Q_1 + Q_2) = B$. The subspaces Q_1, Q_2 are in general position if and only if the factorspace $B/(Q_1 \cap Q_2)$ is the direct sum of the subspaces $Q_1/(Q_1 \cap Q_2)$, $Q_2/(Q_1 \cap Q_2)$.*

6   Separability of a System of Convex Cones

Definition 6.1 A system $K_0, K_1, \ldots, K_s$ of convex cones with common apex x_0 in B is said to be *separated* (cf. [3], [4]) if there exists a hyperplane Γ through x_0 that separates one of the cones from the intersection of the others (i.e., for an index i, the cone K_i is situated in one of the closed half-spaces with boundary Γ, and the intersection of the other cones is situated in the second half-space).

The following theorem gives an answer to the question posed at the end of Section 4.

Theorem 6.1 *Let $\Omega_0, \Omega_1, \ldots, \Omega_s$ be sets in a Banach space B with a common point x_0. Let further $K_0, K_1, \ldots, K_s$ be tents of these sets at the point x_0. We suppose that each cone K_i is standard. Also, we suppose that the system $\mathrm{aff}\,K_0, \mathrm{aff}\,K_1, \ldots, \mathrm{aff}\,K_s$*

possesses the property of general intersection. Then, if the cones $K_0, K_1, \ldots, K_s$ are not separated, then the cone $K_0 \cap \ldots \cap K_s$ is a tent of the set $\Omega_0 \cap \ldots \cap \Omega_s$ at the point x_0.

Theorem 6.2 *Let $\Omega_0, \Omega_1, \ldots, \Omega_s$ be sets in a Banach space B with a common point x_0, and let $K_0, K_1, \ldots, K_s$ be the tents of these sets at the point x_0. We suppose that the conditions of Theorem 6.1 on the system of these cones are satisfied and at least one of the cones does not coincide with its affine hull (i.e., $K_i \neq \mathrm{aff}\, K_i$ for an index i). Under these conditions, if the cones $K_0, K_1, \ldots, K_s$ are not separated, then there exists a point $x' \in \Omega_0 \cap \ldots \cap \Omega_s$ that is distinct from x_0. In other words, the separability of the cones $K_0, K_1, \ldots, K_s$ is a necessary condition for the validity of the relation (3.1).*

The theorems stated give necessary criteria for the solution of the abstract intersection problem. Consequently, they contain necessary criteria for the solutions of different extremal problems. These criteria are formulated in geometrical form, i.e., as requirement of separability of a system of convex cones. In order to reformulate these necessary contitions in algebraic form (as a system of equalities and inequalities), we need an algebraic condition of separability of a system of convex cones with common apex.

Let first s $= 1$, and let the convex cones K_0, K_1 with common apex x_0 possess the separability property, i.e., there exists a hyperplane Γ that separates K_0 and K_1. Let us denote by f_0 a nonzero continuous functional on B with the kernel $\Gamma - x_0$ that is nonpositive on the half-space containing $K_0 - x_0$, and let us denote by $f_1 = -f_0$ the functional that is nonpositive on the other half-space (containing the cone $K_1 - x_0$). Then, the following conditions are satisfied:

(C1) $f_0 + f_1 = 0$;

(C2) $f_0(x - x_0) \leq 0$ for any point $x \in K_0$, and
$\qquad f_1(x - x_0) \leq 0$ for any point $x \in K_1$.

The following definition and theorem extend these visual reasonings to the general case s ≥ 1.

Definition 6.2 *Let K be a convex cone with apex x_0 in a Banach space B. We denote by $D(K)$ its polar cone, i.e., the cone in the conjugate space B' consisting of all linear continuous functionals $a \in B'$ such that $a(x - x_0) \leq 0$ for any $x \in K$.*

Theorem 6.3 *Let $K_0, K_1, \ldots, K_s$ be a system of standard convex cones with common apex x_0 in a Banach space B. We suppose that the planes $\mathrm{aff}\, K_0, \mathrm{aff}\, K_1, \ldots, \mathrm{aff}\, K_s$*

possess the property of general intersection. For the separability of the system $K_0, K_1, \ldots, K_s$, it is necessary and sufficient the existence of vectors $a_i \in D(K_i)$, $i = 0, \ldots, s$, not all equal to zero, such that

$$(6.1) \qquad a_0 + a_1 + \ldots + a_s = 0.$$

According to the terminology of Dubovitski and Miljutin, the relation (6.1) is called the *Euler equation*.

A comparison of Theorems 6.2 and 6.3 give us the following necessary condition for the solution of the abstract intersection problem.

Theorem 6.4 *Let $\Omega_0, \Omega_1, \ldots, \Omega_s$ be sets in a Banach space B with a common point x_0, and let $K_0, K_1, \ldots, K_s$ be the tents of these sets at the point x_0. We suppose that the conditions of Theorem 6.1 are satisfied and that at least one of the cones $K_0, K_1, \ldots, K_s$ does not coincide with its affine hull. For the validity of equation (3.1), it is necessary the existence of vectors $a_i \in D(K_i)$, $i = 0, \ldots, s$, not be equal to zero, such that the Euler equation (6.1) is true.*

As a consequence, we obtain the following necessary condition for the solution of the abstract extremal problem.

Theorem 6.5 *Let f be a smooth scalar function that is defined on the set $\Omega = \Omega_1 \cap \ldots \cap \Omega_s$, and let $K_1, \ldots, K_s$ be the tents of the sets $\Omega_1, \ldots, \Omega_s$ at a point $x_0 \in \Omega$. We suppose that the cones $K_1, \ldots, K_s$ are standard, that the planes aff $K_1, \ldots,$ aff K_s possess the property of general intersection, and that $\frac{\partial f(x_0)}{\partial x} \neq 0$. If x_0 is a minimum point of the function f on the set Ω, then there exist vectors $a_1 \in D(K_1), \ldots, a_s \in D(K_s)$, such that*

$$\frac{\partial f(x_0)}{\partial x} + a_1 + \ldots + a_s = 0.$$

Sufficient conditions may be obtained with the help of the following theorem.

Theorem 6.6 *Let $K_0, K_1, \ldots, K_s$ be convex cones with common apex x_0 in a Banach space B, and let $\Omega_0, \Omega_1, \ldots, \Omega_s$ be sets such that the inclusions $\Omega_i \subset K_i$, $i = 0, \ldots, s$, are satisfied. If there exist vectors $a_0 \in D(K_0)$, $a_1 \in D(K_1), \ldots, a_s \in D(K_s)$ such that $a_0 \neq 0$ and the Euler equation (6.1) holds, then*

$$(int\, \Omega_0) \cap \Omega_1 \cap \ldots \cap \Omega_s = \emptyset.$$

Remark 6.1 If the conditions of Theorem 6.3 are satisfied, then instead of the relation $a_0 \neq 0$ we may require the following: the vectors $a_0, a_1, \ldots, a_s$ are not all equal to zero and the cones $K_1, \ldots, K_s$ are not separated.

Remark 6.2 Let us suppose that the cones $K_0, K_1, \ldots, K_s$, except maybe K_s, are bodies, i.e., $int\, K_i \neq \emptyset$ for $i = 0, 1, \ldots, s-1$. Then, the property of general intersection holds trivially. So, Theorem 6.5 takes place in this case. Thus, we do not need the separation theory of convex cones if the cones $K_0, K_1, \ldots, K_s$, except maybe one, are bodies (the proofs are simpler in this case). This is just the Dubovitski-Miljutin method. The tent method is more general and does not require the cones to be bodies.

Remark 6.3 In the author's first proof of the maximum principle [5], the construction of a tent of the controllability domain Ω_1 for the system $\dot{x} = g(x, u)$, $u \in U$ (cf. Example 2.2) was given. This construction was the central point of the proof of the maximum principle.

We recall that the first statement of the maximum principle was given by Gamkrelidze, who had established (generalizing the famous Legendre theorem) a sufficient condition for a sort of weak optimality problem. Then Pontryagin, Gamkrelidze, and the author verified that the Gamkrelidze condition allows one to obtain the synthesis of optimal trajectories in the Bushaw problem [7] and in similar problems as well. Then, Pontryagin proposed that the Gamkrelidze condition be called *Maximum Principle* and that it be formulated as a *sufficient* condition of optimality in general case [8]. This was the only contribution of Pontryagin to the discovery of the maximum principle. After that, Gamkrelidze established that the maximum principle is a *necessary and sufficient* condition of optimality for linear controlled systems [9]. Finally, while preparing the complete account [10] of the article [8], this author understood that the maximum principle is not a sufficient condition, but only a *necessary* condition of optimality. This author gave the first proof of the maximum principle [5] in this correct statement as a *necessary* condition of optimality in the general, nonlinear case.

Pontryagin was the Department Chairman in the Steklov Mathematical Institute and he could insist on his interests. So, this author had to use the title "Pontryagin's maximum principle" in his paper [5]. This is why today mathematics and engineering researchers know the main optimization criteriom as the *Pontryagin maximum principle*.

References

[1] Boltyanski, V.G. *The method of tents in the theory of extremal problems* (Russian). Uspehi Mat. Nauk 30 (1975) no. 3, 3–55.

[2] Dubovitski, A., Miljutin, A. *Problems on extrema under constraints* (Russian). Doklady AN SSSR, 149 (1963), no. 4, 759–762.

[3] Boltyanski, V.G. *Extremal problems and the method of tents* (Russian). Sbornik Trudov VNIISI, Moscow (1989), no. 14, 136–147.

[4] Boltyanski, V.G. *The method of tents and problems of system analysis* (Russian). Math. theory of systems, Moscow, Nauka (1986), 5–24.

[5] Boltyanski, V.G. *The maximum principle in the theory of optimal processes* (Russian). Doklady AN SSSR, 119 (1958), no. 6, 1070–1074.

[6] Ljusternik, L.A., Sobolev, V.I. *A short course on functional analysis* (Russian). Moscow, (1982), 271 pp.

[7] Bushaw, D.W., *Thesis*, Dept. of Math., Princeton Univ. 1952.

[8] Boltyanski, V.G., Gamkzelidze, R.V., Pontryagin, L.S. *Zur Theorie der optimalen Prozesse* (Russian). Doklady AN SSSR, 110 (1956), 7–10.

[9] Gamkzelidze, R.V. *Zur Theory der optimalen Prozesse in linearen Systemen* (Russian). Doklady AN SSSR, 116 (1957), 9–11.

[10] Boltyanski, V.G. Gamkzelidze, R.V. Pontryagin, L.S. *Die Theorie der optimalen Prozesse I. Das Maximumprinzip* (Russian). Izvestija AN SSSR (matem), 24 (1980), no. 1, 3–42.

Author's address

Prof. Dr. V.G. Boltyanski
Scientific Institute of System Research
VNIISI
Pr. 60 Let Octjabrja 9
SU-117312 Moscow, Russia

Pontryagin's Maximum Principle for Multidimensional Control Problems

Rolf Klötzler and Sabine Pickenhain

Abstract

A weak maximum principle is shown for general problems

$$\text{minimize} \quad f(x, w) \quad \text{on } X_0 \times X_1$$

with respect to *linear* state constraints

$$A_0 x = A_1 w$$

in Banach spaces X_0 and local convex topological vector spaces X_1, where $f(x, \cdot)$ is a convex functional on X_1 and A_j are linear and continuous operators from X_j to a Hilbert space X $(j = 0, 1)$. The proved theorem is applied to Dieudonné–Rashevsky–type and relaxed control problems.

1. Introduction

In the past there where many efforts to extend Pontryagin's maximum principle (PMP) of optimal control theory to the case of multiple integral problems. The investigations are separated into two directions:

1. The large theory of optimal control with distributed parameters, in which one of the independent variables plays a distinctive leading role.

2. The theory of Dieudonné–Rashevsky–type control problems, where the independent variables have an equal rank.

The second kind of problems are the topic of this paper. A basic problem can
be formulated in the following way:

$$(1) \qquad \text{minimize} \quad \int_\Omega r(t, x(t), u(t)) \, dt \, , \quad \Omega \subset \mathbb{E}^m$$

with respect to the *state functions* $x \in W_p^{1,n}(\Omega)$ and the *control functions* $u \in L_p^\nu(\Omega)$
$(p > m)$ fulfilling the *state equations*

$$(2) \qquad x_{t_\alpha}(t) = g_\alpha(t, x(t), u(t)) \quad \text{a.e. on } \Omega \quad (\alpha = 1, \dots, m),$$

the *control restrictions*

$$(3) \qquad u(t) \in V \subseteq \mathbb{E}^\nu \quad \text{a.e. on } \Omega$$

and the *boundary conditions*

$$(4) \qquad x(s) = 0 \quad \text{for } s \in \partial\Omega.$$

This problem (1) – (4) will be denoted by (P).

In 1969 L. Cesari [1] stated a generalized maximum principle for the problem
(P) in the following sense:
Let (x_0, u_0) be an optimal process of (P), then there are multipliers $\lambda_0 \in \mathbb{E}_+$
and $y \in W_1^{1,nm}(\Omega)$, not vanishing simultaneously, such that with the *Pontryagin
function*

$$H(t, \xi, v, \eta, \lambda_0) = -\lambda_0 r(t, \xi, v) + \sum_{\alpha=1}^m \eta^{\alpha T} g_\alpha(t, \xi, v)$$

the *maximum condition*

$$(5) \qquad \max_{v \in V} H(t, x_0(t), v, y(t), \lambda_0) = H(t, x_0(t), u_0(t), y(t), \lambda_0) \quad \text{a.e. on } \Omega$$

and the *canonical equations*

$$(6) \qquad \sum_{\alpha=1}^m y_{t_\alpha}^\alpha(t) = - \operatorname{grad}_\xi H(t, x_0(t), u_0(t), y(t), \lambda_0) \quad \text{a.e. on } \Omega$$

hold.

In 1976 relevant papers followed by R. Klötzler [3] and H. Rund [5]. But we must say that these investigations are insufficient, since they use assumptions on the existence and regularity of the solution of the corresponding *Hamilton – Jacobi – equation*

$$(7) \qquad \sum_{\alpha=1}^{m} S_{t_\alpha}^{\alpha}(t,\xi) \; + \; \max_{v \in V} H(t,\xi,v,\mathrm{grad}_\xi S(t,\xi),1) \; = \; 0,$$

which are is general not fulfilled. Therefore these necessary optimality conditions are in fact *pseudo – necessary* conditions.

Moreover the following example of an unrestricted variational problem shows that a maximum principle as a formal extension of the known (PMP) in one dimensional case in form of (5) and (6) does not hold.

Example:

$$\text{minimize } J(x,u) = \int_\Omega \{u_1^1(t)u_2^2(t) - u_2^1(t)u_1^2(t)\}\, dt, \quad \Omega \subset \mathbb{E}^2,$$

$$(\partial\Omega \quad \text{piecewise smooth}),$$

with respect to $x \in W_p^{1,2}(\Omega)$, $u \in L_p^4(\Omega)$ with

$$x_{t_\alpha}(t) = u_\alpha(t) \quad \text{a.e. on} \quad \Omega\,(\alpha = 1,2),$$

$$x(s) = 0 \quad \text{on} \quad \partial\Omega.$$

It is easy to be seen, that the admissible and optimal process $(x_0 = 0, u_0 = 0)$ does not satisfy the maximum condition (5).

The purpose of our investigation is now to demonstrate a new approach to necessary optimality conditions without restricting assumptions and without any references to the validity of (7).

2. A general concept for an ϵ – maximum principle

We consider a general optimization problem

$$\text{minimize} \quad f(x,w) \quad \text{on } X_0 \times X_1$$

$$(8) \qquad \text{with respect to } \textit{linear} \text{ state constraints}$$

$$A_0 x = A\,w$$

under the following basic **Assumptions:**

1. Let A_j are linear and continuous operators from X_j to X $(j = 0, 1)$.
2. Assume that X be a Hilbert space, X_0 a Banach space and X_1 a compact and convex subset of a local convex topological vector space Ω_0.
3. There is an optimal solution (x_0, w_0) of (8).
4. For the reel functional f the one side Gâteaux derivative $\partial^+ f(x_0, w_0; \xi, \partial w)$ exists in each direction $(\xi = x - x_0, \partial w = w - w_0)$, with $x \in X_0$ and $w \in X_1$ and can be expressed in the following way:

$$\partial^+ f(x_0, w_0; \xi, \partial w) = \langle f_0(x_0, w_0), \xi \rangle_{X_0} + \langle f_1(x_0, w_0), \partial w \rangle_{X_1}$$

(Here $\langle \cdot, \cdot \rangle_{X_i}$ are the interior products in X_i, $i = 0, 1$.)

5. Assume the convexity of $f(x, \cdot)$ for each $x \in X_0$.
6. If the sequence $\{A_0 \xi_n\}_{n=1}^\infty$, $\xi_n \in X_0$, is weakly convergent to c, then a unique element $\xi \in X_0$ exists with $c = A_0 \xi$ and $\{\xi_n\}_{n=1}^\infty$ is weakly convergent to ξ in X_0.

To prove an ϵ – maximum principle we study the following convex set in $\mathbb{E} \times X$ for given $\epsilon \geq 0$:

$$(9) \qquad \begin{aligned} M_\epsilon := \{ &(f(x_0, w) - f(x_0, w_0) + \langle f_0(x_0, w_0), x - x_0 \rangle_{X_0} + \tau, \\ & A_0(x - x_0) - A_1(w - w_0)) \mid x \in X_0, \ w \in X_1, \ \tau \geq \epsilon \} \end{aligned}$$

and the convex cone generated by this set,

$$(10) \qquad K(M_\epsilon) := \{ \lambda \zeta \mid \lambda \geq 0, \ \zeta \in M_\epsilon \}.$$

Firstly we mention an important **special case:**

$$(11) \qquad \overline{K(M_0)} \neq \mathbb{E} \times X.$$

Then, by a well - known separating theorem in Banach spaces, there is a nonzero vector $(\lambda_0, y) \in \mathbb{E} \times X$ with

$$(12) \qquad \lambda_0 z_0 + \langle y, z \rangle_X \geq 0 \quad \text{for all} \quad (z_0, z) \in K(M_0).$$

From (12) immediately follows

$$(13) \quad \lambda_0 \geq 0, \quad \lambda_0[f(x_0, w) - f(x_0, w_0)] - \langle y, A_1(w - w_0) \rangle_X \geq 0 \quad \text{for all } w \in X_1$$

as well as

$$(14) \quad \begin{aligned} A_0 x_0 &= A_1 w_1, \\ 0 &= \lambda_0 \langle f_0(x_0, w_0), x - x_0 \rangle_{X_0} - \langle y, A_0 x - x_0 \rangle_X \quad \text{for all } x \in X_0. \end{aligned}$$

We can interpret $(13) - (14)$ as a general form of (PMP) for convex problems in function spaces.

In the **general case** we shall prove the following

Assertion. *If the problem* (8) *fulfills the basic assumptions* 1. – 6. *then*

$$(15) \qquad \overline{K(M_\epsilon)} \neq \mathbb{E} \times X \quad \text{for each} \quad \epsilon > 0.$$

Proof. We show that the element $(z_0, 0) \in \mathbb{E} \times X$ with $z_0 < 0$ does not belong to $\overline{K(M_\epsilon)}$. Let $\{(z_0^n, z^n)\}, (z_0^n, z^n) \in K(M_\epsilon)$, with $z^n \to 0$ and $z_0^n \to z_0$, be an arbitrary sequence in $K(M_\epsilon)$. Then

$$(16) \qquad z^n = \lambda^n(A_0 \xi^n - A_1(w^n - w_0)) \in X$$

and

$$(17) \qquad z_0^n = \lambda^n[f(x_0, w^n) - f(x_0, w_0) + \langle f_0(x_0, w_0), \xi^n \rangle_{X_0} + \tau^n]$$

are valid with $\xi^n \in X_0$, $w^n \in X_1$, $\tau^n \geq \epsilon$ and $\lambda^n \geq 0$. In consequence of the convexity of $f(x, \cdot)$ and assumption 4 it holds

$$(18) \qquad z_0^n \geq \lambda^n[\langle f_0(x_0, w_0), \xi^n \rangle_{X_0} + \langle f_1(x_0, w_0), w^n - w_0 \rangle_{X_1} + \epsilon].$$

Taking assumption 2 into account we can conclude that $\{w^n\}$ has a convergent subsequence $\{w^{n'}\}$ (in the topology of X_1) with the limit w.
a) The sequence $\{\lambda^{n'}\}$ is bounded and has therefore a convergent subsequence which is again denoted by $\{\lambda^{n'}\}$.

1. Let $\lim_{n'} \lambda^{n'} = \lambda > 0$. Then from (16) and $\{w^{n'}\} \to w$, $\{z^{n'}\} \to \mathbf{0}$ it follows that $\{A_0 \xi^{n'}\}$ is weakly convergent. By assumption 6 there is a unique element $\xi \in X_0$ with $\{\xi^{n'}\} \rightharpoonup \xi$, $\{A_0 \xi^{n'}\} \rightharpoonup A_0 \xi$ and

$$(19) \qquad\qquad A_0 \xi - A_1(w - w_0) = 0.$$

Therefore (x, w) is admissible to (8). From (17) we get

$$(20) \qquad
\begin{aligned}
z_0 = \lim_{n' \to \infty} z_0^{n'} &\geq \lambda[\langle f_0(x_0, w_0), \xi\rangle + \langle f_1(x_0, w_0), (w - w_0)\rangle + \epsilon\,] \\
&\geq \lambda[\partial^+ f(x_0, w_0; \xi, w - w_0) + \epsilon] \geq \lambda\epsilon
\end{aligned}$$

since in virtue of the optimality of (x_0, w_0) to (8) and (19) as well as the convexity of X_i, $i = 0, 1$ the one side Gâteaux derivative $\partial^+ f(x_0, w_0; \xi, w - w_0)$ is non negative and therefore $z_0 > 0$.

2. Let $\lim_{n'} \lambda^{n'} = 0$. Then it follows $\lim_{n' \to \infty} \lambda^{n'} A_1(w^{n'} - w_0) = 0$ and from (16) and $\{z^{n'}\} \to \mathbf{0}$ we conclude the weak convergence of $\{A_0 \hat{\xi}^{n'}\} \to \mathbf{0}$ in X, with $\hat{\xi}^{n'} := \lambda^{n'} \xi^{n'}$. Applying assumption 6 it holds $\{\hat{\xi}^{n'}\} \rightharpoonup \Theta$ and from (18) we thus obtain

$$z_0 = \lim_{n' \to \infty} z_0^{n'} \geq [\langle f_0(x_0, w_0), \Theta\rangle_{X_0} + \langle f_1(x_0, w_0), 0\rangle_{X_1} = 0.$$

b) If $\{\lambda^n\}$ is unbounded then there exists a subsequence $\{\lambda^{n'}\}$ with $\lim_{n' \to \infty} \lambda^{n'} = \infty$ and with similar arguments as in 1. we obtain $\lim_{n' \to \infty} z_0^{n'} = \infty$. (Remark that this part of the proof only works with $\epsilon > 0$.)

In consequence of the assertion and its proof it follows that for each fixed $\epsilon > 0$ and $a < 0$ there is a radius $\rho > 0$, such that for the ball $B_\rho(a, 0)$ around $(a, 0)$ in $\mathbb{E} \times X$ it holds $\bar{B} \cap \overline{K(M_\epsilon)} = \emptyset$. Therefore a supporting plane through the origin $(0, 0)$ of $\mathbb{E} \times X$ exists and has a normal parallel to the vector $\zeta - (a, 0)$, where $\zeta \in \overline{K(M_\epsilon)}$ is the point of shortest distance from $(a, 0)$ to $\overline{K(M_\epsilon)}$. Using the assertion with the normal $(\lambda_0, y) := \zeta - (a, 0)$ we find that $\lambda_0 > 0$ and

$$(21) \qquad
\begin{aligned}
\lambda_0 z_0 + \langle y, z\rangle \geq 0 \quad &\text{for all } (z_0, z) \in \overline{K(M_\epsilon)} \\
&\text{and especially for all } (z_0, z) \in M_\epsilon.
\end{aligned}$$

Without loss of generality we can assume now $\lambda_0 = 1$. Under consideration of (9) and (21) we get the following

Theorem 1. *If (x_0, w_0) is an optimal process of (8) then for each $\epsilon > 0$ there is a vector $y \in X$ (depending on ϵ) with*

$$(22) \qquad \epsilon + [f(x_0, w) - f(x_0, w_0)] - \langle y, A_1(w - w_0)\rangle_X \geq 0 \quad \text{for all } w \in X_1,$$

as well as

$$(23) \qquad A_0 x_0 = A_1 w_0, \quad 0 = \langle f_0(x_0, w_0), \xi\rangle_{X_0} - \langle y, A_0\xi\rangle_X \quad \text{for all } \xi \in X_0.$$

We may interpret (22) and (23) as a weak variant of Pontryagin's maximum principle for convex–linear problems in function-spaces.

3. Applications

1. Let us consider the problem (P) under the following conditions.
A. Assume $r \in C^1(\Omega \times \mathbb{R}^n \times V)$, $g \in C^{0,nm}(\Omega \times \mathbb{R}^n \times V)$, convexity of $r(t, \xi, \cdot)$ for all $(t, \xi) \in \Omega \times \mathbb{R}^n$ and linearity of $g(t, \cdot, \cdot)$ for all $t \in \Omega$.
B. The function $r(\cdot, x, u)$ is summable for each admissible (x, u) to (P), $r_\xi(\cdot, x_0, u_0)$ and $r(\cdot, x_0, u)$ are summable for all $u \in X_1$, with

$$X_1 := \{u \in L_p^\nu(\Omega)|\ u(t) \in V \text{ a.e. on } \Omega \}.$$

C. Let X_1 be a convex and compact subset of $[L_p^\nu(\Omega), \sigma_w]$, where σ_w is the weak topology in $L_p^\nu(\Omega)$, X_0 the Sobolew space $\mathring{W}_p^{1,n}(\Omega)$ with $(p > m)$ and $X = L_2^{nm}(\Omega)$.

With A. – C. the assumptions of the Theorem 1 are satisfied for (P) and Theorem 1 reads as follows:
Theorem 2. *If (x_0, u_0) is an optimal process to (P), then for each $\epsilon > 0$ there is a multiplier $y \in X$ (depending on ϵ), such that the*

integrated ϵ- maximum condition

$$(24) \qquad \epsilon + \int_{\Omega} [\, H(t, x_0(t)u_0(t), y(t), 1) - H(t, x_0(t), u(t), y(t), 1) \,]dt \geq 0$$

$$\text{for all } u \in X_1,$$

and the canonical equations

$$(25) \qquad \sum_{\alpha=1}^{m} y_{t_\alpha}^\alpha = -\mathrm{grad}_\xi H(\cdot, x_0, u_0, y, 1) \quad \text{in distributional sense}$$

are fulfilled.

Remark. Suppose $r(t, \xi, \cdot)$ is not convex, then we replace r in (P) by $\tilde{r}$ for a sufficiently large constant c,

$$\tilde{r}(t, \xi, v) = r(t\,\xi, v) + c|v - u_0(t)|^2$$

and Theorem 2 holds too, if we replace r in the Pontryagin function by $\tilde{r}$. This result is obviously, since the optimal solution for (P) is also an optimal solution of $(\tilde{P})$, in which r is replaced by $\tilde{r}$.

2. Let us consider the following relaxed or generalized problem $(\bar{P})$ to (P)

$$(\bar{1}) \quad \text{minimize } \bar{J}(x, \mu) := \int_{\Omega} \int_{V} r(t, x(t), v)\, d\mu_t(v)dt, \quad \Omega \subset \mathbb{R}^m, \quad m \geq 1,$$

with respect to the *state functions* $x \in W_p^{1,n}(\Omega)$ and *generalized controls* $\mu = \{\mu_t \,|\, t \in \Omega\} \in \mathcal{M}_V$, fulfilling the *state equations*

$$(\bar{2}) \qquad x_{t_\alpha}(t) = \int_{V} g_\alpha(t, v)\, d\mu_t(v) \quad \text{a.e. on } \Omega, \quad (\alpha = 1, \ldots, m),$$

generalized control restrictions

$$(\bar{3}) \qquad \mathrm{supp}\, \mu_t \subseteq V \subseteq \mathbb{E}^\nu \quad \text{a.e. on } \Omega,$$

and the *boundary conditions*

$$(\bar{4}) \qquad\qquad x(s) = C \quad \text{for } s \in \partial\Omega.$$

Remark. The set μ of generalized controls was introduced in the one dimensional control theory by Gamkrelidze [2]. The definition of generalized controls in multidimensional control problems is the formal extensions of this definition to the multidimensional case.

We consider $(\bar{P})$ under the following conditions.

α. Assume $r \in C^1(\Omega \times \mathbb{R}^n \times V)$, $g \in C^{0,nm}(\Omega \times V)$.

β. Let V be compact. Then following [4], $X_1 := [\mathcal{M}_V, \sigma^*]$ is a convex and compact topological vector space and $\{\mu^n\}$ converges to μ in this topology, if

$$\lim_{n\to\infty} \int_\Omega \int_V f(t,v)\,d\mu_t^n(v)\,dt = \int_\Omega \int_V f(t,v)\,d\mu_t(v)\,dt$$

$$\text{for all} \quad f \in C(\bar{\Omega} \times V).$$

Further on let X_0 be the Sobolew space $\overset{\circ}{W}_p^n(\Omega)$ with $(p > m)$ and $X = L_2^{nm}$.

γ. If (x_0, μ_0) is an optimal process to $(\bar{P})$ then there exists

$$\Phi(t) := grad_\xi\,[\int_V r(t,\xi,v)\,d\mu_{t0}(v)]_{\xi = x_0(t)} \quad \text{with } \Phi \text{ summable on } \Omega.$$

With these assumptions $\alpha. - \gamma.$ Theorem 1 is applicable to $(\bar{P})$ and can be formulated in the following way:

Theorem 3. *Let (x_0, μ_0) an optimal process to $(\bar{P})$. Then for each $\epsilon > 0$ there is a multiplier $y \in X$ (depending on ϵ), such that the integrated ϵ-maximum condition*

$$(26) \qquad \epsilon + \int_\Omega \int_V H(t, x_0(t), v, y(t), 1)\,d[\mu_{t0}(v) - \mu_t(v)]dt \geq 0$$

$$\text{for all } \mu \in \mathcal{M}_V$$

and the canonical equations

$$(27) \qquad \sum_{\alpha=1}^{m} y_{t_\alpha}^{\alpha} = -\mathrm{grad}_\xi \int_V H(\cdot, x_0, v, y, 1) d\mu_{\cdot 0}(v) \quad \text{in distributional sense}$$

are fulfilled.

References

1. L. Cesari, *Optimization with partial differential equations in Dieudonné-Rashevsky form and conjugate problems.* Arch.Rat.Mech. Anal. 33 (1969), 339 - 357.

2. R.V. Gamkrelidze, *Principles of Optimal Control Theory.* Plenum Press, New York and London, 1978.

3. R. Klötzler, *On Pontrjagin´s maximum principle for multiple integrals.* Beiträge zur Analysis 8 (1976), 67 - 75.

4. H. Kraut und S. Pickenhain, *Erweiterung von mehrdimensionalen Steuerungsproblemen und Dualität.* Optimization 21 (1990), 387 - 397.

5. H. Rund, *Pontrjagin functions for multiple integral control problems.* J. Optim. Theory Appl. 18 (1976), 511 - 520.

Authors' addresses

Rolf Klötzler, Sabine Pickenhain, Fachbereich Mathematik/Informatik
Institut für Mathematik der Universität Leipzig
D – O – 7010 Leipzig, Augustusplatz 10

An Algorithm for Abstract Optimal Control Problems Using Maximum Principles and Applications to a Class of Distributed Parameter Systems

Hans Benker, Michael Handschug

Abstract

Algorithms using maximum principles for computing lumped parameter optimal control problems are extended to an abstract optimal control problem in Hilbert spaces. Furthermore the application to a class of distributed parameter systems is considered.

1 Algorithm for an optimal control problem in Hilbert spaces

In [2], [3] the authors discuss the extension of two algorithms for lumped parameter optimal control problems using maximum principles to the following abstract optimal control problem

$$(1.1) \qquad J(u) = j(x(u), u) \to \min_{u \in U \subset E_2}$$

subject to the operator equation (state equation)

$$(1.2) \quad T(x, u) = 0$$

where $x(u)$ is the assumed unique solution of (1.2) for $u \in U$ (bounded and closed), $j \in E_1 \times E_2 \mapsto \Re^1$, $T \in E_1 \times E_2 \mapsto E_3$ and E_i are linear normed spaces with the norm $\| \cdot \|_i$.

Remarks:

1. The aim of this paper is to give an algorithm which generalizes the ideas of the papers [1], [8] and [9] to the abstract problem (1.1), (1.2) and to consider the application to distributed parameter systems.

2. If we will treat real systems governed by integral or differential equations it is convenient to write the state equation (2) in the form $x = S(x, u)$ or $Ax = F(x, u)$.

3. In the following we must suppose that the spaces $E_i = H_i$ are Hilbert spaces and j, T are Lipschitz-continuously Fréchet-differentiable in both arguments in order to ensure the convergence properties of the given algorithm. However the existence of a Fréchet-derivative in a Hilbert space (e.g. L^2) places very heavy demands. In nonlinear applied problems these demands are often not met. Therefore we show in section 2 that we need weaker assumptions if we apply the algorithm to problems in function spaces.

Under the assumption that the spaces $E_i = H_i$ are Hilbert spaces, that j and T are Lipschitz-continuously Fréchet-differentiable in both arguments and that for $\Delta x = x(u + \Delta u) - x(u)$ is satisfied an inequality of the form

$$(1.3) \quad \|\Delta x\|_1 \leq C\|\Delta u\|_2 \quad (C = const. > 0)$$

the following algorithm is obtained by generalizing some ideas of Sakawa and Bonnans [1], [9]. For this method we define the Hamiltonian H by

$$(1.4) \quad H(x, u, p) = j(x, u) + \langle p, T(x, u) \rangle_3$$

and the augmented Hamiltonian H^ε by

$$(1.5) \quad H^\varepsilon(x, u, v, p) = H(x, u, p) + \frac{1}{2\varepsilon}\|u - v\|_2^2$$

where the adjoint variable $p \in H_3$ is the assumed unique solution of the adjoint equation

$$(1.6) \qquad H_x = T_x^\top p + j_x = 0$$

($\langle \cdot, \cdot \rangle_i$ is the inner product in the Hilbert space H_i, B_x denotes the Fréchet derivative of B with respect to x, $B^\top$ the adjoint operator to B.)

Algorithm:

Step 1: Let $u^1 \in U$ and a sequence $\{\varepsilon^k\}$ with $0 < \varepsilon_1 < \varepsilon^k < \varepsilon_o$ be given. Set $k = 1$.

Step 2: Calculate the state vector x^k associated to the control u^k by solving the state equation (1.2) and the adjoint vector p^k by solving the adjoint equation (1.6) for (x^k, u^k).

Step 3: Set $k = k + 1$ and find $u^k \in U$ that minimizes $H^{\varepsilon^k}(x^k, u, u^{k-1}, p^{k-1})$ with respect to $u \in U$ (the existence of a solution is supposed).

Step 4: Stop if a stopping criterion is satisfied. Otherwise go to step 2.

This iterative procedure has the following properties:

Theorem 1: *If we assume that the minimum in step 3 is attained at a control u^k then there exists a constant $K > 0$ such that*

$$(1.7) \qquad J(u^k) - J(u^{k-1}) \leq -\left(\frac{1}{\varepsilon^k} - K\right) \|u^k - u^{k-1}\|_2^2.$$

Thus, the sequence $\left\{J(u^k)\right\}$ is monotone decreasing for $\varepsilon^k > 0$ sufficiently small.

Proof: We have $J(u^k) - J(u^{k-1}) = H(x^k, u^k, p^{k-1}) - H(x^{k-1}, u^{k-1}, p^{k-1}) - \langle p^{k-1}, T(x^k, u^k) - T(x^{k-1}, u^{k-1})\rangle_3$ and with $\Delta x^k = x^k - x^{k-1}$, $\Delta u^k = u^k - u^{k-1}$ it holds $H(x^k, u^k, p^{k-1}) - H(x^{k-1}, u^{k-1}, p^{k-1}) = H(x^k, u^k, p^{k-1}) - H(x^k, u^k - \Delta u^k, p^{k-1}) + H(x^{k-1} + \Delta x^k, u^{k-1}, p^{k-1}) - H(x^{k-1}, u^{k-1}, p^{k-1})$. By using the estimates $H(x^k, u^k, p^{k-1}) - H(x^k, u^k - \Delta u^k, p^{k-1}) \leq \langle H_u(x^k, u^k, p^{k-1}), \Delta u^k\rangle_2 + C_1\|\Delta u^k\|_2^2$ and $H(x^{k-1} + \Delta x^k, u^{k-1}, p^{k-1}) - H(x^{k-1}, u^{k-1}, p^{k-1}) \leq \langle H_x(x^{k-1}, u^{k-1}, p^{k-1}), \Delta x^k\rangle_1 + C_2\|\Delta x^k\|_1^2$
$= C_2\|\Delta x^k\|_1^2$ we obtain $J(u^k) - J(u^{k-1}) \leq \langle H_u(x^k, u^k, p^{k-1}), \Delta u^k\rangle_2 + C_1\|\Delta u^k\|_2^2 + C_2\|\Delta x^k\|_1^2$.

By making use of the augmented Hamiltonian we can write this inequality as

$$J(u^k) - J(u^{k-1}) \leq \langle H_u^{\varepsilon^k}(x^k, u^k, u^{k-1}, p^{k-1}), \Delta u^k \rangle_2 - \tfrac{1}{\varepsilon^k}\|\Delta u^k\|_2^2 + C_1\|\Delta u^k\|_2^2 + C_2\|\Delta x^k\|_1^2.$$

Thus by applying the relations $\langle H_u^{\varepsilon^k}(x^k, u^k, u^{k-1}, p^{k-1}), \Delta u^k \rangle_2 \leq 0$ and $\|\Delta x^k\|_1 \leq C\|\Delta u^k\|_2$

we obtain the estimation $J(u^k) - J(u^{k-1}) \leq -\tfrac{1}{\varepsilon^k}\|\Delta u^k\|_2^2 + K\|\Delta u^k\|_2^2$

and the proof is terminated. $\hspace{2cm}$ q.e.d.

Theorem 2: *Let the cost functional J be bounded from below then there exists $\varepsilon_o > 0$ such that, if $0 < \varepsilon^k < \varepsilon_o$, any sequence $\{u^k\}$ generated by the previous algorithm has the following properties*

$$(1.8) \quad \|u^k - u^{k-1}\|_2 \to 0 \qquad for \ k \to \infty,$$

$$(1.9) \quad \left\|u^k - P_U\left(u^k - \varepsilon^{k+1} J'(u^k)\right)\right\|_2 \to 0 \qquad for \ k \to \infty$$

where P_U is the projection on U (assumed to be convex) and $J'(u)$ denotes the gradient of J.

Proof: In virtue of theorem 1 there exists a constant $L > 0$ such that $J(u^k) - J(u^{k-1}) \leq -L\|u^k - u^{k-1}\|_2^2 \leq 0$. Therefore it follows $J(u^k) \to J^o$ and $J(u^k) - J(u^{k-1}) \to 0$ for $k \to \infty$ and we obtain $\|u^k - u^{k-1}\|_2^2 \leq \tfrac{1}{L}\left(J(u^{k-1}) - J(u^k)\right) \to 0$ for $k \to \infty$.
Now we consider the second assertion: The minimization in step 3 yields the necessary optimality condition
$\langle u^k - u^{k-1} + \varepsilon^k H_u(x^k, u^k, p^{k-1}), v - u^k \rangle_2 \geq 0 \quad \forall v \in U$ or equivalently
$u^k = P_U\left(u^{k-1} - \varepsilon^k H_u(x^k, u^k, p^{k-1})\right)$. With $J'(u^{k-1}) = H_u(x^{k-1}, u^{k-1}, p^{k-1})$ we can write
$u^{k-1} - P_U\left(u^{k-1} - \varepsilon^k J'(u^{k-1})\right) = u^{k-1} - u^k + P_U\left(u^{k-1} - \varepsilon^k H_u(x^k, u^k, p^{k-1})\right) - P_U\left(u^{k-1} - \varepsilon^k H_u(x^{k-1}, u^{k-1}, p^{k-1})\right)$ and obtain $\left\|u^{k-1} - P_U\left(u^{k-1} - \varepsilon^k J'(u^{k-1})\right)\right\|_2 \leq \|u^{k-1} - u^k\|_2 + \varepsilon^k\|H_u(x^k, u^k, p^{k-1}) - H_u(x^{k-1}, u^{k-1}, p^{k-1})\|_2 \to 0$ for $k \to \infty$. $\hspace{1cm}$ q.e.d.

Remarks:

1. The theorems 1 and 2 remain valid if in step 3 of the algorithm we replace $H^{\varepsilon^k}(x^k, u, u^{k-1}, p^{k-1})$ by $H^{\varepsilon^k}(x^{k-1}, u, u^{k-1}, p^{k-1})$ which simplifies the numerical calculation. If x and u in (1.1) and (1.2) are separated, i.e. $T(x,u) = A(x) + B(u)$ and $j(x,u) = j_1(x) + j_2(u)$, then both versions are identical.

2. If we make the additional assumption that the functional $J(u)$ is l.s.c. and convex, that U is convex and that $\varepsilon^k \geq \varepsilon_1 > 0$ then there exists a subsequence of $\{u^k\}$ which converges weakly to an optimal control u^o. The proof of this proposition

is given in [1]. The convexity and l.s.c. of $J(u)$ for instance are ensured if the problem (1.1), (1.2) has the form

$$(1.10) \quad J(u) = j(x(u), u) \to \min_{u \in U}$$

subject to the state equation

$$(1.11) \quad Ax + Bu + f = 0$$

where $j(x, u)$ is l.s.c. and convex with respect to (x, u) and the operators A and B are linear (with A^{-1} and B bounded).

3. The practical choice of the sequence $\{\varepsilon^k\}$ is discussed in chapter 3.

Analogously to [1] we can prove the equivalence to the gradient projection method

$$(1.12) \quad u^k = P_U \left(u^{k-1} - \lambda J'(u^{k-1}) \right)$$

for a particular case:

Theorem 3: *If the optimal control problem (1.1), (1.2) has the form*

$$(1.13) \quad J(u) = j(x) + \frac{K}{2}\|u\|_2^2 \to \min_{u \in U} \qquad (K \geq 0)$$

$$(1.14) \quad \text{subject to} \qquad A(x) + Bu + f = 0$$

where B is a linear operator $\in H_2 \to H_3$, then the given algorithm yields the gradient projection method (1.12) in the space H_2.

Proof: The given algorithm can be written as $u^k = P_U \left(u^{k-1} - \varepsilon^k H_u(x^k, u^k, p^{k-1}) \right)$ (see proof of theorem 2), i.e. u^k is characterized by $\langle u^{k-1} - \varepsilon^k H_u(x^k, u^k, p^{k-1}) - u^k, v - u^k \rangle_2 \leq 0$. With $H_u = Ku + B^T p$ we obtain $\langle (1 + \varepsilon^k K)u^k - (1 + \varepsilon^k K)u^{k-1} + \varepsilon^k(Ku^{k-1} + B^T p^{k-1}), v - u^k \rangle_2 \geq 0$, i.e. $\langle u^k - u^{k-1} + \frac{\varepsilon^k}{(1+\varepsilon^k K)} J'(u^{k-1}), v - u^k \rangle_2 \geq 0$ where $J'(u) = H_u = Ku + B^T p$. This is the property of the control u^k obtained by the gradient projection method
$u^k = P_U(u^{k-1} - \frac{\varepsilon^k}{(1+\varepsilon^k K)} J'(u^{k-1}))$. \hfill q.e.d.

By using theorem 3 we can show under some additional assumptions that the sequence $\{u^k\}$ generated by the given algorithm converges strongly in geometric progression to the unique optimal control u^o:

Theorem 4: *Let be for the problem (1.13), (1.14) $j(x)$ convex with a Lipschitz-continuous gradient, $K > 0$, U convex and the operator A linear. Furthermore we assume that $\gamma^k = \frac{\varepsilon^k}{(1+\varepsilon^k K)}$ satisfies the inequality $0 < \gamma^k < \min\{\frac{\varepsilon_o}{(1+\varepsilon_o K)}, \frac{2}{L}\}$ where L ($> K$) denotes the Lipschitz constant of $J'(u)$. Then we have the strong convergence for the sequence $\{u^k\}$:*

$$(1.15) \quad u^k \to u^o \qquad with \ J(u^o) = \min_{u \in U} J(u)$$

which converges in geometric progression, i.e. $\|u^{k+1} - u^o\|_2 \le q\|u^k - u^o\|_2$ ($0 < q < 1$).

Proof: Under the given hypothesis the functional $J(u)$ in (1.13) is strongly convex, i.e. we have $J(\lambda u^1 + (1 - \lambda)u^2) \le \lambda J(u^1) + (1 - \lambda)J(u^2) - \frac{K}{2}\lambda(1 - \lambda)\|u^1 - u^2\|_2^2$ for $\lambda \in [0, 1]$. By virtue of theorem 3 the given algorithm is equivalent to the gradient projection method and we can use the results of [7]: $\|u^{k+1} - u^o\|_2^2 \le \|u^k - u^o\|_2^2 - \gamma(2 - \gamma L)\langle J'(u^k) - J'(u^o), u^k - u^o\rangle_2 \le \|u^k - u^o\|_2^2 - \gamma(2 - \gamma L)(J(u^k) - J(u^o))$ for the projection method $u^{k+1} = P_U(u^k - \gamma J'(u^k))$ with $\gamma = \frac{\varepsilon^k}{(1+\varepsilon^k K)}$, which implies $J(u^k) \to J(u^o)$ for $k \to \infty$. Using the strong convexity of $J(u)$ we obtain the inequality $\|u^{k+1} - u^o\|_2^2 \le q\|u^k - u^o\|_2^2$ where $0 < q = (1 - \gamma K(2 - \gamma L)) < 1$, which implies that the sequence $\{u^k\}$ converges in geometric progression.   q.e.d.

Now we consider the following particular case of the problem (1.13), (1.14) in order to obtain the strong convergence of the generated sequence $\{(x^k, u^k)\}$ without convexity assumptions on the functional:

$$(1.16) \quad J(u) = \mu j(x) \to \min_{u \in U \subset H_2}$$

with a linear state equation of the form

$$(1.17) \quad Ax + Bu + f = 0$$

where A^{-1} and B are bounded and $\mu > 0$ is a given sufficiently small parameter.

We require the following hypothesis: The solution of the minimization problem in step 3 of the algorithm

$$(1.18) \quad u(p) = \arg\min_{u \in U} \left(\langle p, Bu\rangle_3 + \frac{1}{2\varepsilon}\|u - v\|_2^2\right)$$

is Lipschitz-continuous with respect to p, i.e. $\exists L > 0$ such that

$$(1.19) \quad \|u(p^1) - u(p^2)\|_2 \le L\|p^1 - p^2\|_3$$

Applying an idea of Popov [8] we have proved the strong convergence of the generated sequence $\{u^k\}$ (i.e. $u^k \to u^o$) for an arbitrary functional $j(x)$ in our paper [3].

The parameter μ in the cost functional (1.16) obviously has no influence on the solution of the problem (1.16), (1.17) but it plays an important role for the convergence of the given algorithm. The generated sequence $\{u^k\}$ is also independent of μ, since the optimization problem (1.18) can be written in the following form using the adjoint equation $A^{\mathsf{T}}p = \mu j_x(x^k)$:

$$
\begin{aligned}
u^k &= \arg\min_{u\in U}\left(\langle A^{T-1}\mu j_x(x^{k-1}), Bu\rangle_3 + \frac{1}{2\varepsilon^k}\|u - u^{k-1}\|_2^2\right) \\
&= \arg\min_{u\in U}\left(\langle A^{T-1}j_x(x^{k-1}), Bu\rangle_3 + \frac{1}{2\varrho^k}\|u - u^{k-1}\|_2^2\right)
\end{aligned}
$$

where $\varrho^k = \mu\varepsilon^k$. That means that the parameter μ has no influence on the algorithm. We only have to determine the sequence $\{\varrho^k\}$ with $0 < \mu\varepsilon_1 < \varrho^k < \mu\varepsilon_o$ and $\liminf_{k\to\infty}\varrho^k > 0$, i.e. ϱ^k is proportional to ε^k.

Remarks:

1. Under the given assumptions the associated sequences $\{x^k\}$ and $\{p^k\}$ are also convergent, i.e. $x^k \to x^o$ and $p^k \to p^o$, and the relations

$$
\begin{aligned}
(1.20)\quad Ax^o + Bu^o + f &= 0 \\
(1.21)\qquad\qquad A^{\mathsf{T}}p^o &= \mu j_x(x^o) \\
(1.22)\qquad\qquad u^o &= \arg\min_{u\in U}\langle p^o, Bu\rangle_3
\end{aligned}
$$

   hold, which form for the given problem (1.16), (1.17) a necessary and, if $j(x)$ is convex, also a sufficient optimality condition.

2. If the operator A in the state equation (1.17) is nonlinear, then we must suppose that A has a Fréchet-derivative A_x with a continuous inverse A_x^{-1} and that his inverse A^{-1} is Lipschitz-continuous in order to ensure the convergence of $\{u^k\}$.

2 Application to a distributed parameter system in Lebesgue spaces

We apply the algorithm of section 1 to a distributed parameter system governed by the Volterra functional-operator equation (see [6])

$$
(2.1)\quad x(t) = f(t, A(x)(t), u(t)), \qquad t = (t_1,\ldots,t_n) \in T \subset \Re^n
$$

where $A(\cdot) \in L_m^1 \mapsto L_l^1$ is a regular bounded linear operator, $u \in D_U = \{u \in L_s^\infty \mid u(t) \in U \subset \Re^s, U \text{ bounded}\}$ and $f(\cdot,\cdot,\cdot) \in T \times \Re^l \times \Re^s \mapsto \Re^m$. $L_m^p(T)$ denotes the Lebesgue space of the vector functions $x(t)$ with the norm $\|x\|_{p,m} = (\int_T |x(t)|^p dt)^{1/p}$ for $1 \le p < \infty$ and $\|x\|_{\infty,m} = \operatorname{ess\,sup}_{t \in T} |x(t)|$ where $|\cdot|$ is the Euclidean norm.

The requirements imposed on f (condition of Caratheodory-type) are given next: $f(t,x,u)$ is differentiable with respect to x for all u and almost all t and with respect to u for all x and almost all t, and together with its derivatives it is measurable in t for all (x,u) and Lipschitz-continuous in (x,u) for almost all t.

In [6] the existence of a unique solution $x_u \in L_m^\infty$ for each $u \in D_U$ is proved and the following problems of mathematical physics are transformed to equations of the class (2.1):

- the Cauchy problem and the characteristic problem for hyperbolic equations and systems,

- mixed problems for nonlinear integro-differential equations,

- the Goursat problem and

- the first boundary value problem for a semilinear parabolic equation.

The optimal control problem consists of finding a control $u \in D_u$ that minimizes the functional

$$(2.2) \qquad J(u) = G(x_u(t)) = \int_T g(x(t))dt$$

subject to the state equation (2.1), where it is supposed that the functional G has a Lipschitz-continuous Fréchet-derivative which is regular at each point $x \in L_m^\infty$, i.e., as an element of $L_m^{\infty\prime}$ it is identified with the function $g_x(x(t))$. For the optimal control problem (2.1), (2.2) the following maximum principle is given by Plotnikov and Sumin [6]:

Let be $u^o \in D_U$ an optimal control and $x^o \in L_m^\infty$ the associated solution of the state equation (2.1). Then there exists a function $p^o \in L_m^\infty$ satisfying the adjoint equation

$$(2.3) \qquad p^o(t) - A^\mathsf{T}\left(f_x^\mathsf{T}(t, A(x^o), u^o)p^o\right)(t) = -g_x(x^o(t))$$

such that

$$(2.4) \qquad p^{o\mathsf{T}}(t)f(t, A(x^o)(t), u^o(t)) = \sup_{v \in U} p^{o\mathsf{T}}(t)f(t, A(x^o)(t), v)$$

for almost all $t \in T$.

Defining the Hamiltonian H by

$$(2.5) \qquad H(x, u, p, t) = p^\top(t)\left(x(t) - f(t, A(x)(t), u(t))\right) + g(x(t))$$

and the augmented Hamiltonian H^ε by

$$(2.6) \qquad H^\varepsilon(x, u, v, p, t) = H(x, u, p, t) + \frac{1}{2\varepsilon}|u(t) - v(t)|^2$$

we can apply the algorithm from section 1. Now the condition (1.3) is needed in the form

$$(2.7) \qquad |\Delta x(t)| \leq C|\Delta u(t)|$$

and is proved under some conditions on the operator A in [6].

Lemma: *Under the given assumptions the following inequality holds*

$$(2.8) \qquad J(u^k) - J(u^{k-1}) \leq -\left(\frac{1}{\varepsilon^k} - K\right) \int_T |\Delta u^k(t)|^2 dt \quad (K > 0).$$

i.e. the sequence $\left\{J(u^k)\right\}$ generated by the algorithm is monotone decreasing for $\varepsilon^k > 0$ sufficiently small.

Proof: With $J(u^k) - J(u^{k-1}) = \int_T \left(H(x^k, u^k, p^{k-1}, t) - H(x^{k-1}, u^{k-1}, p^{k-1}, t)\right) dt$ the proof is the same as in theorem 1 if we replace the Fréchet-derivatives by the partial derivatives.

Remarks:

1. By applying the inequality (2.8) the property (1.8) of theorem 2 is now obtained in the norm of $L^2(T)$: $\|u^k - u^{k-1}\|_{L^2} \to 0 \quad for \ k \to \infty$.

2. The given Lemma shows that the main property (2.8) of the algorithm can be obtained without the explicit requirement of the Fréchet-derivability of the state equation if we consider control problems in certain spaces of bounded functions such as L_m^∞.

3. Furthermore we don't use the metric of a Hilbert space (i.e. L_m^2) how it is convenient and customary for the gradient projection and conditional gradient methods. Therefore we don't need the very heavy demands on the state equation that are necessary for the existence of the gradient in L_m^2. In nonlinear problems these demands are often not met.

3 Some computational results

The given algorithm is applicable to systems with lumped or distributed parameters which can be solved step by step like initial value problems because one must simultaneously solve in step 3 the state equation (1.2) and minimize the augmented Hamiltonian (1.5). Such systems are e.g. evolution equations or some hyperbolical systems. We remarked that the algorithm can be modified, if in step 3 we first minimize the augmented Hamiltonian and then solve the state equation. This modified algorithm has the advantage that it is also applicable to boundary value problems and that one can use standard programs for solving differential equations.

In Sakawa, Shindo [9] the sequence $\{\varepsilon^k\}$ is generated in the following way: if $J(u^k) < J(u^{k-1})$ let $\varepsilon^{k+1} := K_1\varepsilon^k$, $K_1 > 1$, otherwise let $\varepsilon^k := K_2\varepsilon^k$, $0 < K_2 < 1$ and repeat step 3. K_1 and K_2 are constants (suitable values are $K_1 = 1.2$, $K_2 = 0.5$). It is no problem to start with a large ε (about 100) because after some steps it will be small enough. Furthermore for large ε the algorithm is similar to the Chernousko algorithm [4]. If ε is too small the cost functional decreases very slowly for many steps. Numerical tests show that it is typical that ε^k increases for some steps, then decreases and so on and it does not tend to zero.

We have also tested another step-size-strategy: find the optimal ε^k in every step, that means $\varepsilon^k = \arg\min_{\varepsilon\in[0,\infty)} J(u^k(\varepsilon))$. In practice this can be done by a quadratic interpolation method: Choose ε_1, ε_2 (e.g. $\varepsilon_1 = \varepsilon^{k-1}$, $\varepsilon_2 = 2\varepsilon_1$) and calculate $J(u^k(\varepsilon_1))$ and $J(u^k(\varepsilon_2))$. From the last step is known $J(u^k(0)) = J(u^{k-1})$. Now we can interpolate these values by a quadratic function. The optimal ε^k will be approximated by the minimum of this function. Numerical tests have shown, that the second step-size-strategy takes about the double amount of time for one step but less then half of the number of steps. Because $J(u^k(\varepsilon_1))$ and $J(u^k(\varepsilon_2))$ can be computed independently it is possible to get a speed up with a parallel computer.

Both step-size-strategies are also applicable to the gradient projection method.

We have tested the algorithm numerically for various hyperbolic systems with various boundary conditions:

$$1a) \qquad x_{ts}(t,s) = x_t(t,s) + u(t,s)$$

$$1b) \qquad x_{ts}(t,s) = x(t,s)^2 u(t,s) + f(t,s)$$

$$(t,s) \in G = [0,1] \times [0,1]$$

$$U = \{u|\ |u(t,s)| \leq 1 \quad \forall (t,s) \in G\}$$

$$2a) \qquad x(0,s) = x(s,0) = 0 \qquad s \in [0,1]$$

$$2b) \quad x(0,s) = -x(s,0) = \sin(4\pi s)/9, \qquad s \in [0,1]$$

$$3a) \qquad J(u) = \int \int_G u(t,s)^2 dt ds + \mu(x(1,1) - 1)^2$$

3b) $\quad J(u) = \int \int_G u(t,s)x(t,s)dtds + \mu(x(1,1)-1)^2$

3c) $\quad\quad J(u) = \mu \int \int_G (x(t,s) - r(t,s))^2 dtds$

where $r(t,s)$ and $f(t,s)$ are given functions. In all cases the algorithm converges to a stationary solution. If $r(t,s)$ in 3c) is the solution of the state equation for a given $u^\star \in U$ the optimal solution is obviously $u^o \equiv u^\star$, $x^o \equiv r$. For this singular case the Chernousko-algorithm does not converge, but the algorithm given in section 1 converges.

Figure 1 shows the generated sequences $\{\varepsilon^k\}$ and $\{ln(J(u^k))\}$ for the test example 1a), 2b) and 3c) with $u^1 \equiv 0$, $\varepsilon^1 = 1$, $K_1 = 1.2$, $K_2 = 0.5$, $\mu = 1$ and $u^\star(t,s) = \text{sig}(t - 0.5)(s - 0.5)$, a discretization of the domain G in 50×50 parts and the first step-size-strategy. Figure 2 shows the same with the second step-size-strategy.

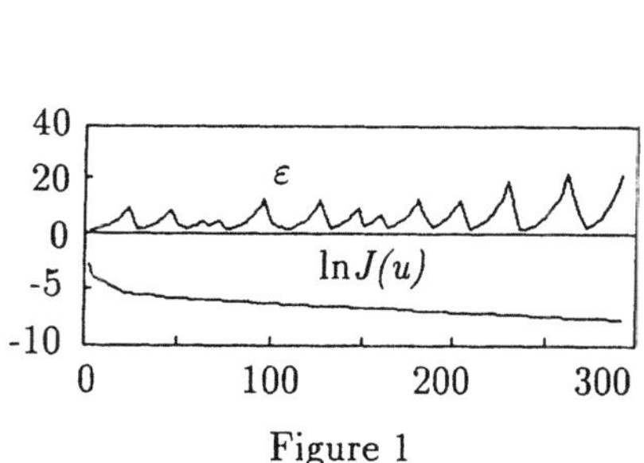

Figure 1

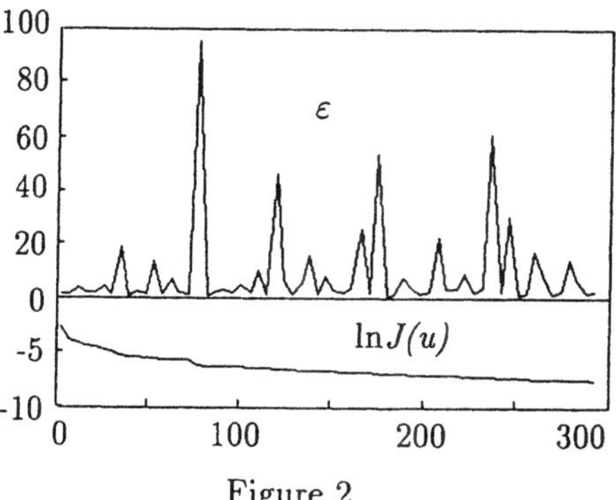

Figure 2

References

[1] Bonnans, J.F. (1986) *On an algorithm for optimal control using Pontryagin's maximum principle*. SIAM J. Control and Opt. 24/3, 579 – 588

[2] Benker, H., Handschug, M. (1990) *Numerical methods for distributed optimal control problems*. Preprint-Reihe Mathematik Univ. Greifswald, Nr. 27, 3 – 5

[3] Benker, H., Handschug, M. (1991) *Algorithms for general optimal control problems using maximum principles*. (in preparation)

[4] Chernousko, F.L., Lyubushin, A.A. (1982) *Method of successive approximations for solution of optimal control problems*. Opt. Contr. Appl. and Methods, vol. 3, 101 – 114

[5] Kazemi-Dehkordi, M.A. (1988) *A method of successive approximation for optimal control of distributed parameter systems*. J. Math. Anal. Appl. 133/2, 484 – 497

[6] Plotnikov, V.I., Sumin, V.I. (1981) *Optimization of distributed parameter systems in Lebesgue-spaces*. (in russian) Sibirskij Mat. Žurn. 6, 142 – 161

[7] Poljak, B.T. (1983) *Introduction to optimization*. (in russian) Moscow, Nauka

[8] Popov, V.A. (1989) *Convergence of the successive approximation method for some optimal control problems*. (in russian) Izvestia Vyšich Učebnych Savedenija, Mat., No. 4, 55 – 61

[9] Sakawa, Y., Shindo, Y. (1980) *On global convergence of an algorithm for optimal control*. IEEE Trans. Autom. Comput. 25, 1149 – 1153

[10] Sakawa, Y., Stachurski (1989) *Convergence properties of an algorithm for solving non-differentiable optimal control problems*. Num. Funct. Anal. and Opt. 10, 765 – 786

Authors' address

Prof. Hans Benker and Dr. Michael Handschug
Technische Hochschule Merseburg
Fachbereich Mathematik und Informatik
Geusaer Straße
D-O-4200 Merseburg
Germany

International Series of Numerical Mathematics, Vol. 111, ©1993 Birkhäuser Verlag Basel

Convexification of Control Problems in Evolution Equations

Winfried Kampowsky and Uldis Raitums

Abstract

This paper deals with optimal control problems governed by nonlinear parabolic differential equations of divergence type including integral constraints. The parabolic initial-boundary value problems are considered as evolution equations in Banach spaces which can be solved within the framework of monotone operators. Using a method of convexification it is possible to pass over to an extended, socalled convexificated control problem. As a first result, optimality conditions in form of a strong minimum principle can be derived for the original control problem without any assumption of convexity and differentiability with respect to the controls.

1 The Problem of Optimal Control

Let S be the time interval $[0, T]$ and let Ω be a bounded domain of $\mathbf{R}^n$ of class $\mathbb{C}^{0,1}$. $\partial\Omega$ denotes the boundary of Ω, and let $\partial_1\Omega$, $\partial_2\Omega \subseteq \partial\Omega$ be measurable sets with $\partial_1\Omega \cup \partial_2\Omega = \partial\Omega$ and $\partial_1\Omega \cap \partial_2\Omega = \emptyset$.

We consider the following parabolic differential equation of second order of divergence type

$$(1.1) \quad \frac{\partial x}{\partial t}(t, s) - \sum_{i=1}^{n} \frac{\partial}{\partial s_i} a_i(t, s, x(t, s), \operatorname{grad}_s x(t, s), u_\Sigma(t, s))$$

$$+ a_0(t, s, x(t, s), \operatorname{grad}_s x(t, s), u_\Sigma(t, s)) = c(t, s), \quad (t, s) \in S \times \Omega = \Sigma,$$

the initial condition

$$(1.2) \quad x(0, s) = x_0(s), \quad s \in \Omega,$$

a homogeneous boundary condition of *Dirichlet* type on the boundary part $\partial_1\Omega$

$$(1.3) \qquad x(t,s) = 0, \quad t \in S, \quad s \in \partial_1\Omega,$$

and a boundary condition of *Neumann* type on the boundary part $\partial_2\Omega$

$$(1.4) \qquad \frac{\partial x}{\partial \nu_A}(t,s) + a_\Gamma(t,s,x(t,s),u_\Gamma(t,s)) = d(t,s), \quad (t,s) \in S \times \partial_2\Omega = \Gamma,$$

where $\frac{\partial x}{\partial \nu_A}(t,s)$ denotes the derivative in direction of the outer conormal. [1]
The vector $u = (u_\Sigma, u_\Gamma)$ of admissible control functions belongs to the set U_0 given by

$$(1.5) \qquad U_0 = \Big\{ u = (u_\Sigma, u_\Gamma) \ : \ u_\Sigma : \Sigma \to \mathbf{R}^{r_1} \text{ is measurable and } u_\Sigma(t,s) \in D_\Sigma \text{ a.e.},$$
$$u_\Gamma : \Gamma \to \mathbf{R}^{r_2} \text{ is measurable and } u_\Gamma(t,s) \in D_\Gamma \text{ a.e.}\Big\},$$

and $D_\Sigma \subset \mathbf{R}^{r_1}$ and $D_\Gamma \subset \mathbf{R}^{r_2}$ are given bounded sets.
We consider the following cost functional of *Lagrange* type

$$(1.6) \qquad F_0(x(.), u_\Sigma(.), u_\Gamma(.))$$

$$= \int_\Sigma f_0(t,s,x(t,s),\mathrm{grad}_s x(t,s), u_\Sigma(t,s))\, ds\, dt$$

$$+ \int_\Gamma g_0(t,s,x(t,s),u_\Gamma(t,s))\, ds\, dt,$$

and inequality constraints of the same integral type

$$(1.7) \qquad F_j(x(.), u_\Sigma(.), u_\Gamma(.))$$

$$= \int_\Sigma f_j(t,s,x(t,s),\mathrm{grad}_s x(t,s), u_\Sigma(t,s))\, ds\, dt$$

$$+ \int_\Gamma g_j(t,s,x(t,s),u_\Gamma(t,s))\, ds\, dt \leq 0, \quad j = 1,...,m.$$

---

[1] Let us recall that

$$\frac{\partial x}{\partial \nu_A}(t,s) = \sum_{i=1}^{n} a_i(t,s,x(t,s),\mathrm{grad}_s x(t,s), u_\Sigma(t,s))\, \cos(\nu_s; s_i),$$

where ν_s is the unit vector of the outer normal in $s \in \partial_2\Omega$.

Our problem of optimal control is to minimize $F_0(x(.), u(.))$ where $(x(.), u(.))$ fulfils (1.1) – (1.4) and (1.7), and $u = (u_\Sigma, u_\Gamma) \in U_0$.

Typical problems of this class are control problems governed by the wellknown heat equation or, more general, the diffusion equation.

Let us formulate the functionally-analytical generalization.

We define the following *Banach* spaces.

$$V = \{v \in W_p^{(1)}(\Omega) : v|_{\partial_1 \Omega} \equiv 0\}, \quad H = L_2(\Omega), \quad X = L_p(S; V),$$

where $1 < p < \infty$. $\langle\langle .; . \rangle\rangle$ denotes the bilinear form on $X^* \times X$.

Let the abstract *Sobolev* space W consist of all $x \in X$ with a derivative $x' \in X^*$ in the sense of V^*-valued distributions.

Let $U = [L_\infty(\Sigma)]^{r_1} \times [L_\infty(\Gamma)]^{r_2}$ be the space of all control functions $u = (u_\Sigma, u_\Gamma)$.

Defining the operator $A : X \times U \longrightarrow X^*$ by

$$(1.8) \qquad \langle\langle A(x, u); y \rangle\rangle$$

$$= \int_\Sigma \{ \sum_{i=1}^n a_i(t, s, x(t, s), \mathrm{grad}_s x(t, s), u_\Sigma(t, s)) \frac{\partial y}{\partial s_i}(t, s)$$

$$+ a_0(t, s, x(t, s), \mathrm{grad}_s x(t, s), u_\Sigma(t, s)) \, y(t, s) \} \, ds \, dt$$

$$+ \int_\Gamma a_\Gamma(t, s, x(t, s), u_\Gamma(t, s)) \, y(t, s) \, ds \, dt, \quad x, y \in X, \quad u \in U,$$

defining the right hand side $b \in X^*$ by

$$(1.9) \qquad \langle\langle b; y \rangle\rangle = \int_\Sigma c(t, s) \, y(t, s) \, ds \, dt + \int_\Gamma d(t, s) \, y(t, s) \, ds \, dt, \quad y \in X,$$

and assuming that the initial data

$$(1.10) \qquad x_0 \in H,$$

we consider the evolution equation

$$(1.11) \qquad x' + A(x, u) = b, \quad x(0) = x_0, \quad x \in W,$$

as the functionally-analytical generalization of the initial-boundary value problem (1.1) – (1.4).

Therefore, our generalized problem of optimal control is to minimize $F_0(x, u)$ where (x, u) fulfills (1.11) and (1.7), and $u = (u_\Sigma, u_\Gamma) \in U_0$, i.e.

$$(1.12) \qquad F_0(x, u) = \min!$$

$$x' + A(x, u) = b, \quad x(0) = x_0, \quad x \in W,$$

$$F_j(x, u) \le 0, \quad j = 1, ..., m, \quad u \in U_0.$$

Remark 1: Obviously, the definition of the generalized control problem (1.12) is correct if the coefficient functions of the operator $A : X \times U \to X^*$ a_i, $i = 0, 1, ..., n$, define mappings from $[L_p(\Sigma)]^{n+1} \times [L_\infty(\Sigma)]^{r_1}$ into $L_q(\Sigma)$, where $\frac{1}{p} + \frac{1}{q} = 1$, and a_Γ defines a mapping from $L_p(\Gamma) \times [L_\infty(\Gamma)]^{r_2}$ into $L_q(\Gamma)$, if $c \in L_q(\Sigma)$, and $d \in L_q(\Gamma)$, if the integrand functions of the cost functional and of the integral functionals of the inequality constraints $F_j : X \times U \to \mathbf{R}$, $j = 0, 1, ..., m$, f_j map from $[L_p(\Sigma)]^{n+1} \times [L_\infty(\Sigma)]^{r_1}$ into $L_1(\Sigma)$ and g_j map from $L_p(\Gamma) \times [L_\infty(\Gamma)]^{r_2}$ into $L_1(\Gamma)$.
Following the theory of *Nemytskij* operators, compare [4], [7], or [11], we assume that the coefficient functions a_i, $i = 0, 1, ..., n$, , a_Γ and the integrand functions f_j, g_j, $j = 0, 1, ..., m$, satisfy the corresponding *Caratheodory* conditions and growth conditions. Therefore, we can assume that the operator A and the functionals F_j are well-defined, and moreover, they are continuous.

2   Convexification of the Control Problem

Let us consider the following convexification of the control problem (1.12).

$$(2.1) \qquad \sum_{l=1}^{k} \lambda_l F_0(x, u_l) = \min!$$

$$x' + \sum_{l=1}^{k} \lambda_l A(x, u_l) = b, \quad x(0) = x_0, \quad x \in W,$$

$$\sum_{l=1}^{k} \lambda_l F_j(x, u_l) + \left(\sum_{l=2}^{k} \lambda_l\right)^2 \le 0, \quad j = 1, ..., m$$

$$u_l \in U_0, \quad \lambda_l \ge 0, \quad l = 1, ..., k, \quad \sum_{l=1}^{k} \lambda_l = 1, \quad k = 1, 2, ... $$

This procedure to extend a given control problem including the technical basis of this convexification is similiar to the method developed by *Raitums* [9], [10] for control problems governed by elliptic differential equations.

Clearly, the original problem (1.12) is a subproblem of this socalled convexificated problem (2.1). Yet, conversly, every cost functional value of the convexificated problem can be approximated by a sequence of cost functional values of the original problem. The following theorem holds.

Theorem 1: Let the following assumption of solvability of the evolution equation (1.11) be fulfilled:

There exists a neighbourhood of the right hand side $b \in X^*$ such that the evolution equation

$$x' + A(x, u) = \bar{b}, \quad x(0) = x_0, \quad x \in W,$$

has an unique solution $x = x(u, \bar{b})$ for all $u \in U_0$ and $\bar{b}$ of the neighbourhood. Moreover,

$$x(u, \bar{b}) \to x(u, b) \quad \text{in } X \quad \text{as} \quad \bar{b} \to b \quad \text{in } X^*,$$

uniformly with respect to $u \in U_0$. Then for an admissible solution $(x, u_1, ..., u_k, \lambda_1, ..., \lambda_k)$ of (2.1) there exists a sequence of admissible solutions (x^ν, u^ν) of (1.12) such that

$$F_0(x^\nu, u^\nu) \longrightarrow \sum_{l=1}^{k} \lambda_l F_0(x, u_l) \quad \text{as} \quad \nu \to \infty.$$

Proof: Let $(\sum_{l=2}^{k} \lambda_l)^2 > 0$, in the other case is nothing to show.

Let $(x, u_1, ..., u_k, \lambda_1, ..., \lambda_k)$ be an admissible solution of the control problem (2.1) where $u_l = (u_{l,\Sigma}, u_{l,\Gamma})$, $l = 1, ..., k$.

Applying a basic technical lemma of convexification, see *Kampowsky, Raitums* [6], we can found sequences of functions $\alpha_l^\nu \in L_2(\Sigma)$, $l = 1, ..., k$, with values 0 and 1, only, and with $\sum_{l=1}^{k} \alpha_l^\nu(t, s) \equiv 1$ on Σ, and $\beta_l^\nu \in L_2(\Gamma)$, $l = 1, ..., k$, with values 0 and 1, only, and with $\sum_{l=1}^{k} \beta_l^\nu(t, s) \equiv 1$ on Γ, such that

$$(2.2) \qquad \alpha_l^\nu \rightharpoonup \lambda_l \quad \text{in } L_2(\Sigma), \quad \beta_l^\nu \rightharpoonup \lambda_l \quad \text{in } L_2(\Gamma) \quad \text{as} \quad \nu \to \infty, \quad l = 1, ..., k.$$

It follows directly that for given functions $c_l \in L_q(\Sigma)$, $d_l \in L_q(\Gamma)$, $l = 1, ..., k$,

$$(2.3) \qquad \sum_{l=1}^{k} \alpha_l^\nu c_l \rightharpoonup \sum_{l=1}^{k} \lambda_l c_l \quad \text{in } L_q(\Sigma) \quad \text{as} \quad \nu \to \infty,$$

$$\sum_{l=1}^{k} \beta_l^{\nu} d_l \rightharpoonup \sum_{l=1}^{k} \lambda_l d_l \quad \text{in } L_q(\Gamma) \quad \text{as} \quad \nu \to \infty.$$

Setting

$$c_l(t,s) = a_0(t,s,x(t,s),\mathrm{grad}_s(t,s),u_{l,\Sigma}(t,s)),$$

$$d_l(t,s) = a_\Gamma(t,s,x(t,s),u_{l,\Gamma}(t,s)),$$

and defining new admissible control functions u_Σ^{ν} on Σ and u_Γ^{ν} on Γ by

$$u_\Sigma^{\nu}(t,s) = u_{l,\Sigma}(t,s) \quad \text{iff} \quad \alpha_l^{\nu}(t,s) = 1,$$

$$u_\Gamma^{\nu}(t,s) = u_{l,\Gamma}(t,s) \quad \text{iff} \quad \beta_l^{\nu}(t,s) = 1,$$

we get

$$\sum_{l=1}^{k} \alpha_l^{\nu} a_0(.,x,\mathrm{grad}_s x, u_{l,\Sigma}) = a_0(.,x,\mathrm{grad}_s x, u_\Sigma^{\nu})$$

$$\rightharpoonup \sum_{l=1}^{k} \lambda_l a_0(.,x,\mathrm{grad}_s x, u_{l,\Sigma}) \quad \text{in } L_q(\Sigma),$$

$$\sum_{l=1}^{k} \beta_l^{\nu} a_\Gamma(.,x,u_{l,\Gamma}) = a_\Gamma(.,x,u_\Gamma^{\nu}) \rightharpoonup \sum_{l=1}^{k} \lambda_l a_\Gamma(.,x,u_{l,\Gamma}) \quad \text{in } L_q(\Gamma).$$

We do this for all the other coefficient functions a_i, $i = 1,...,n$, and for all integrand functions f_j, $j = 0,1,...,m$, respectively, for all integrand functions g_j, $j = 0,1,...,m$. In others words, we can assume a sequence of admissible control functions $u^{\nu} = (u_\Sigma^{\nu}, u_\Gamma^{\nu})$ such that

$$(2.4) \qquad A(x,u^{\nu}) \rightharpoonup \sum_{l=1}^{k} \lambda_l A(x,u_l) \quad \text{in } X^* \quad \text{as} \quad \nu \to \infty,$$

$$(2.5) \qquad F_j(x,u^{\nu}) \to \sum_{l=1}^{k} \lambda_l F_j(x,u_l) \quad \text{as} \quad \nu \to \infty, \quad j = 0,1,...,m.$$

Yet, it is very essential that the construction of the sequences α_l^ν and β_l^ν can be done in such a way that there also exist sequences of functions $\bar{a}_i^\nu \in L_q(\Sigma)$, $i = 0, 1, ..., n$, and $\bar{a}_\Gamma^\nu \in L_q(\Gamma)$ such that

$$\bar{a}_i^\nu \to \sum_{l=1}^{k} \lambda_l \, a_i(., x, \, \mathrm{grad}_s x, u_{l,\Sigma}) \quad \text{in } L_q(\Sigma) \quad \text{as} \quad \nu \to \infty, \quad i = 0, 1, ..., n,$$

$$\bar{a}_\Gamma^\nu \to \sum_{l=1}^{k} \lambda_l \, a_\Gamma(., x, u_{l,\Gamma}) \quad \text{in } L_q(\Gamma) \quad \text{as} \quad \nu \to \infty,$$

and, that is the point, such that [2]

$$\langle a_0(., x, \mathrm{grad}_s x, u_\Sigma^\nu) - \bar{a}_0^\nu; y \rangle_\Sigma + \sum_{i=1}^{n} \langle a_i(., x, \mathrm{grad}_s x, u_\Sigma^\nu) - \bar{a}_i^\nu; \frac{\partial y}{\partial s_i} \rangle_\Sigma$$

$$+ \langle a_\Gamma(., x, u_\Gamma^\nu) - \bar{a}_\Gamma^\nu; y \rangle_\Gamma = 0 \qquad \text{for all } y \in X \quad \text{and} \quad \nu = 1, 2, ... \, .$$

In other words, defining $\bar{a}^\nu \in X^*$ in the usual way by $(\bar{a}_0^\nu, \bar{a}_1^\nu, ..., \bar{a}_n^\nu, \bar{a}_\Gamma^\nu)$, we can assume that

$$(2.6) \qquad \bar{a}^\nu \to \sum_{l=1}^{k} \lambda_l \, A(x, u_l) \quad \text{in } X^* \quad \text{as} \quad \nu \to \infty,$$

$$(2.7) \qquad \langle\langle A(x, u^\nu) - \bar{a}^\nu; y \rangle\rangle = 0 \qquad \text{for all } y \in X \quad \text{and} \quad \nu = 1, 2, ... \, .$$

From the assumption of the theorem we can suppose the unique solution x^ν of the evolution equation

$$(2.8) \qquad (x^\nu)' + A(x^\nu, u^\nu) = b, \quad x^\nu(0) = x_0, \quad x^\nu \in W, \quad \nu = 1, 2, ... \, .$$

Using the relation (2.7) we can transform the evolution equation

$$x' + \sum_{l=1}^{k} \lambda_l \, A(x, u_l) = b, \quad x(0) = x_0,$$

to

$$(2.9) \qquad x' + A(x, u^\nu) = b + \bar{a}^\nu - \sum_{l=1}^{k} \lambda_l \, A(x, u_l), \quad x(0) = x_0.$$

---

[2]Let $\langle .; . \rangle_\Sigma$ and $\langle .; . \rangle_\Gamma$ denote the usual bilinear forms between $L_q(\Sigma)$ and $L_p(\Sigma)$, respectively, between $L_q(\Gamma)$ and $L_p(\Gamma)$.

Obviously, $x^\nu = x(u^\nu, b)$ and $x = x(u^\nu, b + \bar{a}^\nu - \sum_{l=1}^k \lambda_l A(x, u_l))$, therefore, from (2.7) and the assumption of the theorem we obtain

$$(2.10) \quad x^\nu \to x \quad \text{in } X \quad \text{as} \quad \nu \to \infty.$$

Let us show that

$$(2.11) \quad F_j(x^\nu, u^\nu) \to \sum_{l=1}^k \lambda_l F_j(x, u_l) \quad \text{as} \quad \nu \to \infty, \quad j = 0, 1, ..., m.$$

Applying (2.5) we have only to show that

$$(2.12) \quad F_j(x^\nu, u^\nu) - F_j(x, u^\nu) \to 0 \quad \text{as} \quad \nu \to \infty, \quad j = 0, 1, ..., m.$$

This follows from the estimation

$$|F_j(x^\nu, u^\nu) - F_j(x, u^\nu)|$$

$$\leq \sum_{l=1}^k \int_\Sigma |f_j(t, s, x^\nu(t, s), \mathrm{grad}_s x \nu(t, s), u_{l,\Sigma}(t, s))$$

$$- f_j(t, s, x(t, s), grad_s x(t, s), u_{l,\Sigma}(t, s))|\, ds\, dt$$

$$+ \sum_{l=1}^k \int_\Gamma |g_j(t, s, x^\nu(t, s), u_{l,\Gamma}(t, s)) - g_j(t, s, x(t, s), u_{l,\Gamma}(t, s))|\, ds\, dt\,,$$

and from the continuity of the *Nemytskij* operators defined by f_j and g_j, see remark 1. In all, noticing (2.8), (2.11), and that $(\sum_{l=2}^k \lambda_l)^2 > 0$ holds, the proof is complete. •

Corollary: If the assumptions of theorem 1 are fulfilled, then the infima of the original problem (1.12) and of the convexificated problem (2.1) are the same.

Remark 2:   The assumptions of theorem 1 are satisfied if we consider the evolution equation (1.11) using the theory of monotone operators, and if we assume, for instance, that $A(., u) : X \to X^*$ is generalized uniformly monotone, i.e.

$$(2.13) \quad \langle\langle A(x, u) - A(y, u); x - y \rangle\rangle + \lambda \|x - y\|^2_{L_2(S;H)} \geq \rho(\|x - y\|_X)$$

$$\text{for all } x, y \in X, \ u \in U_0,$$

where $\rho : [0, \infty[\to [0, \infty[$ is a strictly monotone increasing function with $\rho(0) = 0$, and $\lambda \geq 0$ and ρ are independend of $u \in U_0$.

The proof of this assertion is standard in the framework of evolution equations with monotone operators, see [8], [3], compare also [6].

3 Optimality Conditions

A first result using the connection between the original problem (1.12) and the convexificated problem (2.1) concerns optimality conditions. The result on necessary optimality conditions is the following one.

Theorem 2: Let $(x^*, u^*) \in W \times U_c$ be optimal for problem (1.12) and let the following assumptions be fulfilled: [3]

The operator $A(., u) : X \to X^*$ is generalized strongly monotone, i.e.

$$(3.1) \qquad \langle\langle A(x, u) - A(y, u); x - y \rangle\rangle + \lambda \|x - y\|^2_{L_2(S;H)} \geq m \|x - y\|^2_X$$

$$\text{for all } x,\, y \in X,\ u \in U_0,$$

with constants $\lambda \geq 0$ and $m > 0$ independend of $u \in U_0$.

For each $u \in U_0$ the operator $A(., u) : X \to X^*$ is *Gâteaux* differentiable in a neighbourhood of x^*, and the *Gâteaux* derivative $A_x(., u)$ is pointwise continuous in x^*.

For each $u \in U_0$ the functionals $F_j(., u) : X \to \mathbf{R}$, $j = 0, 1, ..., m$, are *Fréchet* differentiable in a neighbourhood of x^*, and the *Fréchet* derivative $F_{j,x}$ is pointwise continuous in x^*.

Then the following minimum principle is valid:

There exist multipliers $\eta_j \geq 0$, $j = 0, 1, ..., m$, not all zero, such that

$$(3.2) \qquad \sum_{j=0}^{m} \eta_j\, F_j(x^*, u^*) - \langle\langle A(x^*, u^*); y \rangle\rangle$$

$$= \min\Big\{ \sum_{j=0}^{m} \eta_j\, F_j(x^*, u) - \langle\langle A(x^*, u); y \rangle\rangle \mid u \in U_0 \Big\},$$

where y is the unique solution of the adjoint evolution equation

$$(3.3) \qquad -y' + A_x^*(x^*, u^*)\, y = \sum_{j=0}^{m} \eta_j\, F_{j,x}(x^*, u^*), \quad y(T) = o, \quad y \in W,$$

and the following slack conditions hold:

$$(3.4.) \qquad \eta_j\, F_j(x^*, u^*) = 0, \quad j = 1, ..., m.$$

---

[3]In this paragraph we suppose $p = q = 2$.

Sketch of the proof: Applying theorem 1 including the corollary and remark 2 $(x^*, u^*, 1)$ is an optimal solution of the convexificated problem (2.1). Following a general concept to derive necessary optimality conditions, due to *Bittner* [1], [2], for control problems with evolution equations see *Kampowsky* [5], compare also [6], we define an abstract set of admissible controls by

$$(3.5) \qquad \vec{U}_0 = \Big\{ \, \vec{u} = (u_1, ..., u_k, \lambda_1, ..., \lambda_k) \; : \; u_l = (u_{l,\Sigma}, u_{l,\Gamma}) \in U_0 \,, $$
$$\lambda_l \geq 0 \,, \; l = 1, ..., k \,,$$
$$\textstyle\sum_{l=1}^{k} \lambda_l = 1 \,, \; k = 1, 2, ... \Big\} \,,$$

and the problem (2.1) can be rewritten to

$$(3.6) \qquad \mathcal{F}_0(x, \vec{u}) = \min!$$

$$x' + \mathcal{A}(x, \vec{u}) = b \,, \quad x(0) = x_0 \,, \quad x \in W \,,$$

$$\mathcal{F}_j(x, \vec{u}) \leq 0 \,, \quad j = 1, ..., m \,, \quad \vec{u} \in \vec{U}_0 \,,$$

where for $\vec{u} = (u_1, ..., u_k, \lambda_1, ..., \lambda_k) \in \vec{U}_0$:

$$\mathcal{F}_0(x, \vec{u}) = \sum_{l=1}^{k} \lambda_l \, F_0(x, u_l) \,,$$

$$\mathcal{A}(x, \vec{u}) = \sum_{l=1}^{k} \lambda_l \, A(x, u_l) \,,$$

$$\mathcal{F}_j(x, \vec{u}) = \sum_{l=1}^{k} \lambda_l \, F_j(x, u_l) + \Big(\sum_{l=2}^{k} \lambda_l\Big)^2 \,, \quad j = 1, ..., m \,.$$

We apply directly the concept given in [5], and consider a sequence $\{\vec{u}^\mu\}$ of socalled varied controls with respect to the optimal solution $(x^*, \vec{u}^*)$ with $\vec{u}^* = (u^*, 1)$ given by

$$(3.7) \qquad \vec{u}^\mu = (u^*, u_1, ..., u_k, 1 - \frac{1}{\mu}, \frac{1}{\mu}\lambda_1, ..., \frac{1}{\mu}\lambda_k) \,, \quad \mu = 1, 2, ... \,,$$

where $\vec{u} = (u_1, ..., u_k, \lambda_1, ..., \lambda_k) \in \vec{U}_0$ is fixed for the moment. The existence of the socalled directional limits

$$\Delta(\vec{u}) \, \mathcal{F}_j = \lim_{\mu \to \infty} \mu \, (\mathcal{F}_j(x^*, \vec{u}^\mu) - \mathcal{F}_j(x^*, \vec{u}^*))$$

$$= \sum_{l=1}^{k} \lambda_l \, F_j(x^*, u_l) - F_j(x^*, u^*), \quad j = 0, 1, ..., m,$$

$$\Delta(\vec{u}) \, \mathcal{A} = \lim_{\mu \to \infty} \mu \left(\mathcal{A}(x^*, \vec{u}^{\mu}) - \mathcal{A}(x^*, \vec{u}^*) \right)$$

$$= \sum_{l=1}^{k} \lambda_l \, A(x^*, u_l) - A(x^*, u^*),$$

is evident.

Let $x^{\mu}(\vec{u})$ be the unique solution of the evolution equation

$$(3.8) \qquad (x^{\mu}(\vec{u}))' + \mathcal{A}(x^{\mu}(\vec{u}), \vec{u}^{\mu}) = b, \quad x^{\mu}(\vec{u})(0) = x_0, \quad x^{\mu}(\vec{u}) \in W, \quad \mu = 1, 2, ...,$$

and let $\Delta(\vec{u}) \, x$ be the unique solution of the variational evolution equation

$$(3.9) \qquad (\Delta(\vec{u}) \, x)' + A_x(x^*, u^*) \, \Delta(\vec{u}) \, x = -\Delta(\vec{u}) \, \mathcal{A}, \quad \Delta(\vec{u}) \, x(0) = o, \quad \Delta(\vec{u}) \, x \in W.$$

It can be shown that

$$(3.10) \qquad \Delta(\vec{u}) \, x = \lim_{\mu \to \infty} \mu \left(x^{\mu}(\vec{u}) - x^* \right) \quad \text{in } W.$$

We define the sets

$$(3.11) \qquad K = \Big\{ (\xi_0, \xi_1, ..., \xi_m) \; : \;$$
$$\xi_j = \Delta(\vec{u}) \, \mathcal{F}_j + \langle\langle F_{j,x}(x^*, u^*); \Delta(\vec{u}) \, x \rangle\rangle, \; j = 0, 1, ..., m, \; \vec{u} \in \vec{U}_0 \Big\},$$

$$L = \Big\{ (\xi_0, \xi_1, ..., \xi_m) \; : \; \xi_0 < 0, \; \xi_j < -F_j(x^*, u^*), \; j = 1, ..., m \Big\}.$$

It follows from the optimality of $(x^*, \vec{u}^*)$ that $K \cap L = \emptyset$. Further, K and L are convex sets, therefore, applying a separation theorem, there exist multipliers η_j, $j = 0, 1, ..., m$, not all zero, and $\varepsilon \in \mathbf{R}$, such that

$$(3.12) \qquad \sum_{j=0}^{m} \eta_j \, \xi_j \geq \varepsilon \quad \text{for all } (\xi_0, \xi_1, ..., \xi_m) \in K,$$

$$\sum_{j=0}^{m} \eta_j \, \xi_j \leq \varepsilon \quad \text{for all } (\xi_0, \xi_1, ..., \xi_m) \in L.$$

Discussing the details of the relations (3.12) we obtain $\eta_j \geq 0$, $j = 0, 1, ..., m$, and $\varepsilon = 0$, moreover,

$$(3.13) \qquad \sum_{j=0}^{m} \eta_j \, \Delta(\vec{u}) \, \mathcal{F}_j + \langle\langle \sum_{j=0}^{m} \eta_j \, F_{j,x}(x^*, u^*); \Delta(\vec{u}) \, x \rangle\rangle \geq 0 \quad \text{for all } \vec{u} \in \vec{U}_0,$$

and the slack conditions (3.4).

Finally, defining the adjoint evolution equation (3.3) and using the variational evolution equation (3.9) the variational inequality (3.13) can be rewritten to

$$(3.14) \qquad \sum_{j=0}^{m} \eta_j \, \Delta(\vec{u}) \, \mathcal{F}_j - \langle\langle \Delta(\vec{u}) \, \mathcal{A}; y \rangle\rangle \geq 0 \quad \text{for all } \vec{u} \in \vec{U}_0.$$

The condition (3.2) is equal to (3.14), considered only for $\vec{u} = (u, 1)$, $u \in U_0$. •

Suppose that the functionals F_j and the operator A are separated in x and u, i.e.

$$(3.15) \qquad F_j(x, u) = F_j^1(x) + F_j^2(u), \quad j = 0, 1, ..., m,$$

$$A(x, u) = A^1(x) + A^2(u),$$

and suppose that the functional parts F_j^1 are convex and the operator part A^1 is linear, then the minimum principle given in theorem 2 under the assumption $\eta_0 > 0$ is sufficient, too.

The exact formulation is the following one.

Theorem 3: Let the operator part $A^1 : X \to X^*$ be linear and generalized strongly monotone.

Let the functional parts $F_j^1 : X \to \mathbf{R}$, $j = 0, 1, ..., m$, be convex and continuously *Frèchet* differentiable.

Let the following minimum principle be fulfilled:

There exist multipliers $\eta_0 > 0$, $\eta_j \geq 0$, $j = 1, ..., m$, such that

$$(3.16) \qquad \sum_{j=0}^{m} \eta_j \, F_j^2(u^*) - \langle\langle A^2(u^*); y \rangle\rangle$$

$$= \min \left\{ \sum_{j=0}^{m} \eta_j \, F_j^2(u) - \langle\langle A^2(u); y \rangle\rangle \mid u \in U_0 \right\},$$

$$(3.17) \qquad -y' + (A^1)^* \, y = \sum_{j=0}^{m} \eta_j \, F_{j,x}^1(x^*), \quad y(T) = o, \quad y \in W,$$

$$(3.18) \quad \eta_j \left(F_j^1(x^*) + F_j^2(u^*) \right) = 0 , \quad j = 1, ..., m .$$

Then (x^*, u^*) is optimal for problem (1.12) with (3.15).

References:

1. L. Bittner, *Necessary optimality conditions for a model of optimal control processes.* Banach Center Publ. **1**(1975), 25–32.

2. L. Bittner, *Ein Modell für eine Klasse von Aufgaben optimaler Steuerung.* ZAMM **58**(1978), 251–260.

3. H. Gajewski, H. Gröger und K. Zacharias, *Nichtlineare Operatorgleichungen und Operatordifferentialgleichungen.* Akademie-Verlag, Berlin, 1974.

4. H. Goldberg, W. Kampowsky und F. Tröltzsch, *On Nemytskij operators in L_p-spaces of abstract functions.* Math. Nachr. **155**(1992), 127–140.

5. W. Kampowsky, *Optimal control in nonlinear evolution equations with constraints.* ZAMM **71**(1991) 7/8, 277–288.

6. W. Kampowsky and U. Raitums, *Convexification of parabolic control problems.* SPP der DFG *Anwendungsbezogene Optimierung und Steuerung*, Report No. 349, 1991.

7. M.A. Krasnoselskij, *Integral operators in spaces of summable functions* (Russian). Publ. house Nauka, Moscow, 1967.

8. J.L. Lions, *Quelques méthodes de résolution des problèmes aux limites non linéaires.* Dunod Gauthier-Villars, Paris, 1969.

9. U. Raitums, *Maximum principle in optimal control problems governed by elliptic equations* (Russian). Z. Anal. u. Anwend. **5**(1986), 291-306.

10. U. Raitums, *Optimal control problems governed by elliptic equations* (Russian). Publ. house Sinatne, Riga, 1989.

11. M.M. Vainberg, *Variational methods for the study of nonlinear operators.* Holden-Day, Inc., San Francisco, London, Amsterdam, 1964.

Authors' addresses

Winfried Kampowsky
Fachhochschule Stralsund, Fachbereich Elektrotechnik,
Große Parower Straße 133, D-O-2300 Stralsund, Germany.

Uldis Raitums
Latvian State University, Research Institute of Mathematics and Computer Science,
Boulevard Rainis 29, Riga, Latvia.

Semidiscrete Ritz–Galerkin Approximation of Nonlinear Parabolic Boundary Control Problems

Fredi Tröltzsch

Abstract

In the paper a class of optimal control problems for a parabolic equation with nonlinear boundary condition and constraints on the control and the state is considered. Related to this problem a corresponding approximate one is defined, where the equation of state is tackled by a semidiscrete Ritz– Galerkin method and the set of admissible controls is discretized. It is shown that the optimal controls of the approximate problems converge strongly in L_2–sense to the solution of the original one, provided that the discretization parameter tends to zero and certain sufficient second order optimality conditions are satisfied by the exact optimal control.

1   Introduction

The subject of this paper is an analysis of convergence for numerical approximations of nonlinear parabolic boundary control problems by certain numerical techniques. We consider a semidiscrete Ritz–Galerkin scheme as a prototype for the numerical treatment of the state–equation — in our case this is a parabolic initial boundary value problem with nonlinear boundary condition. Results on the convergence of numerical methods for distributed control systems have already been obtained by Alt and Mackenroth (1989), Knowles (1982), Lasiecka (1980), Malanowski (1981) or Tröltzsch (1987) for linear equations of state and convex objectives. The present work can be considered as a natural extension of these results to the nonlinear case, which is technically much more involved. Our approach can be related to three origins:

The first is the general theory of second order conditions for programming problems in Banach spaces going back to early papers by Ioffe (1979) and Maurer (1981). The extension to parabolic control systems was performed in Goldberg and Tröltzsch (1989,1991).

A second basis, the key to extend those results for the control of ordinary differential equations to our problems, is the geometric theory for parabolic equations with inhomogeneous boundary conditions. The third and decisive origin is a general method to show strong convergence of optimal solutions of approximated programming problems in Banach spaces owing to Alt (1984).

We are going to investigate the following simplified *model problem*:

Minimize

$$(1.1) \qquad F(w,u) = \int_\Omega (w(T,\xi) - q(\xi))^2 d\xi + \lambda \int_0^T \int_\Gamma u(t,\xi)^2 dS_\xi dt$$

subject to

$$(1.2) \qquad \begin{aligned} \frac{\partial w}{\partial t}(t,\xi) &= \Delta_\xi w(t,\xi) - w(t,\xi) && \text{in } \Omega \\ w(0,\xi) &= 0 && \text{in } \Omega \\ \frac{\partial w}{\partial n}(t,\xi) &= b(w(t,\xi), u(t,\xi)) && \text{on } \Gamma, \end{aligned}$$

$0 < t \leq T$, and to the constraints

$$(1.3) \quad |u(t,\xi)| \leq 1$$

a.e. on $[0,T] \times \Gamma$, where the state-function $w \in C([0,T], W_p^\sigma(\Omega))$ is defined as *mild solution* of (1.2) (cf. the definition in section 2) and the *control* u is taken from $L_\infty((0,T) \times \Gamma)$. We confine ourselves to this type of problems without state-constraints. A more general class with inhomogeneous initial condition and additional constraints on the state is discussed in Tröltzsch (1991).

In this setting, the following quantities are given:
$\Omega \subset I\!R^n$, $n \geq 2$, is a bounded domain with C^∞–boundary Γ , $T > 0$ a fixed time, and $q \in L_\infty(\Omega)$ is a real-valued function. We assume for simplicity that $b = b(w,u)$ has the form

$$b(w,u) = b_1(w) + b_2(w)u,$$

where $b_i \in C^1(I\!R), i = 1,2$. For the more general case $b = b(t,\xi,w,u)$ we refer the reader to Tröltzsch (1991). By $\partial/\partial n$ the outward normal derivative at Γ is denoted, dS_ξ is the surface measure on Γ. Throughout the paper we shall use the following *notations*:

$$
\begin{aligned}
\textbf{Spaces:} \quad & X_\infty = C([0,T], C(\Gamma)) = C([0,T] \times \Gamma), \\
& X_p = U_p = L_p((0,T), L_p(\Gamma)),\ 1 \le p < \infty, \quad X_{C,2} = C([0,T], L_2(\Gamma)) \\
\textbf{Norms:} \quad & \|u\|_p = \Big(\int_0^T \int_\Gamma |u(t,\xi)|^p\, dS_\xi dt\Big)^{1/p}, \quad \|x\|_{C,2} = \max_{t\in[0,T]} \Big(\int_\Gamma |x(t,\xi)|^2\, dS_\xi\Big)^{1/2} \\
& \|(x,u)\|_{\alpha,\beta} = \max\{\|x\|_\alpha, \|u\|_\beta\}, \quad \|(x,u)\|_\alpha = \|(x,u)\|_{\alpha,\alpha}.
\end{aligned}
$$

Moreover, we shall denote by $\|.\|_{s,\Omega}$ the usual norm of $H^s(\Omega)$.

Pairings: For suitable (possibly vector-valued) functions x,y we define

$$
(x,y)_\Omega = \int_\Omega x(\xi)y(\xi)\,d\xi, \quad (x,y)_\Gamma = \int_\Gamma x(\xi)y(\xi)\,dS_\xi.
$$

2 Semigroup approach to the control problem

The concept of mild solutions to the state equation (1.2) requires the definition of certain linear operators:

For $1 < r < \infty$, $A_r : L_r(\Omega) \supset D(A) \to L_r(\Omega)$ is defined by

$$
A_r w = -\Delta w + w \quad \text{on} \quad D(A_r) = \{w \in W_r^2(\Omega) : \frac{\partial w}{\partial n} = 0 \quad \text{on } \Gamma\}.
$$

$-A_r$ is known to be the infinitesimal generator of a strongly continuous and analytic semigroup $\{S_r(t)\}$ of linear continuous operators in $L_r(\Omega), t \ge 0$. The *Neumann operator* N_r is defined by $N_r : g \mapsto w$, where $\Delta w - w = 0$ in Ω, $\partial w/\partial n = g$ on Γ and $g \in L_r(\Gamma)$. N_r is continuous from $L_r(\Gamma)$ to $W_r^s(\Omega)$ for all $s < 1 + 1/r$. Now we fix once and for all p and σ by $p > n + 1$,

$$
(2.1) \quad \frac{n}{p} < \sigma < 1 + 1/p
$$

and put $A := A_p, S(t) := S_p(t), N := N_p$. Moreover, a Nemytskij operator B is defined formally by $B(x,u)(t,\xi) = b(x(t,\xi), u(t,\xi))$.

According to the assumptions imposed on b this operator is twice continuously differentiable from $X_\infty \times U_p$ to $L_p((0,T) \times \Gamma)$, hence also from $X_\infty \times U_p$ to $L_p((0,T) \times \Gamma) = L_p((0,T), L_p(\Gamma)) = U_p$. This is the way we are considering B. By τ the trace operator will be denoted.

Definition: Any $w \in C([0,T], W_p^\sigma(\Omega))$ satisfying the *Bochner integral equation*

$$(2.2) \qquad w(t) = \int_0^t AS(t-s)NB(\tau w, u)(s)\,ds,$$

$t \in [0,T]$, is said to be a *mild solution* of (1.2). Note that (2.1) ensures the continuity of $w = w(t,\xi)$, hence $\tau w \in X_\infty$. A control u is said to be *admissible* if it belongs to the set

$$U^{ad} = \{u \in L_\infty((0,T) \times \Gamma) : |u(t,\xi)| \le 1\}.$$

By means of arguments from the geometric theory of parabolic equations it can be shown that there is a sufficiently small $T > 0$ such that for all admissible controls a mild solution $w = w(u)$ exists on $[0,T]$. Moreover, w is unique on its interval of existence and there is a constant $R > 0$ such that

$$(2.3) \qquad |w(t,\xi)| \le R \quad \text{on } [0,T] \times \bar{\Omega}$$

for all $w = w(u)$ belonging to any admissible control u. In all what follows we shall assume that $T > 0$ is chosen in this way. The estimate (2.3) has a simple but important consequence: The function b is uniformly Lipschitz and bounded on $[-R, R] \times [-1, 1]$. Therefore we can suppose without limitation of generality (possibly after cutting of and a smooth re–definition outside of $[-R, R] \times [-1, 1]$ that b is in addition to the former assumptions *uniformly Lipschitz and bounded* for all $(w, u) \in \mathbb{R}^2$. Thus in particular

$$(2.4) \qquad\qquad |b(w,u)| \le b_M$$

$$(2.5) \qquad |b(w_1, u_1) - b(w_2, u_2)| \le \lambda_b \max\{|u_1 - u_2|, |w_1 - w_2|\}.$$

The proof of existence and uniqueness of w is closely connected with estimates of the norm of $AS(t)N$. It holds

$$(2.6) \qquad \|A_r S_r(t) N_r\|_{L_r(\Gamma) \to W_r^\sigma(\Omega)} \le ct^{-(1-(\sigma'-\sigma)/2)}$$

for all $0 < \sigma < \sigma' < 1 + 1/r$ (cf. Amann (1988)). Related to this kernel $AS(t)N$ the following linear control operators occur:

$$(Lz)(t) = \int_0^t AS(t-s)Nz(s)\,ds, \quad \Lambda z = (Lz)(T) = \int_0^T AS(T-s)Nz(s)\,ds.$$

$(Kz)(t) = (\tau Lz)(t)$. It follows from (2.6) and $p > n + 1$ that these operators are continuous between the following spaces: $K : U_p \to X_\infty$, $L : U_p \to C([0,T], C(\bar{\Omega}))$, $\Lambda : U_p \to C(\bar{\Omega})$.

3 Known results on necessary and sufficient optimality conditions

Let M be the set of all $(x,u) \in X_\infty \times U^{ad}$ satisfying all constraints of (P). Each element (x,u) of M is said to be *feasible* , M is the *feasible set*. If $(x^o,u^o) \in M$ and $f(x^o,u^o) \leq f(x,u)$ for all $(x,u) \in M$, then (x^o,u^o) is said to be an *optimal pair* and u^o an *optimal control*. If this holds for all $(x,u) \in M \cap \{(x,u) : \|u - u^o\|_p < \epsilon\}, \epsilon > 0$, then u^o or (x^o,u^o) is called *locally optimal*.

In all what follows we suppose that (x^o,u^o) is locally optimal. The partial Fréchet derivatives of B at (x^o,u^o) are denoted by B_x and B_u, hence $B'(x^o,u^o)(x,u) = B_x x + B_u u$.

Definition: The set $M(x^o,u^o)$ consisting of all elements $(x - x^o, z)$ such that $z = \lambda(u - u^o)$, $u \in U^{ad}$, $\lambda \geq 0$, and

$$x - x^o = K(B_x(x - x^o) + B_u z)$$

(*linearized equation of state*) is said to be the *linearizing cone* at (x^o,u^o).
The *Lagrange function* $\mathcal{L}$ is defined by

$$\mathcal{L}(x,u;y) = f(x,u) + \int_0^T (x(t) - (KB(x,u))(t),\, y(t))_\Gamma \, dt.$$

Theorem 1 (Goldberg and Tröltzsch (1991)) *Suppose that (x^o,u^o) is locally optimal for (P). Then there exists a Lagrange multiplier $y \in L_\infty((0,T) \times \Gamma)$ such that*

$$(3.1) \qquad \mathcal{L}_x(x^o,u^o;y) \;=\; 0$$
$$(3.2) \qquad \mathcal{L}_u(x^o,u^o;y)(u - u^o) \;\geq\; 0 \quad \forall u \in U^{ad}.$$

Theorem 2 (Goldberg and Tröltzsch (1991)) *Let the feasible pair (x^o,u^o) satisfy the first order necessary conditions (3.1-2), where $y \in L_\infty((0,T) \times \Gamma)$. Suppose further that (x^o,u^o) fulfils the following second order sufficient optimality condition:*

There is a $\delta > 0$ such that

$$(3.3) \qquad \mathcal{L}''(x^o,u^o;y) \geq \delta \|(x - x^o, u - u^o)\|_2^2$$

for all $(x,u) \in M(x^o,u^o)$. Then there exist $\epsilon > 0$ and $\delta_1 > 0$ such that

$$(3.4) \qquad f(x,u) - f(x^o,u^o) \geq \delta_1 \|(x - x^o, u - u^o)\|_2^2,$$

for all $(x, u) \in M$ with $\|u - u^o\|_p < \epsilon$, hence (x^o, u^o) is locally optimal in the sense of $X_\infty \times U_p$.

4 The Ritz – Galerkin approximation

Let $V_h \subset H^1(\Omega)$ be a family of finite-dimensional subspaces depending on a discretization parameter $h > 0$ and enjoying the following properties: There is a constant c, independent of h and s, such that

$$(4.1) \qquad \|w - P_h w\|_{0,\Omega} + h \|w - P_h w\|_{1,\Omega} \leq c h^s \|w\|_{s,\Omega}$$

for all $1 \leq s \leq 2$ and for all $w \in H^s(\Omega)$, where $P_h : H^1(\Omega) \to V_h$ denotes the L_2-projector onto V_h. Moreover, the *inverse estimate*

$$(4.2) \qquad \|w\|_{1,\Omega} \leq c h^{-1} \|w\|_{0,\Omega}$$

is assumed to hold for all $w \in V_h$ and all $h > 0$, where $c > 0$ is independent from h. The spaces V_h of piecewise linear splines on sufficiently regular meshes on Ω comply with these requirements, see Ciarlet (1978).

The *approximate control problem* is to *minimize* $F(w_h, u)$ subject to $w_h : [0, T] \to V_h$,

$$\frac{d}{dt}(w_h(t)\,,\,v)_\Omega + (\nabla w_h(t)\,,\,\nabla v)_\Omega + (w_h(t)\,,\,v)_\Omega \;=\; (B(\tau w_h, u)(t)\,,\,v)_\Gamma$$

$$(4.3) \qquad\qquad\qquad\qquad\qquad\qquad\qquad\qquad w_h(0) \;=\; 0,$$

for all $v \in V_h$ and almost all $t \in [0, T]$, and to the constraints

$$(4.4) \qquad u \in U_h^{ad}, \quad \|u - u^o\|_p \leq \epsilon.$$

Here ϵ indicates the diameter of the region, where u^o is optimal and convergence of u_h to u^o can be expected at all. In real computations the restriction (4.4) should be substituted by a trust region, where we are sure to have a unique locally optimal control.

The system (4.3) is uniquely solvable for all $u \in U_h^{ad}$, as $b = b(w, u)$ is uniformly bounded and Lipschitz (according to (2.4), (2.5)).

Let $g \in L_2((0, T), L_2(\Gamma)) = U_2$ be given . Then the linear system (4.3) with right hand side $(g(t)\,,\,v)_\Gamma$ substituted for $(B(\tau w_h, u), v)_\Gamma$ has a unique solution w_h, too. Completely analogous to L, K, and Λ we define

$$L_h : \quad U_2 \to C([0,T], H^1(\Omega)), \quad L_h : \quad g \mapsto w_h$$
$$K_h : \quad U_2 \to C([0,T], L_2(\Gamma)), \quad K_h : \quad g \mapsto \tau w_h$$
$$\Lambda_h : \quad U_2 \to L_2(\Omega), \qquad\qquad \Lambda_h : \quad g \mapsto w_h(T).$$

After setting $x_h(t) := \tau w_h(t)$ the approximate control problem admits the form

$$(P_h^\epsilon) \qquad f_h(x,u) = \min!$$
$$x = K_h B(x,u)$$
$$u \in U_h^{ad}, \quad \|u - u^\circ\|_p \le \epsilon,$$

with

$$f_h(x,u) = \|q - \Lambda_h B(x,u)\|^2_{L_2(\Omega)} + \lambda \|u\|^2_{U_2}.$$

By means of estimates for the linear version of (4.3) owing to Lasiecka (1986) it was shown by Tröltzsch (1988) that

$$\|w(t) - w_h(t)\|_{1,\Omega} \le ch^{1/2-\varepsilon},$$

where w solves (2.2), w_h solves (4.3), and c is independent from u but depending in general on ε.

5 Strong convergence of approximating controls

In this section we shall prove strong convergence of optimal controls of (P_h^ϵ) for $h \downarrow 0$ under natural assumptions specified below.

We shall use the following notations: The distance of $A_1 \subset U_p$ to $A_2 \subset U_p$ is

$$d_p(A_1, A_2) = \sup_{v \in A_1} (\inf_{z \in A_2} \|v - z\|_p).$$

By $a_U(h), a_f(h)$, and $a_K(h)$ positive functions are denoted, tending to zero for $h \downarrow 0$ and r_j^A is the j-th order remainder term of a Taylor expansion of a mapping A at the point (x°, u°), λ_K and λ_f are positive constants. We suppose that U_h^{ad} satisfies the approximation condition

(i) $d_p(U_h^{ad}, U^{ad}) < a_U(h), \quad d_p(u^0, U_h^{ad}) < a_U(h)$

Moreover, the following properties hold true for our problem:

(ii) B is twice continuously Fréchet differentiable from $X_\infty \times U_p$ to U_p and

$$(5.1) \quad \|r_1^B(v)\|_2/\|v\|_2 \to 0, \quad \|r_2^\mathcal{L}(v)\|_2/\|v\|_2^2 \to 0,$$

if $\|v\|_{\infty,p} \to 0$.

This is a more or less immediate consequence of the linearity of b with respect to u. If b is fully nonlinear, then (ii) can only hold for $p = \infty$.

(iii) Let u, u_h be arbitrary elements of $U^{ad} \cup U_h^{ad}$. Then the equations $x = KB(x,u)$, $x_h = K_h B(x_h, u_h)$ possess exactly one solution $x, x_h \in X_\infty$ and $X_{C,2}$, respectively. Furthermore,

$$(5.2) \quad \|x - x_h\|_{C,2} \leq a_K(h) + \lambda_K \|u - u_h\|_p.$$

(iv) $|f_h(x_1, u_1) - f(x_2, u_2)| \leq a_f(h) + \lambda_f \, max\{\|x_1 - x_2\|_{C,2}, \|u_1 - u_2\|_p\}$
for all $u_i \in U^{ad} \cup U_h^{ad}$ and all corresponding x_i. This follows from (iii).

(v) $|\mathcal{L}''(x^o, u^o; y)[v_1, v_2]| \leq \lambda_\mathcal{L} \|v_1\|_2 \cdot \|v_2\|_2$
for all $v_1, v_2 \in X_\infty \times U_p$.

For a detailed discussion of (iii) the reader is referred to Tröltzsch (1991). In view of the estimate $\|w(t) - w_h(t)\|_{1,\Omega} \leq ch^{1/2-\epsilon}$ it can be shown that $a_K(h) = O(h^{1/2-\epsilon})$. Clearly, the feasible set of (P_h^ϵ) is non-empty for all sufficiently small h. In all what follows let (x_h, u_h) be a globally optimal solution of (P_h^ϵ). It exists due to the linearity of b with respect to u.

Lemma 1 *It holds*

$$f_h(x_h, u_h) - f(x^o, u^o) \leq \alpha_1(h)$$

for all sufficiently small h, where $\alpha_1(h) = a_f(h) + c_K^1 a_K(h) + c_U^1 a_U(h)$ and c_K^1, c_U^1 do not depend on h.

Proof: We approximate u^o by $\tilde{u}_h \in U_h^{ad}$ according to (i). If h is small enough, then $\|\tilde{u}_h - u^o\|_p \leq \epsilon$. Let $\tilde{x}_h$ be the corresponding state. Then

$$\begin{aligned}
f_h(x_h, u_h) &\leq f_h(\tilde{x}_h, \tilde{u}_h) = f(x^o, u^o) + f_h(\tilde{x}_h, \tilde{u}_h) - f(x^o, u^o) \\
&\leq f(x^o, u^o) + a_f(h) + \lambda_f \max\{\|\tilde{x}_h - x^o\|_{C,2}, \|\tilde{u}_h - u^o\|_p\} \\
&\leq f(x^o, u^o) + a_f(h) + \lambda_f(c_K a_K(h) + c_U a_U(h)) \\
&= f(x^o, u^o) + a_f(h) + c_K^1 a_K(h) + c_U^1 a_U(h)
\end{aligned}$$

by (i), (iii), and (iv). $\qquad\qquad\qquad\qquad\qquad\qquad\qquad\qquad\qquad\qquad\square$

The more difficult part in the proof of strong convergence of optimal controls is to establish a useful lower estimate for $f_h(x_h, u_h) - f(x^o, u^o)$. This estimate employs the second order condition, which is formulated for the linearized cone $M(x^o, u^o)$.

Lemma 2 *Let $\epsilon > 0$ be sufficiently small. Suppose that (x^o, u^o) is feasible for (P) and fulfils the first order necessary conditions (3.1-2) as well as the second order sufficient condition (3.3). Then there is a $\bar\delta > 0$ such that for all sufficiently small $h > 0$*

$$f_h(x_h, u_h) - f(x^o, u^o) \geq \bar\delta \, \|(x_h - x^o, u_h - u^o)\|_2^2 - \alpha_2(h),$$

where $\alpha_2(h) = c_f^2 a_f(h) + c_K^2 a_K(h) + c_U^2 a_U(h)$, and c_f^2, c_K^2, c_U^2 are independent from h

Proof: Let $\alpha(h)$ denote a generic function of the form $\alpha(h) = c_1 a_f(h) + c_2 a_K(h) + c_4 a_U(h)$ (with generic constants c_i). We approximate u_h by $\tilde u_h \in U^{ad}$ such that $\|u_h - \tilde u_h\|_p < \alpha_U(h)$. If h is sufficiently small, then $\|\tilde u_h - u^o\|_p < 2\varepsilon$. Let the auxiliary state $\bar x_h$ be defined by $\bar x_h = KB(\bar x_h, \tilde u_h)$. This is the exact state belonging to the approximate $\tilde u_h$. In the sequel we shall write $\mathcal{L}(x, u) = \mathcal{L}(x, u; y)$ as y remains fixed. Moreover, we put $\bar v_h = (\bar x_h, \tilde u_h), v^o = (x^o, u^o), v_h = (x_h, u_h)$. As (x_h, u_h) is feasible for (P_h^ϵ)

$$\begin{aligned}
f_h(x_h, u_h) &\geq f_h(x_h, u_h) + (x_h - K_h B(x_h, u_h), y)_\Gamma \\
&\geq f(\bar x_h, \tilde u_h) + (\bar x_h - KB(\bar x_h, \tilde u_h), y)_\Gamma - \alpha(h).
\end{aligned}$$

This is a conclusion of (iii) and (iv). Hence

$$\begin{aligned}
f_h(x_h, u_h) &\geq \mathcal{L}(\bar x_h, \tilde u_h) - \alpha(h) \\
&= \mathcal{L}(v^o) + \mathcal{L}'(v^o)(\bar v_h - v^o) + \frac{1}{2}\mathcal{L}''(v^o)[\bar v_h - v^o, \bar v_h - v^o] \\
&\quad + r_2^{\mathcal{L}}(\bar v_h - v^o) - \alpha(h).
\end{aligned}$$

It holds $\mathcal{L}(v^o) = f(v^o)$ and

$$\mathcal{L}'(v^o)(\bar v_h - v^o) = \mathcal{L}_x(v^o)(\bar x_h - x^o) + \mathcal{L}_u(v^o)(\tilde u_h - u^o) \geq 0$$

in view of the first order conditions (3.1)-(3.2). Thus

$$f_h(x_h, u_h) - f(x^o, u^o) \geq \frac{1}{2}\mathcal{L}''(v^o)[\bar v_h - v^o, \bar v_h - v^o] + r_2^{\mathcal{L}}(\bar v_h - v^o) - \alpha(h).$$

We have $\bar x_h = KB(\bar x_h, \tilde u_h)$. Linearizing at (x^o, u^o)

$$\bar x_h - x^o = K(B_x(x^o, u^o)(\bar x_h - x^o) + B_u(x^o, u^o)(\tilde u_h - u^o) + r_1^B(\bar v_h - v^o)).$$

Thus

$$\bar{x}_h - x^o = (I - KB_x(x^o, u^o))^{-1}KB_u(x^o, u^o)(\bar{u}_h - u^o) + (I - KB_x(x^o, u^o))^{-1}Kr_1^B$$
$$= \tilde{x}_h - x^o + R.$$

By definition,

$$\tilde{x}_h - x^o = K(B_x(x^o, u^o)(\tilde{x}_h - x^o) + B_u(x^o, u^o)(\bar{u}_h - u^o)),$$

hence $(\tilde{x}_h - x^o, \bar{u}_h - u^o)$ belongs to the linearized set, where the sufficient condition (3.3) applies. Moreover, $\|R\|_2 \leq c\|r_1^B\|_2$, as K and $(I - KB_x)^{-1}$ are continuous from X_2 to X_2 (this can be shown by means of (2.6)). We put $\tilde{v}_h = (\tilde{x}_h, \bar{u}_h)$. Then

$$(5.3) \qquad \|\tilde{v}_h - \bar{v}_h\|_2 \leq c\|r_1^B(\bar{v}_h - v^o)\|_2,$$

and

$$\begin{aligned}
f_h(v_h) - f(v^o) &\geq \frac{1}{2}\mathcal{L}''(v^o)[\tilde{v}_h - v^o, \tilde{v}_h - v^o] + \frac{1}{2}\mathcal{L}''(v^o)[\bar{v}_h - \tilde{v}_h, \bar{v}_h - \tilde{v}_h] \\
&\quad + \mathcal{L}''(v^o)[\tilde{v}_h - v^o, \bar{v}_h - \tilde{v}_h] + r_2^{\mathcal{L}} - \alpha(h) \\
&\geq \frac{\delta}{2}\|\tilde{v}_h - v^o\|_2^2 - c(\|\bar{v}_h - \tilde{v}_h\|_2^2 + \|\tilde{v}_h - v_h^o\|_2\|\bar{v}_h - \tilde{v}_h\|_2) \\
&\quad + r_2^{\mathcal{L}} - \alpha(h),
\end{aligned}$$

where $r_2^{\mathcal{L}} := r_2^{\mathcal{L}}(\bar{v}_h - v^o)$.

Here we employed the second order condition (3.3) and the estimate (v). Re-substituting $\bar{v}_h$ for $\tilde{v}_h$ we arrive after some formal calculations by means of (5.3) at

$$\begin{aligned}
f_h(v_h) - f(v^o) &\geq \|\bar{v}_h - v^o\|_2^2\left\{\frac{\delta}{2} - c\left(\frac{\|r_1^B\|_2^2}{\|\bar{v}_h - v^o\|_2^2} + \frac{\|r_1^B\|_2}{\|\bar{v}_h - v^o\|_2}\right) - \frac{|r_2^{\mathcal{L}}|}{\|\bar{v}_h - v^o\|_2^2}\right\} \\
&\quad - c\|r_1^B\|_2^2 - 2ca(h)\|r_1^B\|_2 - ca(h)^2 - c\|\bar{v}_h - v^o\|_2^2 \cdot a(h) - \alpha(h),
\end{aligned}$$

where $r_1^B = r_1^B(\bar{v}_h - v^o)$. The term after the curled brackets is of the type $\alpha(h)$. If $\epsilon \downarrow 0$, then $\|\bar{u}_h - u^o\|_p \to 0$, hence $\|\bar{x}_h - x^o\|_\infty \to 0$, too. Thus $\|\bar{v}_h - v^o\|_{\infty,p} \to 0$ and the term in the brackets tends to $\delta/2$ by (5.1). Therefore, if ε is sufficiently small,

$$f_h(v_h) - f(v^o) \geq \frac{\delta}{4}\|\bar{v}_h - v^o\|_2^2 - \alpha(h) \geq \frac{\delta}{4}\|v_h - v^o\|_2^2 - \tilde{\alpha}(h),$$

implying the statement of the Lemma. $\qquad\qquad\qquad\qquad\qquad\qquad\qquad\qquad\qquad$ $\square$

Lemma 1 and Lemma 2 yield the

Theorem 3 *Let (x^o, u^o) be a locally optimal for the control problem (P) satisfying the sufficient second order condition (3.3). Let a sequence of (globally) optimal solutions (x_h, u_h) of (P_h^ϵ) be given.*

If $\epsilon > 0$ is sufficiently small, then for all sufficiently small $h > 0$ the estimate

$$(5.4) \qquad \|(x_h - x^o, u_h - u^o)\|_2^2 \leq c\,\alpha(h)$$

takes place, where $\alpha(h) = \alpha_1(h) + \alpha_2(h) = c_U \alpha_U(h) + c_f \alpha_f(h) + c_K \alpha_K(h)$ and c_U, c_f, c_K are positive constants not depending on h.

References

[1] Alt, H.W. (1984), On the approximation of infinite optimization problems with an application to optimal control problems. Appl. Math. Opt. 12, 15-27.

[2] Alt, H.W. and U. Mackenroth (1989), Convergence of finite element approximations to state constrained convex parabolic boundary control problems. SIAM J. Contr. Opt. 27, 718-736.

[3] Amann (1988), H., Parabolic evolution equations with nonlinear boundary conditions. Journal of Differential Equations 72, 201-269.

[4] Ciarlet, P. (1978), The finite element method for elliptic problems. North Holland, Amsterdam.

[5] Goldberg, H., and F. Tröltzsch (1989), Second order optimality conditions for a class of control problems governed by nonlinear integral equations with application to parabolic boundary control. Optimization 20, 687-698.

[6] Goldberg, H., and F. Tröltzsch (1991), Second order sufficient optimality conditions for a class of nonlinear parabolic boundary control problems, appears in SIAM J. Contr. Opt.

[7] Ioffe, A.D. (1979), Necessary and sufficient conditions for a local minimum 3, Second order conditions and augmented duality. SIAM J.Control Opt. 17, 266-288.

[8] Knowles, G. (1982), Finite element approximation of parabolic time optimal control problems. SIAM J. Control Opt. 20, 414-427.

[9] Lasiecka, I. (1980), Boundary control of parabolic systems, finite element approximation Appl. Math. Optim. 6, 31-62.

[10] Lasiecka, I. (1986), Ritz–Galerkin approximation of abstract parabolic boundary value problems with rough boundary data — L_p-theory. Math. of Comp. 47, 55–75.

[11] Malanowski, K. (1981), Convergence of approximations vs. regularity of solutions for convex, control-constrained optimal control problems. Appl. Math. Opt. 8, 69-95.

[12] Maurer, H. (1981), First and second order sufficient optimality conditions in mathematical programming and optimal control. Math. Programming Study. 14, 163-177.

[13] Tröltzsch, F. (1987), Semidiscrete finite element approximation of parabolic boundary control problems – convergence of switching points. In: Optimal control of partial differential equations II. Int. Ser. Num. Math., Vol. 78, 219 - 232, Birkhäuser, Basel.

[14] Tröltzsch, F. (1988), On convergence of semidiscrete Ritz-Galerkin schemes applied to the boundary control of parabolic equations with non-linear boundary conditions. Appears in ZAMM 1992.

[15] Tröltzsch, F. (1991), Semidiscrete Ritz–Galerkin Approximation of Nonlinear Parabolic Boundary Control Problems — Strong Convergence of Optimal Controls. Submitted to Appl. Math. Opt.

Author's address

Prof. Dr. F. Tröltzsch
Technische Universität Chemnitz
Fachbereich Mathematik
PSF 964
D -O–9010 Chemnitz
troeltz@mira.mathematik.tu-chemnitz.de

Iterative Methods for Optimal Control Processes governed by Integral Equations

Werner H. Schmidt

Abstract

We formulate a method of successive approximations for control processes described by Volterra or Fredholm integral equations using necessary optimality conditions. Assumptions are given under which the iterative methods converge.

1 Introduction

Krylov and Chernousko applied the maximum principles for differential processes in order to construct a sequence of controls: Start with a dispatcher control u_{old}, calculate the corresponding state x and the solution of the adjoint equation. From the maximum principle we get a new control u_{new}. Replace u_{old} by u_{new} and iterate the procedure. In the linear case u_{new} is the optimal control. However in many nonlinear examples the controls do not converge in the sense of functional. Therefore modifications of the iterative method were given by the inventors themselves [5]. The idea is to restrict the difference between u_{old} and u_{new}. If the control set U is convex u_{new} can be replaced by a convex combination of u_{new} and u_{old}. Another possibility is to replace u_{old} by u_{new} only in a small time–interval , on this idea based the proof of the maximum principle. Chernousko and Lyubushin gave some modifications of the method for differential processes. Using a lemma of Rozonoer [7] they proved convergence under strong assumptions. We apply such a modified method for integral processes. The stage effect is to fulfil the fundamental inequality of theorem 1 of the paper.

2 Necessary optimality conditions

For some kinds of integral processes maximum principles were derived in [8],[9] using general results of Bittner [1], Kampowsky [4], Schmidt [10] for abstract control problems. Side conditions and isoperimetric conditions we shall take as penalty terms. Therefore we consider simple problems with a functional and a process equation:

Let be E and S Banach spaces, $J = [0,T]$ a fixed interval, $U \subseteq S$ a control set. Let denote g a functional on $J \times E \times U$ and f an E–valued mapping on $J \times J \times E \times U$. Assume the continuity of g and f and the existence of continuous partial Frèchet–derivatives g_e, f_e with respect to $e \in E$. Discontinuities of $f(s,t,e,u)$ and $f_e(s,t,e,u)$ are allowed if $s = t$ (Volterra integral equation).

Denote $C(J,E)$ the space of all E–valued functions on J and $PCL(J,E^*)$ the set of all piecewise continuous functions on J into the dual space E^*; $h \in C(J,E)$.

The problem under consideration is:
Minimize

$$(2.1) \qquad \mathcal{F}(u) = \int_0^T g(t,x(t),u(t))dt$$

subject to

$$(2.2) \qquad x(s) = h(s) + \int_0^T f(s,t,x(t),u(t))dt \ , \quad s \in J$$

$$(2.3) \qquad x \in C(J,E), \quad u \in PCL(J,U)$$

Necessary optimality conditions for $\bar{x}, \bar{u}$ to be an optimal solution of (2.1)–(2.3) are proved in [8], see lemma 1. In the Fredholm case is assumed the integral operator which belongs to the linearized equation of (2.2) with the arguments $\bar{x}(t), \bar{u}(t)$ has an inverse one on $C(J,E)$.
$\bar{\Psi} \in PCL(J,E^*)$ denotes the solution of the adjoint equation

$$(2.4) \qquad \Psi(t) = \int_0^T \Psi(s)f_e(s,t,\bar{x}(t),\bar{u}(t))ds - g_e(t,\bar{x}(t),\bar{u}(t)) \ , \quad t \in J \ .$$

Define

$$(2.5) \qquad H(t,e,\Psi(\cdot),u) = \int_0^T \Psi(s)f(s,t,e,u)ds - g(t,e,u)$$

on $J \times E \times PCL(J,E^*) \times U$. Keep in mind $f(s,t,e,u) = 0$ and $f_e(s,t,e,u) = 0$ for $t > s$ if (2.2) is a Volterra integral equation.

Lemma 1. *If* $\bar{x}, \bar{u}$ *is an optimal solution of* (2.1), (2.2),(2.3) *then the maximum principle*

$$(2.6) \qquad H(t, \bar{x}(t), \bar{\Psi}(\cdot), \bar{u}(t)) = \max_{v \in U} H(t, \bar{x}(t), \bar{\Psi}(\cdot), v), \quad 0 \leq t \leq T$$

is valid, where $\bar{\Psi}$ *is the solution of* (2.4).

3 Basic lemma

Let be $x, u; \bar{x}, \bar{u}$ admissible solutions of (2.2),(2.3) and $\Delta x = \bar{x} - x$.
We arrange to abbreviate

$$(3.1) \qquad \bar{f}(s,t) = f(s,t,\bar{x}(t),\bar{u}(t)), \qquad \bar{f}_e(s,t) = f_e(s,t,\bar{x}(t),\bar{u}(t))$$

$$(3.2) \qquad \bar{g}(t) = g(t,\bar{x}(t),\bar{u}(t)), \qquad \bar{g}_e(t) = g_e(t,\bar{x}(t),\bar{u}(t))$$

$$(3.3) \qquad \bar{H}(t) = H(t,\bar{x}(t),\bar{\Psi}(\cdot),\bar{u}(t))$$

$$(3.4) \qquad DH(t) = H(t,\bar{x}(t),\bar{\Psi}(\cdot),u(t)) - \bar{H}(t)$$

$$(3.5) \qquad DH_e(t) = H_e(t,\bar{x}(t),\bar{\Psi}(\cdot),u(t)) - \bar{H}_e(t)$$

$$(3.6) \qquad R(t) = \int_0^T \bar{\Psi}(s)[f(s,t,x(t),u(t)) - f(s,t,\bar{x}(t),u(t)) - f_e(s,t,\bar{x}(t),u(t))\Delta x(t)]\,ds$$

$$(3.7) \qquad S(t) = g(t,x(t),u(t)) - g(t,\bar{x}(t),u(t)) - g_e(t,\bar{x}(t),u(t))\Delta x(t)$$

where $s, t \in J$. From these definitions follows directly

Lemma 2. *Let be* $x, u; \bar{x}, \bar{u}$ *admissible solutions of* (2.2), (2.3) *and* $\bar{\Psi}$ *a solution of* (2.4) *corresponding to* $\bar{x}, \bar{u}$. *Then the equation*

$$\mathcal{F}(u) - \mathcal{F}(\bar{u}) = \int_0^T [g(t,x(t),u(t)) - \bar{g}(t)]\,dt =$$

$$= \int_0^T [S(t) - R(t)]\,dt - \int_0^T DH(t)dt - \int_0^T DH_e(t)\Delta x(t)dt$$

holds.

4   Method of successive approximations for integral processes

We formulate one of the modified algorithms given in [2] in order to solve the control problems (2.1)–(2.3) described by Volterra or Fredholm integral equations.

(i) Input N_{max} as the maximal number of subintervals taken into consideration.

(ii) $k := 0$. Start with an admissible control $u_0(\cdot)$.

(iii) Comput a solution x_k of (2.2) and Ψ_k of (2.4) with respect to the arguments x_k, u_k.
Find a control $v_k \in PCL(J, U)$ as a solution of the optimalization problems

$$H(t, x_k(t), \Psi_k(\cdot), v_k(t)) = \max_{v \in U} H(t, x_k(t), \Psi_k(\cdot), v) , \quad t \in J .$$

(iv) Put

$$\Delta H(t) = H(t, x_k(t), \Psi_k(\cdot), v_k(t)) - H(t, x_k(t), \Psi_k(\cdot), u_k(t)) , \quad t \in J$$

and calculate $\mu_k = \int_0^T \Delta H(t) dt$.

(v) $N := 1$

(vi) $h := 2^{-N} T$
Find $\tau \in \{h, 3h, \ldots, (2^N - 1)h\}$ such that

$$\frac{1}{2h} \int_{\tau - h}^{\tau + h} \Delta H(t) dt \geq \frac{1}{T} \mu_k .$$

(vii) Define $u_{\tau, h}(t) = v_k(t)$ for $\tau - h < t \leq \tau + h$ and $u_{\tau, h}(t) = u_k(t)$ else.

(viii) Compute $x_{\tau, h}$ corresponding to $u_{\tau, h}$ as solution of (2.2) and the value $\mathcal{F}(u_{\tau, h})$ of the functional (2.1).

(ix) If $\mathcal{F}(u_{\tau, h}) \leq \mathcal{F}(u_k) - \frac{h}{T} \mu_k$ then
$\{$put $u_{k+1} := u_{\tau, h} ; k := k + 1$ and goto (iii)$\}$ else goto (x).

(x) $N := N + 1$

(xi) If $N \leq N_{max}$ goto (vi) else stop.

Remark: Here it is supposed the process equation (2.2) corresponding to all controls u_k calculated by this iterative method is always solvable, but this is not always true! Also we are not able to guarantee $u_k \in PCL(J, U)$, in concrete processes the optimalization problems of step (iii) will be considered only for a finite number of points $t \in J$.

5 Convergence theorems

To prove convergence in [2] is supposed the trajectories x_k are bounded and all functions and their derivatives fulfil Lipschitz–conditions. Denote $\mathcal{E}$ the set of all points $e \in E$ reachable by the process in any time $t \leq T$ with any admissible control. More precisely we assume the existence of functions $b_1, b_3, c_1 \in C(J \times U \times U, \mathbf{R}_+)$; $b_2, b_4, c_2, d_1, d_2 \in C(J, \mathbf{R}_+)$, $d \in C(J \times J, \mathbf{R}_+)$ and nonnegativ numbers b_1', b_3', c_1' such that

$$(5.1) \qquad \|f_e(s, t, e, u_1) - f_e(s, t, e, u_2)\| \leq b_1(t, u_1, u_2) \leq b_1' \qquad \text{and}$$

$$b_1(t, u, u) = 0 \quad \text{for all} \quad s, t \in J\, ;\ e \in \mathcal{E}\, ;\ u, u_1, u_2 \in U$$

$$(5.2) \qquad \|f_e(s, t, e_1, u) - f_e(s, t, e_2, u)\| \leq b_2(t)\|e_1 - e_2\|$$

$$(5.3) \qquad \|g_e(t, e, u_1) - g_e(t, e, u_2)\| \leq b_3(t, u_1, u_2) \leq b_3' \quad \text{and} \quad b_3(t, u, u) = 0$$

$$(5.4) \qquad \|g_e(t, e_1, u) - g_e(t, e_2, u)\| \leq b_4(t)\|e_1 - e_2\|$$

$$(5.5) \qquad \|f(s, t, e, u_1) - f(s, t, e, u_2)\| \leq c_1(t, u_1, u_2) \leq c_1' \quad \text{and} \quad c_1(t, u, u) = 0$$

$$(5.6) \qquad \|f(s, t, e_1, u) - f(s, t, e_2, u)\| \leq c_2(t)\|e_1 - e_2\|$$

$$(5.7) \qquad \|f_e(s, t, e, u)\| \leq d(s, t) \leq d_1(s)$$

$$(5.8) \qquad \|g_e(t, e, u)\| \leq d_2(t)$$

for all $s, t \in J\, ;\ e_1, e_2 \in \mathcal{E}\, ;\ u, u_1, u_2 \in U$.
If (2.2) is a Fredholm integral equation we assume

$$(5.9) \qquad \max_t \int_0^T d(s, t)ds < 1, \qquad \int_0^T c_2(t)dt < 1\, .$$

Lemma 3. *Assume* (5.1)–(5.4). *Then for all admissible control functions*
$u = u(\cdot)$, $u^0 = u^0(\cdot)$, *trajectories* $x = x(\cdot)$, $x^0 = x^0(\cdot)$ *and the solution* $\Psi^0 = \Psi^0(\cdot)$ *of the adjoint equation* (2.4) *corresponding to* u^0, x^0 *the inequality*

$$\mathcal{F}(u) - \mathcal{F}(u^0) \leq -\int_0^T \Delta H(t)dt+$$

$$+\int \Phi_0^T \int_0^T \|\Psi^0(s)\|ds\, [b_1(t, u(t), u^0(t)) + b_2(t)\|\Delta x(t)\|]\, \|\Delta x(t)\|dt+$$

$$+\int_0^T [b_3(t, u(t), u^0(t)) + b_4(t)\|\Delta x(t)\|]\, \|\Delta x(t)\|dt$$

holds.

The proof follows immediately from lemma 2.

Lemma 4. *Let be* u, u^0 *admissible control functions,* $\Delta x = x - x^0$ *and* (2.2) *a Volterra integral equation. Assume* (5.5),(5.6). *Then it is*

$$(5.10) \qquad \|\Delta x(s)\| \leq \exp(\int_0^T c_2(w)dw) \int_0^s c_1(t, u(t), u^0(t))dt , \quad s \in J .$$

The proof is a consequence of the Gronwall lemma.

Lemma 5. *The assumptions* (5.5),(5.6),(5.9) *let be fulfilled. Then we get the following inequalities if* (2.2) *is a Fredholm integral equation:*

$$\|\Delta x(s)\| \leq \int_0^T c_2(t)\|\Delta x(t)\|dt + \int_0^T c_1(t, u(t), u^0(t))dt$$

and

$$(5.11) \qquad \|\Delta x(s)\| \left(1 - \int_0^T c_2(t)dt\right) \leq \int_0^T c_1(t, u(t), u^0(t))dt \leq c_1'T, \quad s \in J .$$

Now we return to the method of successive approximations. In this method formulated above the control u_k coincides with $u_{\tau,h}$ defined by step (vii) outside of $]\tau - h, \tau + h]$. Therefore $b_i(t, u_{\tau,h}(t) , u_k(t)) = 0 , i = 1, 3$ and $c_1(t, u_{\tau,h}(t) , u_k(t)) = 0$ if $t \notin]\tau - h, \tau + h]$.

Lemma 6. *If $u = u_{\tau,h}$, $u^0 = u_k$, $\Delta x = x_{\tau,h} - x_k$ then there exist a constant K_1 independent of τ, h and u_k such that*

$$(5.12) \qquad \|\Delta x(s)\| \leq K_1 h \qquad \text{for all} \quad s \in J.$$

Proof: With $\quad K_1 = 2c_1' \exp(\int_0^T c_2(w)du) \quad$ in the Volterra case and

$$K_1 = c_1' T(1 - \int_0^T c_2(t)dt)^{-1} \quad \text{in the Fredholm case}$$

the inequality (5.12) follows from (5.10) and (5.11), respectively.

Now we try to estimate $\|\Psi(s)\|$.

Lemma 7. *Assume (5.7),(5.9). There is a nonnegativ number K_2 such that $\|\Psi(t)\| \leq K_2$ for all $t \in J$, where K_2 does not depend on controls u and state x to which Ψ belongs as solution of (2.4).*

Proof: Consider a Volterra integral equation (2.2). Then

$$\|\Psi(t)\| \leq \int_t^T \|\Psi(s)\|d_1(s)ds + d_2(t) , \ t \in J .$$

From Gronwall's lemma we obtain

$$\|\Psi(t)\| \leq d_2(t) + \int_t^T d_1(s)d_2(s) \exp(\int_s^s d_1(w)dw)ds .$$

Take K_2 as the maximum of the right–hand side. If (2.2) is a Fredholm integral equation it follows from (2.4)

$$(5.13) \qquad \|\Psi(t)\| \leq \int_0^T \|\Psi(s)\|d(s,t)ds + d_2(t) , \ t \in J .$$

The integral equation

$$z(t) = \int_0^T d(s,t)z(s)ds + d_2(t) , \ t \in J$$

has a unique solution $z \in C(J, \mathbf{R})$, that follows from (5.9) and Banach's fixed point theorem. The integral operator P defined by

$$(Pz)(t) = \int_0^T d(s,t)z(s)ds \, , \ t \in J$$

on $C(J, E)$ is positiv as a consequence of $d(s,t) \geq 0$. Neumann's operator serie $\sum_{k=0}^{\infty} P^k$ converges strongly. The solution z of $z = Pz + d$ can be represented as $z = d + Rd$, where $R = \sum_{k=0}^{\infty} P^k$ is positiv. Then

$$z(s) = d_2(s) + \int_0^T R(s,t)d_2(t)dt \, , \ s \in J$$

with a resolvent operator R on $J \times J$, $R(s,t) \geq 0$ for all $s,t \in J$. Let be

$$r(t) = \int_0^T y(s)d(s,t)ds + d_2(t) - y(t) \quad \text{with} \quad y(t) = \|\Psi(t)\|$$

then $r(t) \geq 0$, $t \in J$, $r \in PCL(J, \mathbf{R})$. The integral equation

$$y(t) = \int_0^T y(s)d(s,t)ds + [d_2(t) - r(t)] \, , \ t \in J$$

has in $PCL(J, \mathbf{R})$ a unique solution y, too, and it is

$$y(t) = [d_2(t) - r(t)] + \int_0^T R(s,t)[d_2(s) - r(s)] \, ds \leq d_2(t) + \int_0^T R(s,t)d_2(s)ds = z(t) \, ,$$

$t \in J$. Lemma is satisfied taking K_2 as the maximum norm of z.

Theorem 1: *Assume* (5.1)–(5.9). *Let be* u, u^0 *admissible controls which coincide outside of an interval* $(\tau - h, \tau + h]$. *Then there exist a number* $C > 0$ *independent of* τ, h, u *and* u^0 *such that*

$$(5.14) \qquad \mathcal{F}(u) - \mathcal{F}(u^0) \leq Ch^2 - \int_0^T \Delta H(t)dt \, .$$

Proof: Combining the results of the lemmata 3–7 we obtain inequality (5.14). **Remark:** Assume (5.1)–(5.9). Then (5.14) is valid for all $u = u_{k+1}$, $u^0 = u_k$. C is independent of k.

Lemma 8. *If g is bounded on $J \times \mathcal{E} \times \mathcal{U}$ and (5.1)–(5.9) are assumed, there is always an integer $N = N(k)$ such that the steps (iv)–(xi) of the iterative method can be realized.*

Proof: The numbers μ_k defined in step (iv) are uniformly bounded, it exists a real number $\delta > 0$ such that $\mu_k \leq T^2 \delta$ for all k. Without loss of generality $C > \delta$, where C is chosen according theorem 1. Then $0 \leq \mu_k \leq CT^2$. For every k we take $M = M(k)$ as the integer for which $2^{-M} \leq \frac{\mu_k}{CT^2} \leq 2^{1-M}$. Put $h = 2^{-M}T$ and find $\tau, u_{\tau,h}$ from the steps (vi),(vii). Then

$$\mathcal{F}(u_{\tau,h}) - \mathcal{F}(u_k) \leq Ch^2 - \frac{2h}{T}\mu_k \leq -\frac{h}{T}\mu_k \leq \frac{-1}{2CTT}\mu_k^2 \,.$$

Take $N = N(k)$ to be the smallest integer M which satisfies (ix).

We want to show that the controls u_k computed by the iterative method fulfil the necessary optimality condition asymptoticly. Obviously, $\mu_k = 0$ implies

$$H(t, x_k(t), \Psi_k(\cdot), u_k(t)) = \max_{v \in U} H(t, x_k(t), \Psi_k(\cdot), v) \,, \quad t \in J \,.$$

Theorem 2. *Suppose $\inf\{\mathcal{F}(u) \mid u \text{ admissible}\} > -\infty$ and assume (5.1)–(5.9). Then*

$$\lim_{k \to \infty} \mu_k = 0 \,.$$

Proof: Defining $\beta = \frac{1}{2CTT}$ and using the proof of lemma 8 we obtain

$$\beta \sum_{k=0}^{m} \mu_k^2 \leq \mathcal{F}(u_0) - \mathcal{F}(u_m) \leq \mathcal{F}(u_0) - \inf \mathcal{F}(u) < \infty \quad \text{for all positiv integers } m \,.$$

Since $\beta > 0$ the serie $\sum_{k=0}^{\infty} \mu_k^2$ converges, therefore $\lim_{k \to \infty} \mu_k = 0$.

Consider control processes (2.1)–(2.3) separated in e and u, which have the special view

$$(5.15) \qquad f(s,t,e,u) = A(s,t)e + B(s,t,u)$$

$$(5.16) \qquad g(t,e,u) = a(t,e) + b(t,u) \,; \quad s,t \in J, \ e \in E, \ u \in U$$

where $a(t, e)$ is assumed to be convex in $e \in E$ for every $t \in J$.

Assume (5.1)–(5.9), that means

$$(5.4') \qquad \|a_e(t, e_1) - a_e(t, e_2)\| \le b_4(t)\|e_1 - e_2\|$$

$$(5.5') \qquad \|B(s, t, u_1) - B(s, t, u_2)\| \le c_1(t, u_1, u_2) \le c_1'$$

$$(5.8') \qquad \|a(t, e) + b(t, u)\| \le d_3(t) \, , \ \|a_e(t, e)\| \le d_2(t)$$

for all $s, t \in J$, $e \in \mathcal{E}$, $u \in U$,
and, if (2.2) is a Fredholm integral equation,

$$(5.9') \qquad \max_t \int_0^T \|A(s, t)\| ds < 1 \, , \qquad \max_s \int_0^T \|A(s, t)\| dt < 1 \, .$$

By applying the method of successive iteration to such processes we obtain controls converging in the sense of functional:

Theorem 3. *Let u_k be computed by the iterative method for problem (2.1)–(2.3) with specifications (5.15),(5.16), suppose (5.4')–(5.9')and $\inf \mathcal{F}(u) > -\infty$. There exists $L > 0$ such that*

$$(5.17) \qquad \mathcal{F}(u_k) - \inf \mathcal{F}(u) \le \frac{L}{k} \qquad \text{for all } k.$$

Proof: If we put $b_k = \mathcal{F}(u_k) - \inf \mathcal{F}(u)$, lemma 8 and the proof of theorem 2, respectively, imply $b_{k+1} - b_k = \mathcal{F}(u_{k+1}) - \mathcal{F}(u_k) \le -\beta \mu_k^2$. Concider an arbitrary sequence of controls $\{\tilde{u}_j\}$ with $\lim_{j \to \infty} \mathcal{F}(\tilde{u}_j) = \inf \mathcal{F}(u)$.
Since the process is separated lemma 2 and the convexity of $a(t, e)$ give the inequality

$$\mathcal{F}(\tilde{u}_v) - \mathcal{F}(u_k) \ge \int_0^T [H(t, x_k(t), \Psi_k(\cdot), u_k(t)) - H(t, x_k(t), \Psi_k(t), \tilde{u}_v(t))] \, dt \ge -\mu_k$$

for all v. It follows $b_k \le \mu_k$ and $b_{k+1} - b_k \le \beta b_k^2$ for all positive integers k. From a lemma of Luybushin [6] follows $b_k \le \frac{L}{k}$ for all k, where $L = \max\{b_1, \frac{2}{\beta}\}$. Indeed, define $L_n = nb_n$ for all n. Then $L_{n+1} \le L_n[1 + (1 - \beta L_n)\frac{1}{n} - \frac{\beta}{n^2}L_n]$. Since $n + 1 \le 2n$ and $\beta L_n > 0$ it is $L_{n+1} \le 2L_n$ for $n \ge 1$. The case $\beta L_n \le 1$ is trivial.
Suppose $1 - \beta L_j \le 0$ for all $j = 1, \ldots, n$ then $b_1 > L_2 > \ldots > L_n > L_{n+1}$. In the complementary case $1 - \beta_j \le 0$ for $2 \le k \le j \le n$ and $1 - \beta L_{k-1} > 0$ we obtain

$L_k > L_{k+1} > \ldots > L_n > L_{n+1}$ and $L_{k-1} < \frac{1}{\beta}$. From $L_k < 2L_{k-1}$ follows $L_j < \frac{2}{\beta}$ for $j = k-1, \ldots, n+1$.

Remark: The method of successive approximations described here looks like the method of conditioned gradients in nonlinear programming. The convergence theorems have similarities with those known in programming.

Example: To minimize is

$$\mathcal{F}(u) = \int_0^2 \left[x_1(t)u_1(t) + x_2(t)u_2(t) + \alpha x_1(t)x_2(t) \right] dt$$

subject to

$$x_1(s) = -1 + \int_0^s \left[tx_1(t)u_1(t) + x_2(t) \right] dt$$

$$x_2(s) = 1 + \int_0^s \left[x_2(t)u_1(t) - u_2(t) \right] dt$$

$$0 \leq s \leq 2, \ |u_i(t)| \leq 1, \ i = 1, 2.$$

The iterative method was realized on a computer EC1035. The system of integral (differential) equations was solved by means of the procedure FA6C (Runge–Kutta–method of order 7 with Bulirsch–control for the stepsize) for some chosen real α. The optimal controls are bang–bang, they have only one switching. For instance we found

α	$u_1 = 1$	$u_1 = -1$	$u_2 = 1$	$u_2 = -1$	$\mathcal{F}(u)$
-5	$0 \leq t \leq 1.97$	$1.97 < t \leq 2$	–	$0 \leq t \leq 2$	-374
0	$0 \leq t \leq 1.53$	$1.53 < t \leq 2$	–	$0 \leq t \leq 2$	-9.4
5	–	$0 \leq t \leq 2$	$0 \leq t \leq 1.81$	$1.81 < t \leq 2$	0.8

The results were verificated by solving the so–called Π–system.

6 Control processes with functionals of the generalized Mayer type

Let φ be a continuous functional on E^n; $\varphi = \varphi(e_1, \ldots, e_n)$, $e_1, \ldots, e_n \in E$. Assume the existence of continuous partial Frèchet–derivatives φ_{e_i} with respect to all $e_i \in$

E, $i = 1, \ldots, n$. Given are timepoints $0 \le t_1 < \ldots < t_n \le T$. What happens if the functional

$$(6.1) \qquad \mathcal{F}(u) = \varphi(x(t_1), \ldots, x(t_n))$$

is to minimize subject to (2.2),(2.3) without changing the assumptions concerning f? We obtain the adjoint equation corresponding to $\bar{x}, \bar{u}$ as

$$(6.2) \quad \Psi(t) = \int_0^T \Psi(s) f_e(s, t, \bar{x}(t), \bar{u}(t)) ds - \sum_{i=1}^n \varphi_{e_i}(\bar{x}(t_1), \ldots, \bar{x}(t_n)) f_e(t_i, t, \bar{x}(t), \bar{u}(t)), \quad t \in J$$

and the H–function on $J \times E^{n+1} \times PCL(J, E^*) \times U$ as

$$(6.3) \quad H(t, e, e_1, \ldots, e_n, \Psi(\cdot), u) = \int_0^T \Psi(s) f(s, t, e, u) ds - \sum_{i=1}^n \varphi_{e_i}(e_1 \ldots, e_n) f(t_i, t, e, u).$$

In [9] we proved

Lemma 9. *If $\bar{x}$, $\bar{u}$ is an optimal solution of* (6.1),(2.2) (2.3) *and $\bar{\Psi}$ of* (6.2) *then*

$$H(t, \bar{x}(t), \bar{x}(t_1), \ldots, \bar{x}(t_n), \bar{\Psi}(\cdot), \bar{u}(t)) = \max_{v \in U} H(t, \bar{x}(t), \bar{x}(t_1), \ldots, \bar{x}(t_n), \bar{\Psi}(\cdot), v), \quad t \in J.$$

Lemma 10. *Let be* $x, u; \bar{x}, \bar{u}$ *admissible solutions of* (2.2),(2,3). *Define*
$\bar{\varphi}_{e_i} = \varphi_{e_i}(\bar{x}(t_1), \ldots, \bar{x}(t_n))$, $\bar{\varphi} = \varphi(\bar{x}(t_1), \ldots, \bar{x}(t_n))$,

$$(6.4) \quad P(t) = \sum_{i=1}^n \bar{\varphi}_{e_i}[f(t_i, t, x(t), u(t)) - f(t_i, t, \bar{x}(t), u(t)) - f_e(s, t, \bar{x}(t), u(t)) \Delta x(t)]$$

$$(6.5) \quad Q = \varphi(x(t_1), \ldots, x(t_n)) - \bar{\varphi} - \sum_{i=1}^n \bar{\varphi}_{e_i} \Delta x(t_i)$$

$$(6.6) \quad DH(t) = H(t, \bar{x}(t), \bar{x}(t_1), \ldots, \bar{x}(t_n), \bar{\Psi}(\cdot), u(t)) - \bar{H}(t), \quad t \in J.$$

Then $\quad \varphi(x(t_1), \ldots, x(t_n)) - \bar{\varphi} = Q - \int_0^T DH(t) dt - \int_0^T DH_e(t) \Delta x(t) dt + \int_0^T [P(t) - R(t)] dt$

The proof is a consequence of the definitions (6.3)–(6.4).

Theorem 4. *If there are real numbers b_5^i, d_i such that*
$$\|\varphi_{e_i}(e_1,\ldots,e_n) - \varphi_{e_i}(e_1',\ldots,e_n')\| \le b_5^i \|e_i - e_i'\| \text{ and } \|\varphi(e_1,\ldots,e_n)\| \le d_i$$
for all $e_i, e_i' \in \mathcal{E}$, $i = 1,\ldots,n$ then

$$\mathcal{F}(u) - \mathcal{F}(u\prime) \le -\int_0^T \Delta H(t)dt +$$

$$+\int_0^T \int_0^T \|\Psi_0(s)\|[b_1(t,u(t),u_0(t)) + b_2(t)\|\Delta x(t)\|] \, \|\Delta x(t)\| ds dt +$$

$$+\sum_{j=1}^n d_j \int_0^T b_1(t,u(t),u_0(t)) \, |\Delta x(t)\| dt + \sum_{j=1}^n d_j \int_0^T b_2(t)\|\Delta x(t)\|^2 dt +$$

$$+\max_j \|\Delta x(t_j)\| \sum_{i=1}^n b_5^i$$

and the corresponding adjoint function Ψ is uniformly bounded. Therefore the inequality of theorem 1 is valid, too. The controls u_k obtained by the method of successive approximations have the property $\lim_{k\to\infty} \mu_k = 0$ given in theorem 2, too.

Theorem 5. *Let the integral process be separated in e and u, see (5.15) and suppose φ is convex, $\inf \mathcal{F}(u) > -\infty$. Then we find $L > 0$ such that*

$$\mathcal{F}(u_k) - \inf \mathcal{F}(u) \le \frac{L}{k} \qquad \text{for all } k.$$

Proof: By means of lemma 10, theorem 4 and the convexity of φ we estimate $b_k \le \mu_c$, where μ_k is defined in (iv). Now we follow the proof of theorem 3.

References

[1] L.Bittner, *Ein Modell für eine Klasse von Aufgaben optimaler Steuerung.* ZAMM 58 (1978), 251–260.

[2] F.L.Chernousko, *A.A.Lyubushin, Method of successive approximations for solution of optimal control problems.* Optimal Control Appl. & Methods. Vol.3 (1982), 101–114.

[3] P.N.Fedorenko, *Approximative solution of optimal control problems* (in Russian). Science. Moscow 1978.

[4] W.Kampowsky, *Optimal Control in Nonlinear Evolution Equations with Constraints.* ZAMM 72 (1991) 7/8. 277–288.

[5] J.A.Krylov, F.L.Chernousko, *An Algorithm of successive approximations in control problems* (in Russian).
Jour. Num. Math./Math. Phys. 1962, 2, Nr.6, 1132–1139.

[6] A.A.Lyubushin, *Modifications and studies of convergence of the method of successive approximations in optimal control problems* (in Russian).
Jour. Num. Math./Math. Phys. 1979, 19, Nr.6, 1414–1421.

[7] L.I..Rozonoer, *The maximum principle of L.S.Pontryagin in the theory of optimal control* (in Russian) I,II. Autom. and Telemech. 20, 10 (1059), 1320–1334 and 20,11 (1959), 1441–1458.

[8] W.H.Schmidt, *Durch Integralgleichungen beschriebene optimale Prozesse mit Neben- bedingungen in Banachräumen—notwendige Optimalitätsbedingungen.* ZAMM 62 (1982), 65-75.

[9] W.H.Schmidt, *Optimalitätsbedingungen für verschiedene Aufgaben von Integralprozessen in Banachräumen und das Iterationsverfahren von Chernousko.* Habilitation. Greifswald 1988.

[10] W.H.Schmidt, *Necessary optimality conditions for abstract control problems with certain constraints.* submitted to ZAMM 1992.

Author's address

Werner H.Schmidt, Fachbereich Mathematik/Informatik, Ernst–Moritz–Arndt-Universität, L.-Jahn-Str. 15a, O-2200 Greifswald, Bundesrepublik Deutschland

Solving Equations –
a Problem of Optimal Control

Leonhard Bittner

Abstract

This paper connects embedding methods for solving equations with control theory. The question how to select the arbitrary functions in homotopy statements motivates new problems of optimal control.

1 Introduction

Homotopy methods connect an equation $F(x) = 0$, to be solved, with another equation $F_o(x) = 0$, easily solvable, by means of a family of equations $H(x,t) = 0, 0 \leq t \leq 1$, such that $H(x,0) = F_o(x), H(x,1) = F(x)$. In this paper we take into account special kinds of families formed with the aid of certain functions v arbitrary apart from few restrictions. We consider these arbitrary functions as steering functions and formulate problems of optimal control.

I)Let X, Y be Banach spaces, $F : X \to Y$ a sufficiently smooth mapping with a (Frechet) derivative $F'(x), x_o$ an approximate solution of the equation $F(x) = 0, t$ a real variable, $v : [0,1] \to R$ a real sufficiently smooth function of t with the values $v(0) = 1, v(1) = 0$, f.e. $v(t) = 1 - t$. Define a differentiable trajectory $x : [0,1] \to X$ starting at $x(0) = x_o$ by means of

$$(1.1) \quad F(x(t)) = v(t)F(x_o) = v(t)F_o$$

or by means of the resulting initial value problem

$$(1.2) \quad F'(x)\dot{x} = \dot{v}(t)F_o = F_o\dot{v}(t), x(0) = x_o.$$

II) Assume $X = Y = R^n, F(x) = (F_1(x), ..., F_n(x))^T \in R^n, F_i'(x) = grad F_i(x)$. Let $v_i : [0, 1] \to R$ $(i = 1, ..., n)$ be real differentiable functions with values $v_i(0) = 1, v_i(1) = 0$. Define a smooth trajectory $x : [0, 1] \to R^n$ starting at $x(0) = x_o$ and satisfying the equations

$$(1.3) \quad F_i(x(t)) = v_i(t)F_i(x_o) \quad (i = 1, ..., n)$$

or satisfying the initial value problem

$$(1.4) \quad F_i'(x(t))\dot{x} = \dot{v}_i(t)F_i(x_o) \quad (i = 1, ..., n), x(0) = x_0$$

With the aid of a diagonal matrix F_o and a vector function v , i.e.

$$(1.5) \quad \mathbf{F_0} = \begin{pmatrix} F_1(x_0) & 0 & \cdots & 0 \\ 0 & F_2(x_0) & \cdots & 0 \\ \cdots\cdots\cdots\cdots\cdots\cdots\cdots\cdots \\ 0 & 0 & \cdots & F_n(x_0) \end{pmatrix}$$

$$v(t) = (v_1(t), v_2(t), \cdots v_n(t))^T$$

the relations (1.3),(1.4) can formally be written in the same manner as (1.1),(1.2).

III) More generally, let X, Y, F be as in example I and let V be a sufficiently smooth mapping $V : X \times [0, 1] \to Y$ with the partial (Frechet) derivatives $V_x(x, t), V_t(x, t)$ and the values

$$(1.6) \quad V(x_o, 0) = F(x_o), V(x, 1) = 0 \quad \text{f.a.} \quad x \in X,$$

for example

$$V(x, t) = v(t)(F(x_o) + M(x) - M(x_o)),$$

where M denotes a smooth mapping from X into Y and v a real function with values $v(0) = 1, v(1) = 0$. Define a smooth trajectory $x : [0, 1] \to X$ by

$$(1.7) \quad F(x(t)) = V(x(t), t), x(0) = x_o$$

or by

$$(1.8) \quad (F'(x) - V_x(x, t))\dot{x} = V_t(x, t), x(0) = x_o$$

IV) This is a particular case of example III. Let F be composed of an affine and a minor nonlinear mapping, i.e.

$$(1.9) \quad F(x) = Ax - b + C(x),$$

where $A : X \to Y$ is linear, bounded, one-to-one and onto, $b \in Y, C : X \to Y$ differentiable and $Ax_0 - b = 0$. Pose

$$(1.10) \quad V(x,t) = v(t)C(x),$$

where v has the previous meaning. Then indeed

$$V(x_o,0) = C(x_o) = Ax_o - b + C(x_o) = F(x_o), V(x,1) = 0.$$

Note that

$$H(x,t) = F(x) - V(x,t) = Ax - b + (1 - v(t))C(x)$$

is a homotopy connecting $Ax - b$ with $F(x)$ and that the initial value problem (1.8) now reads

$$(1.11) \quad (I + (1 - v(t))A^{-1}C'(x))\dot{x} = \dot{v}(t)A^{-1}C(x), x(0) = x_o.$$

Evidently, in each example the value $x(1)$ of the trajectory, if it exists, is a solution of the equation $F(x) = 0$.

2 Existence of a solution

The following statements provide some conditions which guarantee the existence of a solution of the initial value problems up to $t = 1$.

Theorem 1. a) Assume that F is (Frechet) differentiable in the ball $B(x_o,r) = \{x \in X \mid \|x - x_o\| < r\}$ and that the derivative $F'(x)$ has a unique bounded inverse satisfying a Lipschitz condition

$$\|F'^{-1}(x)F_o - F'^{-1}(x_o)F_o\| \le L\|x - x_o\|.$$

Assume that

$$\|F'^{-1}(x_o)F_o\| \int_o^1 \|\dot{v}(s)\| \exp(L \int_s^1 \|\dot{v}(\sigma)\| \, d\sigma) \, ds < r.$$

Then the initial value problem (1.2),(1.4) has a unique solution $x(\cdot)$ in the whole intervall $[0,1]$ and $x(1)$ is a solution of the equation $F(x) = 0$ in $B(x_o, r)$.

b) Assume that F is twice continuously differentiable in the ball $B(x_o, r)$ and that the inverse $F'^{-1}(x_o)$ exists. Pose

$$a = \|F'^{-1}(x_o)F_o\|,$$

$$b = sup\{\|F'^{-1}(x_o)F''(x)\| \mid \|x - x_o\| < r\}$$

and suppose

$$2ab \int_o^1 \|\dot{v}(s)\| \, ds < 1,$$

$$1 - \sqrt{1 - 2ab \int_o^1 \|\dot{v}(s)\| \, ds} < b \, r.$$

Then there exists a solution $x(\cdot)$ of the initial value problem (1.2),(1.4) in the whole unit intervall and $F(x(1)) = 0$ (c.[5], ch.XVIII).

c) Suppose that C is continuously differentiable in a ball $B(x_o, r)$ and $Ax_o - b = 0$. Pose

$$a = \|A^{-1}C(x_o)\|,$$

$$b = sup\{\|A^{-1}C'(x)\| \mid \|x - x_o\| < r\}$$

and suppose that

$$|1 - v(t)|b < 1 \quad \text{f.a.} \quad t \in [0,1],$$

$$\int_o^1 \frac{a \, |\dot{v}(s)|}{1 - |1 - v(s)| \, b} \exp(\int_s^1 \frac{|\dot{v}(\sigma)| \, b}{1 - |1 - v(\sigma)| \, b} d\sigma) ds < r.$$

Then there exists a solution $x(\cdot)$ of the initial value problem (1.11) over $[0,1]$ and $F(x(1)) = 0$. In case of a nonincreasing v the conditions mentioned above simply read

$$b < 1 \quad , \quad a/(1-b) < r.$$

Proof a) The solution $x(\cdot)$ of (1.2) and (1.4) exists for small $t > 0$ and is prolongable beyond any $t < 1$ with $x(t) \in B(x_o, r)$. For as long as $x(t)$ exists and belongs to $B(x_o, r)$ there follows

$$\|x(t) - x_o\| = \left\| \int_o^t F'^{-1}(x(s)) F_o \dot{v}(s) ds \right\|$$

$$\leq \int_o^t \left\| [F_o'^{-1} F_o + (F'^{-1}(x(s)) - F'^{-1}(x_o)) F_o] \dot{v}(s) ds \right\|$$

$$\leq \int_o^t \| F_o'^{-1} F_o \| \|\dot{v}(s)\| ds + \int_o^t L \|x(s) - x_o\| \|\dot{v}(s)\| ds$$

where $F_o' = F'(x_o)$. Hence by Gronwall's lemma

$$\|x(t) - x_o\| \leq \| F_o'^{-1} F_o \| \int_o^t \|\dot{v}(s)\| exp\left(\int_s^t L \|\dot{v}(\sigma)\| d\sigma \right) ds.$$

But the righthand side remains smaller than r for all $t \leq 1$. Thus $x(1)$ is defined up to $t = 1$.

b) Put

$$F_o^{(i)} = F^{(i)}(x_o), \quad F^{(i)} = F^{(i)}(x) \quad (i = 1, 2), \quad 0 < \epsilon \ll 1.$$

If

$$\|x - x_o\| < r \quad \text{and} \quad b \|x - x_o\| < 1$$

holds, then

$$F'^{-1} = (F_o'[I + F_o'^{-1}(F' - F_o')])^{-1}$$

$$= [I + F_o'^{-1}(F' - F_o')]^{-1} F_o'^{-1}$$

exists, since

$$\|F_o'^{-1}(F' - F_o')\| \le b\,\|x - x_0\| < 1.$$

Further

$$\|F'^{-1}F_o\| = \|F_o'^{-1}F_o + F_o'(F_o'^{-1} - F')F'^{-1}F_o\|$$

$$\le a + b\,\|x - x_0\|\,\|F'^{-1}F_o\|,$$

$$\|F'^{-1}F_o\| \le a/(1 - b\,\|x - x_0\|)$$

is valid. Of course, the solution $x(t)$ is defined for sufficiently small positive t and is prolongable beyond t, if $\|x(t) - x_0\| < r$ and $F'^{-1}(x(t))$ exists, which is the case, if $b\,\|x(t) - x_0\| < 1$ besides $\|x(t) - x_0\| < r$. Thus, while $\|x(s) - x_0\| < r$ and $b\,\|x(s) - x_0\| < 1, 0 \le s \le t$, the integration of the differential equation (1.11) yields the integral inequality

$$\|x(t) - x_0\| = \left\| \int_o^t F'^{-1}(x(s))F_o\dot{v}(s)ds \right\|$$

$$< \int_o^t \frac{a\,\|\dot{v}(s)\|}{1 - b\,\|x(s) - x_0\|}\,ds + \epsilon$$

for every positive ϵ. Let w_ϵ be the solution of a corresponding integral equality

$$w_\epsilon(t) = \int_o^t \frac{a\,\|\dot{v}(s)\|}{1 - b\,w_\epsilon(s)}\,ds + \epsilon$$

or the equivalent differential equation

$$\dot{w}_\epsilon = \frac{a\,\|\dot{v}(t)\|}{1 - b\,w_\epsilon}, \quad w_\epsilon(0) = \epsilon,$$

i.e.

$$w_\epsilon(t) = \left[1 - \sqrt{(1 - 2ab \int_o^t \|\dot{v}(s)\|\,ds + b^2\epsilon^2 - 2b\epsilon)}\,\right]/b.$$

For $0 < \epsilon \ll 1$ and all $t \in [0, 1]$ the radicand remains positive, $w_\epsilon(t)$ is well defined and is $\le w_\epsilon(1) < r$ according to the conditions of the theorem. Evidently

$$\|x(s) - x_0\| < w_\epsilon(s)$$

for $s = 0$ and small positive s. If there were a first positive t such that

$$\|x(t) - x_o\| = w_\epsilon(t),$$

the integral inequality would imply

$$\|x(t) - x_o\| < \int_o^t \frac{a\,\|\dot{v}(s)\|}{1 - b\,w_\epsilon(s)}\,ds + \epsilon = w_\epsilon(t),$$

a contradiction. Hence for all $t \in [0,1]$, for which $x(t)$ is defined,

$$\|x(t) - x_o\| < w_\epsilon(t) \le w_\epsilon(1) < r$$

and

$$b\,\|x(t) - x_o\| \le bw_\epsilon(1) < 1,$$

which implies that $x(\cdot)$ is defined up to $t = 1$.

c) According to the assumptions

$$\|A^{-1}C(x)\| = \|A^{-1}C(x_o) + A^{-1}(C(x) - C(x_o))\| \le a + b\,\|x - x_o\|$$

holds in $B(x_o, r)$. Thus integration of the differential equation (1.11) yields

$$\|x(t) - x_o\| \le \int_o^t \|(I + (1 - v(s))A^{-1}C'(x(s)))^{-1}A^{-1}C(x(s))\| \cdot |\dot{v}(s)|\,ds$$

$$\le \int_o^t \frac{a + b\,\|x(s) - x_o\|}{1 - |1 - v(s)|\,b}|\dot{v}(s)|\,ds \quad \text{for} \quad 0 < t \ll 1$$

or, more generally, for all t with $x(s) \in B(x_o, r), 0 \le s \le t$. By Gronwall's lemma and the conditions of theorem 1c

$$\|x(t) - x_o\| < \int_o^t \frac{a\,|\dot{v}(s)|}{1 - |1 - v(s)|\,b}\exp\Big(\int_s^t \frac{b\,|\dot{v}(\sigma))|}{1 - |1 - v(\sigma)|\,b}\,d\sigma\Big)\,ds < r.$$

This means that $x(t)$ cannot end before $t = 1$.

3 How to select the controls v

The functions v,V are arbitrary apart from certain restrictions. A first idea might be to choose them in such a way that the left sides of the inequalities, which they have to fulfil in theorem 1a, 1b or 1c respectively, be as small as possible.

Theorem 2. The left sides of the inequalities, which have to be fulfilled by the (vector) function v in theorem 1a, 1b, 1c, become as small as possible, if (each component v_i of) v is a nonincreasing function of t (independently of i), especially if (each component v_i of) $v(t)$ is equal to $1 - t$. (Here, in case of an n-vector function v it is simply assumed that the norm of an n-vector is the maximum of the absolute values of its components.)

Proof. a) Let $< w_i >$ be an abbreviation for an n-vector with components w_i. Then

$$\| < [\exp L - 1]/L > \| = \| \int_o^t \frac{d}{ds} < \exp{(Lv_i(s))}/L > ds \| \leq$$

$$\leq \int_o^1 \| < \dot{v}_i \exp{(Lv_i)} > \| \, ds \leq \int_o^1 \| < |\dot{v}_i(s)| \exp{(L \int_1^s \dot{v}_i(\sigma) \, d\sigma)} > \| \, ds \leq$$

$$\leq \int_o^1 \| < |\dot{v}_i(s)| \exp{(L \int_s^1 \|\dot{v}(\sigma)\| \, d\sigma)} > \| \, ds = \int_o^1 \|\dot{v}(s)\| \exp{(L \int_s^1 \|\dot{v}(\sigma)\| d\sigma)} \, ds.$$

Because of

$$v_1(t) = .. = v_n(t), \|\dot{v}(t)\| = |\dot{v}_i(t)| = -\dot{v}_i(t), \quad 0 \leq t \leq 1,$$

the last item of the inequalities coincides with the first.

b)Obviously

$$\| < -1 > \| = \|v(1) - v(0)\| = \| \int_o^1 \dot{v}(s) \, ds \| \leq \int_o^1 \|\dot{v}(s)\| \, ds.$$

Equality holds if all components coincide and are nonincreasing.

c) For each function v, which satisfies

$$|1 - v(t)| \, b < 1, \quad 0 \leq t \leq 1,$$

one gets

$$\frac{a}{1-b} = \int_o^1 \frac{a(-\dot{v}(s))}{1-b(1-v(s))} \exp\Big(\int_s^1 \frac{-b\dot{v}(\sigma)}{1-b|1-v(\sigma)|}\,d\sigma\Big)\,ds.$$

$$\leq \int_o^1 \frac{a|\dot{v}(s)|}{1-b|1-v(s)|} \exp\Big(\int_s^1 \frac{b|\dot{v}(\sigma)|}{1-b|1-v(\sigma)|}\,d\sigma\Big)\,ds.$$

Equality takes place, if v is nonincreasing.

Initial value problems

$$\dot{x} = g(x,t), \quad x(0) = x_o, \quad 0 \leq t \leq 1,$$

like those of (1.2), (1.4), (1.8), (1.11) can usually be solved only by numerical methods, f.e. by means of predictor, corrector methods. Let N be a positive integer and $h = 1/N$ the constant step of a predictor method

$$x_{k+1} = x_k + h\sum_{i=o}^{q-2} \gamma_i\, g(x_{k-i},(k-i)h) \quad (k = q-2,\ldots,N-1).$$

Assume for the sake of simplicity that the initial values x_k $(k = 0,\ldots,q-2)$ coincide with the exact values $x(kh)$ of the solution $x(\cdot)$. Then an error estimation shows that

$$\|x(1) - x_N\| \leq c\cdot h^{q-1}\cdot \sup_{o\leq t\leq 1} \|x^{(q)}(t)\|,$$

where c depends on a Lipschitz constant of g with respect to x in a neighbourhood of the solution $x(\cdot)$ and on N, q and γ_i. A similar estimation holds for a corrector method. Suppose that the numerical integration of (1.2), (1.4), (1.8), (1.11) is performed for different controls v, V with the same number N of steps. The final numerical value x_N differs from the root $x(1)$ to be computed according to an estimation mentioned above. The third term of the right side of the estimation is the essential part determined by the applied controls, the second does not depend on them at all and the first is of minor importance. The aim should be to assure that the third term be small. This aim suggests to formulate the following task.

4 New problem of optimal control

Given a dynamical system

$$(4.1) \quad \dot{x} = f(x,v(t),\dot{v}(t)), \quad 0 \leq t \leq 1, \quad x(0) = x_o$$

with a sufficiently smooth mapping f, find such a control v out of a certain set of arbitrary q-times differentiable controls v, which minimizes the functional

$$(4.2) \qquad \Phi(v) = \sup_{o \leq t \leq 1} \|x^{(q)}(t)\|$$

or minimizes a certain mean value

$$\Phi_p(v) = (\int_o^1 \|x^{(q)}(t)\|^p \, dt)^{1/p}.$$

The values $\Phi(v)$ represent some measures of curvature of the trajectory, $\Phi(v)$ is the arc length if $q = 1$. In the sequel we are only concerned with Φ. By means of the total derivatives of f with respect to the differential equation, i.e. by means of

$$f^{(1)}(x, v_o, v_1) = f(x, v_o, v_1),$$

$$f^{(i)}(x, v_o, v_1, \ldots, v_i) = f_x^{(i-1)} f + \sum_{k=o}^{i-1} f_{v_k}^{(i-1)} v_{k+1} \quad (i = 2, \ldots, q)$$

$\Phi(v)$ gets a more convenient appearance

$$\Phi(v) = \sup_{o \leq t \leq 1} \|f^{(q)}(x(t), v(t), \ldots, v^{(q)}(t))\|.$$

Φ can be investigated by its directional derivative. If X is a Hilbert space with scalar product $[\cdot, \cdot]$, then, under appropriate conditions, it is possible to prove that Φ has a directional derivative (c.[4])

$$\Phi'(v, \delta v) = \sup\{[x^{(q)}(t), \delta x^{(q)}(t)]/\Phi(v) \mid t \in T(x)\},$$

where $x = x(\cdot)$ corresponds to $v = v(\cdot)$ according to (4.1) and

$$x^{(q)}(t) = f^{(q)}(x(t), v(t), ..., v^{(q)}(t)),$$

$$\delta x^{(q)}(t) = f_x^{(q)}(x(t), ..., v^{(q)}(t)) \, \delta x(t) + \sum_{k=o}^{q} f_{v_k}^{(q)} \, \delta v^{(k)}(t),$$

$$\delta \dot{x}(t) = f_x(x(t), v(t), \dot{v}(t)) \delta x(t) + f_{v_o} \delta v(t) + f_{v_1} \delta \dot{v}(t),$$

$$\delta x(0) = 0,$$

$$T(x) = \{t \in [0,1] \mid \|x^{(q)}(t)\| = \Phi(v)\}.$$

Necessary for v to achieve the minimum of Φ is the existence of a $t \in T(x)$ for each $\delta x(\cdot)$ corresponding to a feasible increment δv such that

$$[x^{(q)}(t), \delta x^{(q)}(t)] \geq 0.$$

5 Special cases

5.1 Refering to the particular exemples I - IV let us at first consider

$$(5.1) \qquad f(x,v,\dot v) = F'^{-1}(x)F_o\dot v \quad \text{for} \quad F(x) = Ax - b,$$

where A is a linear, bounded, one-to-one mapping of the Banach space X onto the Banach space Y and b an element of Y. No constraints are posed on the (vector) function v besides those mentioned above in section 1. The solution of the initial value problem (4.1) or (1.2) reads as follows

$$x(t) = A^{-1}b + A^{-1}F_o v(t).$$

Thus

$$x^{(q)}(t) = A^{-1}F_o v^{(q)}(t),$$

$$\Phi(v) = \sup_{o \leq t \leq 1} \|A^{-1}F_o v^{(q)}(t)\|.$$

Theorem 3. Φ attains its minimum zero, if v is equal to a (vector) polynom of degree $\leq q - 1$. (Each component of) such a minimizing function reads

$$(5.2) \qquad v(t) = 1 - t + a_2(t^2 - t) + \cdots a_{q-1}(t^{q-1} - t)$$

with arbitrary real coefficients $a_2, \ldots, a_{q-1}$.

Proof. (Each component of) $v(0) = 1, v(1) = 0$. Moreover $v^{(q)}(t) = 0$ f.a. t. Hence $\Phi(v) = 0$.

5.2 Let f,A,b,X,Y be as defined above in 5.1, but

$$(5.3) \qquad F(x) = Ax - b + B(x),$$

where B is a small sufficiently smooth mapping of X into Y, ϵ a small real number and

$$(5.4) \qquad B(x) = \epsilon D(x), \quad \|A^{-1}D^{(k)}(x)\| \leq d_k \quad (k = 1, \ldots, q) \quad \text{f.a.} \quad x$$

We prove that an inequality

$$\|x^{(q)}(t)\| \leq \|A^{-1}F_o v^{(q)}(t)\|/(1 - \epsilon d_1) + \epsilon\omega(\|\dot{v}(t)\|, \ldots, \|v^{(q-1)}(t)\|, \epsilon)$$

is valid, where ω denotes a certain polynomial function of its first $q - 1$ arguments with coefficients depending on ϵ, but bounded for $\epsilon \to 0$. Indeed, successive differentiation of the differential equation

$$(A + \epsilon D')\dot{x} = F_o \dot{v}$$

implies

$$(A + \epsilon D')\ddot{x} = F_o \ddot{v} - \epsilon D'' \dot{x}^2,$$

$$(A + \epsilon D')x^{(k)} = F_o v^{(k)} - \epsilon \sum m(l, n_1, \ldots, n_{l-1})D^{(l)}(\dot{x})^{n_1} \cdots (x^{(l-1)})^{n_{l-1}}$$

where the summation has to be carried out over all nonnegative integers $l, n_1, \ldots, n_{l-1}$ satisfying $2 \leq l \leq k - 1, n_1 + \ldots + n_{l-1} = l$ and the coefficients $m(l, n_1, \ldots)$ denote certain positive numbers. Successive elimination of $x, \dot{x}, \ldots$ yields

$$\|\dot{x}(t)\| \leq \|A^{-1}F_o\dot{v}(t)\|/(1 - \epsilon d_1) \leq \|A^{-1}F_o\| \, \|\dot{v}(t)\|/(1 - \epsilon d_1)$$

$$\|\ddot{x}(t)\| \leq \|A^{-1}F_o\ddot{v}(t)\|/(1 - \epsilon d_1) + \epsilon d_2 \|A^{-1}F_o\|^2 (1 - \epsilon d_1)^{-3}\|\dot{v}(t)\|^2$$

$$\leq \|A^{-1}F_o\| \, \|\ddot{v}(t)\|/(1 - \epsilon d_1) + \epsilon\omega(\|\dot{v}(t)\|, \epsilon),$$

$$\|x^{(k)}(t)\| \leq \|A^{-1}F_o v^{(k)}(t)\|/(1 - \epsilon d_1) + \epsilon \sum m(l, n_1, \ldots, n_{l-1})d_l\|\dot{x}(t)\|^{n_1} \cdots \|x^{(l-1)}(t)\|$$

$$\leq \|A^{-1}F_o v^{(k)}(t)\|/(1 - \epsilon d_1) + \epsilon\omega(\|\dot{v}(t)\|, \ldots, \|v^{(k-1)}(t)\|, \epsilon)$$

Evidently the ω of k arguments is defineable by the ω of fewer arguments and by the values $\|A^{-1}F_o\|/(1 - \epsilon d_1), \|v^{(i)}(t)\|, i < k$, through polynomial operations. This implies

Theorem 4. If v is equal to a (vector) polynom (5.2), $\Phi(v)$ differs from $\inf \Phi$ at most by a value (ϵ).

5.3 Now let us consider

$$(5.5) \qquad f(x, V, \dot{V}) = F'^{-1}(x)V_t$$

as in exemple III, but assuming that V depends only on t. Further let us assume that $\mathcal{F}$ has a unique sufficiently smooth inverse mapping G (at least in a certain neighbourhood of x_o), i.e.

$$x = G(y) \quad \text{iff} \quad F(x) = y.$$

We want to show how the functional Φ has to be calculated. The solution of the initial value problem (1.8) for the given V is

$$x(t) = G(V(t)), \quad 0 \le t \le 1.$$

A formal Taylor expansion

$$x(t + \epsilon) = x(t) + \epsilon \dot{x}(t) + (1/2!)\epsilon^2 \ddot{x}(t) + \ldots =$$

$$= G(V(t + \epsilon)) = G(V(t) + \epsilon \dot{V}(t) + (1/2!)\epsilon^2 \ddot{V}(t) + \ldots) =$$

$$= G(V(t)) + G'(V(t))(\epsilon \dot{V}(t) + (1/2!)\epsilon^2 \ddot{V}(t) + \ldots) + \ldots$$

and comparison with respect to ϵ^q supplies the representation

$$(5.6) \qquad \frac{1}{q!}x^{(q)}(t) = \sum_{l=1}^{q} \frac{1}{l!} \sum \frac{\cdot}{k_1! \cdots k_l!} G^{(l)}(V(t))V^{(k_1)}(t) \cdots V^{(k_l)}(t),$$

where the summation for the inner sum has to carried out over all decompositions $k_1 + \cdots k_l = q$, $k_i = $ integer ≥ 1, taking account of all permutations of items. The values $G^{(l)}(y)w^l$, hence the required values $G^{(l)}(y)w_1 \cdots w_l$ of the derivatives of G, can subsequently be replaced by corresponding values of the derivatives of F. For this purpose put

$$x + \Delta x = G(y + \epsilon w) = x + \epsilon G'(y)w + (1/2!)\epsilon^2 G''(y)w^2 + \ldots.$$

Then

$$y + \epsilon w = F(x + \Delta x) = y + F'(x)[\epsilon G'(y)w + (1/2!)\epsilon^2 G''(y)w^2 \ldots]$$

$$+(1/2!)F''(x)[\epsilon G'(y)w + (1/2!)\epsilon^2 G''(y)w^2 + \ldots]^2 + \ldots \quad .$$

Comparing the coefficients of equal power of ϵ on the left and right side of this equation gives

$$(5.7) \quad w = F'(x)G'(y)w$$

and for $l \geq 2$

$$0 = \frac{1}{l!}F'(x)G^{(l)}(y)w^l \quad +$$

$$+\sum_{i=2}^{l} \frac{1}{i!} \sum \frac{1}{k_1! \cdots k_l!} F^{(i)}(x)(G^{(k_1)}(y)w^{k_1}) \cdots (G^{(k_l)}(y)w^{k_l})$$

where the summation of the inner sum has to be carried out over all decompositions $k_1 + \cdots + k_l = l$, $\quad k_i = \text{integer} \geq 1$. Multiplication by $F'^{-1}(x)$ yields a recursion formula for the subsequent values $G'(y)w, \ldots, (1/l!)G^{(l)}(y)w^l, \ldots$. $G^{(l)}(y)w_1 \ldots w_l$ is representable as an appropriate linear combination of values $G^{(l)}(y)(w_1 + \ldots - \ldots + -w_l)^l$ recursively calculated above.

References

[1] Allgower E.L.,Georg K.(1990), Numerical Continuation Methods. Springer-Verlag , Berlin

[2] Bittner L.(1967), Einige kontinuierliche Analogien von Iterationsverfahren. Funktionalanalysis, Approximationstheorie, Numerische Mathematik, ISNM Vol 7, p.114-135, Birkhäuser-Verlag

[3] Boltjanski W.G.(1971), Mathematische Methoden optimaler Steuerung. Akad. Verlagsgesellschaft, Leipzig

[4] Demjanow V.F.(1975), Einführung in Minimax - Probleme. Akad.Verlagsgesellschaft, Leipzig

[5] Kantorowitsch L.W., Akilow G.P.(1964), Funktionalanalysis in normierten Räumen, Akademie Verlag, Berlin

[6] Leitmann G.(1986), The Calculus of Variations and Optimal Control. Plenum Press, New York

Author's address

Leonhard Bittner, Fachbereich Mathematik/Informatik, Universitt Greifswald, L.-Jahn- Str. 15a, Greifswald O - 2200, BRD

On the Minimax Optimal Control Problem
and Its Variations

Ping Lu and Nguyen X. Vinh

Abstract. This paper considers some analytical properties of the minimax optimal control problem and the closely related problems. In particular, some recent development in necessary conditions is reviewed. The properties of controls, including continuity, singularity and switching are discussed. Examples are presented to signify the discussions.

1. Introduction

A minimax optimal control problem is a nonclassical optimal control problem in the sense that it is not readily in the form of problems of Bolza, Mayer and Lagrange. Interest has remained strong in seeking the solution to the problem since the 1960s and many applications have been found (for example, see Johnson (1967), Powers (1972), Miele *et al* (1982–1989)). Although a transformation first suggested by Warga (1965) easily converts the problem into a problem of Mayer, it was recent time that this technique was independently employed to derive a set of necessary conditions of optimality for the problem in a general context (Vinh *et al* (1987, 1988), Lu *et al* (1988, 1991)). These necessary conditions have rendered a powerful tool to treat minimax type of optimization problems emerging from various disciplines. Despite the progress, many interesting features and extensions of the minimax problem remain to be explored, both analytically and numerically. It is the intent of this paper to discuss some of these aspects. First, the necessary conditions for the minimax problem are reviewed. The connection between the necessary conditions for the minimax problem and those for some closely related problems is addressed. Then some properties of the optimal control such as continuity and number of switchings in these problems are investigated. This is the emphasis of this paper. Examples are presented to demonstrate the theory.

2. Minimax Problem and Its Variations

2.1 Problem Statement. The minimax optimal control problem is formally defined as the following:

$$(2.1) \qquad \min_{u} \ \{ \max_{t_0 \leq t \leq t_f} F(\mathbf{x}(t)) \},$$

$$(2.2) \qquad \dot{\mathbf{x}}(t) = \mathbf{f}(\mathbf{x}(t), \mathbf{u}(t)),$$

$$(2.3) \qquad \mathbf{u}(t) \in \mathbf{U}, \quad \forall t \in [t_0, t_f],$$

$$(2.4) \qquad \mathbf{S}_0(\mathbf{x}(t_0)) = 0,$$

$$(2.5) \qquad \mathbf{S}_f(\mathbf{x}(t_f)) = 0,$$

where $F : \mathbf{R}^n \to R$, and $\mathbf{f} : \mathbf{R}^n \times \mathbf{R}^m \to \mathbf{R}^n$ are C^1. $\mathbf{x} \in \mathbf{R}^n$ and $\mathbf{u} \in \mathbf{R}^m$ are the state vector and control vector, respectively. $\mathbf{U} \subset \mathbf{R}^m$ is a control set. Admissible controls are all piecewise continuous m-dimensional functions that satisfy Eq. (2.3). $\mathbf{S}_0$ and $\mathbf{S}_f$ represent smooth initial and terminal manifolds of dimension $n - s$ and $n - r$. The final time t_f may be prescribed, or implicitly specified by (2.5). The consideration of only autonomous systems is not necessary, but for simplicity.

The natural variations of the performance index (2.1) include:

$$(2.6) \qquad \max_{u} \ \{ \min_{t_0 \leq t \leq t_f} F(\mathbf{x}(t)) \}, \quad \max_{u} \ \{ \max_{t_0 \leq t \leq t_f} F(\mathbf{x}(t)) \}, \quad \min_{u} \ \{ \min_{t_0 \leq t \leq t_f} F(\mathbf{x}(t)) \},$$

We shall call them, accordingly, the maximin, maximax and minimin problems.

2.2 Necessary Conditions. A complete derivation of the necessary conditions for the minimax problem is available in Lu *et al* (1991). We shall reproduce some the results here for the convenience of reference in the subsequent discussions. The minimax problem of (2.1)–(2.5) is transformed into the following equivalent problem of Mayer

$$(2.7) \qquad \min\{x_{n+1}(t_f)\},$$

$$(2.8) \qquad \dot{\mathbf{x}}(t) = \mathbf{f}(\mathbf{x}(t), \mathbf{u}(t)),$$

$$(2.9) \qquad \dot{x}_{n+1} = f_{n+1}(\mathbf{x}(t), \mathbf{u}(t)) \triangleq 0,$$

$$(2.10) \qquad F(\mathbf{x}) - x_{n+1} \leq 0,$$

$$(2.11) \qquad x_{n+1}(t_0) = \text{free},$$

with constraints (2.3)–(2.5). Suppose that the pair of $\mathbf{x}^*(t)$ and $\mathbf{u}^*(t)$, $t_0 \leq t \leq t_f$, is an optimal solution pair hereafter. Suppose that $F(\mathbf{x}^*(t))$ attains its maximum k times $(k \geq 1)$ in the *open* interval (t_0, t_f) at some isolated points t_i, $i = 1, ..., k$, $t_0 < t_1 < ... < t_k < t_f$. We have the necessary conditions stated in the form of a theorem:

Theorem 1

There exists a nonzero, piecewise absolutely continuous function $(\mathbf{p}^T(t), p_{n+1}(t)) = (p_1(t), ..., p_n(t), p_{n+1}(t))$ called the adjoint state vector, such that for the Hamiltonian defined by

$$(2.12) \qquad H(\mathbf{x}, \mathbf{p}, \mathbf{u}) = \mathbf{p}^T \mathbf{f}(\mathbf{x}, \mathbf{u}) + p_{n+1} f_{n+1} = \mathbf{p}^T \mathbf{f}(\mathbf{x}, \mathbf{u}),$$

the following must be satisfied for almost all $t \in [t_0, t_f]$,

$$(2.13) \qquad \dot{\mathbf{x}}^* = \frac{\partial H(\mathbf{x}^*, \mathbf{p}, \mathbf{u}^*)}{\partial \mathbf{p}},$$

$$(2.14) \qquad \dot{\mathbf{p}} = -\frac{\partial H(\mathbf{x}^*, \mathbf{p}, \mathbf{u}^*)}{\partial \mathbf{x}},$$

$$(2.15) \qquad H(\mathbf{x}^*, \mathbf{p}, \mathbf{u}^*) = \sup_{\iota \in U} H(\mathbf{x}^*, \mathbf{p}, \mathbf{u}) = C.$$

where C is a constant. At $t_1, t_2, ..., t_k$, the jump condition holds

$$(2.16) \qquad \mathbf{p}(t_i^+) = \mathbf{p}(t_i^-) + \mu_i \frac{\partial F(\mathbf{x}^*(t_i))}{\partial \mathbf{x}},$$

$$(2.17) \qquad \mu_i \geq 0, \quad i = 1, ..., k, \quad \sum_{i=1}^{k} \mu_i = 1.$$

The transversality conditions at t_0 and t_f are

$$(2.18) \qquad \mathbf{p}(t_0) = \frac{\partial \mathbf{S}_0^T(\mathbf{x}^*(t_0))}{\partial \mathbf{x}}\eta, \quad \mathbf{p}(t_f) = \frac{\partial \mathbf{S}_f^T(\mathbf{x}^*(t_f))}{\partial \mathbf{x}}\beta,$$

where η and β are constant multiplier vectors of appropriate dimensions.

Remarks:

(1) We emphasize that the complete adjoint state to the optimal control problem (1.9)–(1.16) is $(\mathbf{p}^T(t), p_{n+1}(t))$ instead of $\mathbf{p}^T(t)$ alone. p_{n+1} turns out to be piecewise constant. In particular, $p_{n+1}(t) = 0$ for $t \in [t_0, t_1)$ and $p_{n+1}(t) = -1$ for $t \in [t_k, t_f]$.

(2) Condition (2.17) is the distinction between the necessary conditions for a minimax problem and the necessary conditions for a Mayer's problem with a state inequality constraint. When the maximum of $F(\mathbf{x}^*(t))$ is unique ($k = 1$), (2.17) eliminates the need of searching for the multiplier μ_1 since $\mu_1 = 1$ by (2.17).

When $F(\mathbf{x}^*(t))$ attains its maximum in some finite intervals in (t_0, t_f), which is called flat maximum, we need the the definition of the order of $F(\mathbf{x})$ to characterize the necessary conditions. Loosely speaking, if $F^{(q)}(\mathbf{x}, \mathbf{u})$ is the q-th time derivative of $F(\mathbf{x})$ in which components of $\mathbf{u}$ first appear explicitly, q is defined as the order of $F(\mathbf{x})$ (For more rigorous definition, see Lu *et al* (1991)). When $q = 1$ and flat maximum occurs, the necessary conditions for the minimax problem are given by Lu *et al* (1991). We will not repeat them here.

When $q > 1$ and $m = 1$ (scalar control), and $F(\mathbf{x}^*(t))$ has flat maximum over some interval $(t_j, t_j') \subset (t_0, t_f)$, $t_j < t_j'$, $j = 1, ..., l \le k$, then Eqs. (2.13)–(2.16) and (2.17)–(2.18) apply outside (t_j, t_j'). Inside (t_j, t_j'), we have

$$(2.19) \qquad \dot{\mathbf{p}} = -H_x + (F_u^{(q)})^{-1} F_x^{(q)} (\mathbf{f}_u)^T \mathbf{p}$$

The jump condition at t_j is

$$(2.20) \qquad \mathbf{p}(t_j^+) = \mathbf{p}(t_j^-) + \mu_j F_x + [\dot{F}_x, \ddot{F}_x, \ldots, F_x^{(q-1)}]\alpha^j$$

where $(\alpha^j)^T = (\alpha_1^j, \ldots, \alpha_{q-1}^j)^T$ is a multiplier vector, $\dot{F}_x = \partial\dot{F}/\partial\mathbf{x}$, and so on. All multipliers μ_i corresponding to isolated and flat maxima in (t_0, t_f) sum up to one

$$(2.21) \qquad \sum_{j=1}^{k} \mu_j = 1$$

The determination of the type of maximum, isolated or flat, may require an educated guess before applying the necessary conditions. But in some cases, the jump conditions and other relationship will dictate which type of maximum is to occur, as will be seen in Section 4.

Now let us consider the maximin problem. Obviously the solution of the minimax problem for $G(\mathbf{x}) = -F(\mathbf{x})$ is the same as that of the maximin problem for $F(\mathbf{x})$

$$\min_{t_0 \le t \le t_f} \max \ G(\mathbf{x}) = \max_{t_0 \le t \le t_f} \min \ F(\mathbf{x})$$

So the necessary conditions above become the necessary conditions for the maximin problem if the sign in front of F in (2.16) and (2.20) is changed into $(-)$.

For the maximax problem, a transformation similar to (2.7)–(2.11) plus additional intermediate point constraints at t_i, $i = 1, ..., k$, leads to the necessary conditions of the same form as those for the maximin problem (Vinh *et al* (1987)). Likewise, the necessary conditions for the minimin problem are the same as those for the minimax problem.

3. Some Properties of Optimal Control

3.1 Continuity of Control at Jump Points. In this Section we shall discuss the continuity property of the controls based on the above necessary conditions. First, we discuss the effect of the jump condition (2.16). Because of the relationship (2.15), the discontinuity in the adjoint state $\mathbf{p}$ at the jump points t_i, $i = 1, ... k$, may influence the continuity of $\mathbf{u}^*(t)$. To investigate this, we assume that the Hamiltonian $H(\mathbf{x}, \mathbf{p}, \mathbf{u})$ is regular at t_i^-, that is, it admits a unique maximizing $\mathbf{u}^* = \mathbf{u}^*(t_i^-)$ when $\mathbf{x} = \mathbf{x}^*(t_i^-)$ and $\mathbf{p} = \mathbf{p}(t_i^-)$.

Theorem 2

Suppose that the problem is of order one, and $F(\mathbf{x}^*(t))$ attains its isolated maximum at t_i, $i = 1, ... k$, $t_i \in (t_0, t_f)$. Then

(1) For the minimax problem, the control is continuous at t_i.

(2) For the maximax problem, the control is discontinuous at t_i in general.

Proof:

(1) The Hamiltonian is continuous in $[t_0, t_f]$. In particular,

$$(3.1) \qquad \mathbf{p}^T(t_i^-)\mathbf{f}(\mathbf{x}^*(t_i), \mathbf{u}^*(t_i^-)) = \mathbf{p}^T(t_i^+)\mathbf{f}(\mathbf{x}^*(t_i), \mathbf{u}^*(t_i^+))$$

In this discussion we do not distinguish $\mathbf{x}^*(t_i^-)$ from $\mathbf{x}^*(t_i^+)$ because the state vector is absolutely continuous throughout $[t_0, t_f]$. Using (2.16) to replace $\mathbf{p}(t_i^+)$ in (3.1)

$$(3.2) \qquad \mathbf{p}^T(t_i^-)\mathbf{f}(\mathbf{x}^*(t_i), \mathbf{u}^*(t_i^-)) - \mathbf{p}^T(t_i^-)\mathbf{f}(\mathbf{x}^*(t_i), \mathbf{u}^*(t_i^+)) = \mu_i \dot{F}(\mathbf{x}^*(t_i), \mathbf{u}^*(t_i^+))$$

By the optimality condition (2.15), the left hand side of (3.2) should be nonpositive. Therefore,

$$(3.3) \qquad \mu_i \dot{F}^+ \geq 0$$

Since $F(\mathbf{x}^*(t_i))$ is a maximum of $F(\mathbf{x}^*(t))$, it is necessary that

$$(3.4) \qquad \dot{F}^+ \leq 0$$

Combining (3.3) and (3.4), and using the condition $\mu_i \geq 0$ (Eq. (2.17)), we have

$$(3.5) \qquad \mu_i \dot{F}^+ = 0$$

Substituting (3.5) into (3.2) gives

$$(3.6) \qquad H(\mathbf{x}^*(t_i), \mathbf{p}(t_i^-), \mathbf{u}^*(t_i^-)) = H(\mathbf{x}^*(t_i), \mathbf{p}(t_i^-), \mathbf{u}^*(t_i^+))$$

We combine the optimality condition (2.15) and (3.6):

$$(3.7) \quad H(\mathbf{x}^*(t_i), \mathbf{p}(t_i^-), \mathbf{u}^*(t_i^+)) = H(\mathbf{x}^*(t_i), \mathbf{p}(t_i^-), \mathbf{u}^*(t_i^-)) = \sup_{u \in U} H(\mathbf{x}^*(t_i), \mathbf{p}(t_i^-), \mathbf{u})$$

By the regularity of H, we must have

$$(3.8) \qquad \mathbf{u}^*(t_i^-) = \mathbf{u}^*(t_i^+)$$

(2) The jump condition for the maximax problem is

$$(3.9) \qquad \mathbf{p}(t_i^+) = \mathbf{p}(t_i^-) - \mu_i \frac{\partial F(\mathbf{x}^*(t_i))}{\partial \mathbf{x}}$$

By a procedure similar to part (1), we arrive at

$$(3.10) \qquad \mu_i \dot{F}^+ \leq 0$$

Unlike Eq. (3.3), the combination of (3.10) and (3.4) will not necessarily lead to (3.5), and ultimately (3.8). In general, since $\mathbf{p}(t)$ is discontinuous across t_i, an appropriate jump in $\mathbf{u}(t)$ at t_i is required to keep $H(\mathbf{x}, \mathbf{p}, \mathbf{u})$ continuous.

Remarks

(1) H is assumed to be regular at t_i^-. We would have the same results if H is regular at t_i^+. If H is not regular either at t_i^- or t_i^+, the control can be discontinuous at t_i for the minimax problem.

(2) Part (1) in Theorem 2 applies to the maximin problem, and part (2) to the minimin problem as well.

Example 1.

$$(3.11) \qquad \min \ \{ \max_{0 \le t \le t_f} (x_1(t) + x_2(t)) \}$$

$$(3.12) \qquad \dot{x}_1 = x_2,$$

$$(3.13) \qquad \dot{x}_2 = u,$$

$$(3.14) \qquad |u| \le 1,$$

$$(3.15) \qquad x_1(0) = x_2(0) = 2, \quad x_1(t_f) = x_2(t_f) = 0, \quad t_f \text{ free}$$

This minimax problem is shown to have infinitely many solutions (Lu *et al* (1991)). Let $0 < \Delta t \le 2$. Define

$$(3.16) \qquad t_2 = 2 + \Delta t, \quad t_3 = \frac{4 + 2\Delta t}{\Delta t}, \quad t_f = t_3 + \Delta t$$

The application of the necessary conditions (2.13)–(2.18) gives rise to
(3.17)
$$p_1(t) = \begin{cases} -1; & t \in [0,1) \\ 0; & t \in [1,t_f] \end{cases}, \quad p_2(t) = \begin{cases} t-2; & t \in [0,1) \\ 0; & t \in [1,t_f] \end{cases}, \quad p_3(t) = \begin{cases} 0; & t \in [0,1) \\ 1; & t \in (1,t_f] \end{cases}$$

A group of optimal controls is given by

$$(3.18) \qquad u = \begin{cases} -1; & t \in [0,t_2) \\ 0; & t \in [t_2,t_3) \\ 1; & t \in [t_3,t_f] \end{cases}$$

The optimal performance index is

$$(3.19) \qquad \min \ \max_{0 \le t \le t_f} F(x_1(t), x_2(t)) = x_1(1) + x_2(1) = 4.5$$

Note that since the Hamiltonian is regular at $t_1^- = 1^-$ ($p_2(t_1^-) \ne 0$), the control is continuous at t_1 where p_1 and p_2 have a jump. Another observation is that a singularity characterized by $p_1(t) = p_2(t) = 0$ appears in $[1, t_f]$. This is not a contradiction to the necessary conditions which assert the existence of a nonzero adjoint vector, because, as Remark (1) in Section **2** points out, the complete adjoint vector in this problem is (p_1, p_2, p_3).

Figure 1 shows the variations of $F = x_1 + x_2$ and the control u with $\Delta t = 1$.

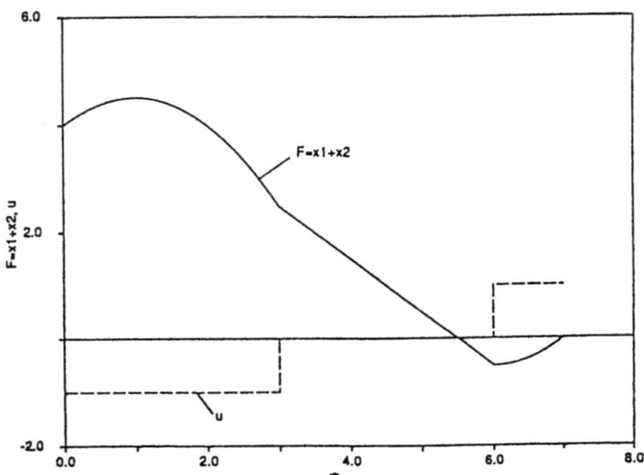

Fig. 1 Variations of $F(\mathbf{x})$ and $u(t)$ for Example 1.

Example 2.

This example illustrates that the control is discontinuous in a maximax problem.

$$(3.20) \qquad \max \quad \{\max_{0 \le t \le 3} x_2(t)\}$$

$$(3.21) \qquad \dot{x}_1 = x_2 + u^2$$

$$(3.22) \qquad \dot{x}_2 = -x_2 + u$$

$$(3.23) \qquad x_1(0) = -5, \ x_2(0) = 0, \ x_1(3) = x_2(3) = 0$$

Applying the necessary conditions (2.13)–(2.17) with (2.16) replaced by (3.9), we obtain

$$(3.24) \qquad p_1 = -0.146144, \quad p_2 = \begin{cases} 0.22641e^t - 0.146144; & t \in [0, 1.5) \\ (0.22641 - e^{-1.5})e^t - 0.146144; & t \in [1.5, 3] \end{cases}$$

The optimal control

$$(3.25) \qquad u = \begin{cases} -0.5(-1.54922e^t + 1); & t \in [0, 1.5) \\ -0.5(-0.02244e^t + 1); & t \in [1.5, 3] \end{cases}$$

The maximum of $x_2(t)$ occurs at $t_1 = 1.5$

$$(3.26) \qquad \max_{0 \leq t \leq 3} x_2(t) = x_2(1.5) = 1.260927$$

Note that the control is discontinuous at t_1, and the discontinuity is given by

$$(3.27) \qquad u(t_1^+) - u(t_1^-) = 3.423$$

Plotted in Fig. 2 are the variations of $x_2(t)$ and the control $u(t)$.

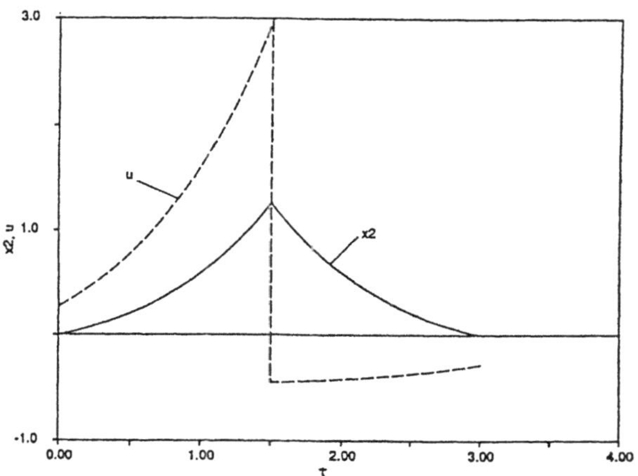

Fig. 2 Variations of $x_2(t)$ and $u(t)$ for Example 2.

3.2 Number of Control Switchings for Linear Systems. When the system
equation (2.2) is linear and time-invariant,

$$(3.28) \qquad \dot{\mathbf{x}} = A\mathbf{x} + B\mathbf{u}$$

$$(3.29) \qquad u_{jmin} \leq u_j \leq u_{jmax}, \quad j = 1, ..., m$$

where $A \in \mathbf{R}^{n \times n}$ and $B = [\mathbf{b}_1, ..., \mathbf{b}_m] \in \mathbf{R}^{n \times m}$, the optimal control for a minimax
problem can be of bang-bang type or singular, as both appear in Example 1. In the
nonsingular case, the control will switch between the control bounds. The maximum
number of switchings of each control component u_j is of interest. The following
theorem is stated for the minimax problem, but is adaptable to the variations of the
minimax problem introduced in Section 2.

Theorem 3

Let $\mathbf{x}^*(t)$ and $\mathbf{u}^*(t)$, $t \in [t_0, t_f]$, be an optimal solution pair to the minimax problem for the system (3.28)–(3.29) and $\mathbf{p}(t)$ is the associated adjoint state. Suppose that all eigenvalues of A are real, and system (3.28) is completely controllable with respect to each of the control component u_j, $j = 1, ..., m$. Define the set

$$(3.30) \qquad T = \{\tau \in [t_0, t_f] \mid F(\mathbf{x}^*(\tau)) < \max_{t_0 \leq t \leq t_f} F(\mathbf{x}^*(t))\}$$

Let $\bar{T}$ be the closure of T. Suppose that $\mathbf{u}^*(t)$ is of bang-bang type in $\bar{T}$. Then the maximum number of switchings of u_j in $\bar{T}$ between u_{jmin} and u_{jmax} is $(k+1)(n-1)$, where k is the number of non-connected subintervals in which $F(\mathbf{x}^*(t))$ attains its maximum.

Proof:

This theorem is a direct extension of Theorem 10 by Pontryagin *et al* (1962). Under the assumptions of Theorem 3, in any one of the $k+1$ intervals in $[t_0, t_f]$ separated by the k maxima of $F(\mathbf{x}^*(t))$, the j-th switching function $\mathbf{p}^T(t)\mathbf{b}_j$ has at most $n-1$ zeros, hence each u_j has at most $n-1$ switchings. The jump conditions (2.16) or (2.20) provide a new initial condition for $\mathbf{p}$ in the next interval, which may result in another $n-1$ switchings of u_j in that interval. Therefore the total switchings of u_j in $\bar{T} \subset [t_0, t_f]$ will be at most $(k+1)(n-1)$.

Remarks:

(1) Unlike the minimum-time problem for a linear system, the controllability and real eigenvalues of A do not guarantee an optimal control of bang-bang type for a minimax problem, as seen from Example 1. But if the optimal control is bang-bang type, Theorem 3 gives the maximum number of control switchings in $\bar{T}$.

(2) If all maxima of $F(\mathbf{x}^*(t))$ are isolated, obviously $\bar{T} = [t_0, t_f]$. Then Theorem 3 gives the maximum number of possible control switchings in $[t_0, t_f]$.

Example 3.

$$(3.31) \qquad min \quad \{\max_{0 \leq t \leq 10} x_1(t)\}$$

$$(3.32) \qquad \dot{x}_1 = x_2 \qquad \dot{x}_2 = x_3 \qquad \dot{x}_3 = u$$

$$(3.33) \qquad |u| \leq u_{max} = 2$$

$$(3.34) \qquad x_1(0) = x_1(10) = 0, \; x_2(0) = -x_2(10) = 1, \; x_3(0) = x_3(10) = 2$$

For this third order example, $x_1(t)$ has a flat maximum. Applying the necessary conditions with (2.19) and (2.20), we have

$$(3.35) \qquad p_1(t) = \begin{cases} -0.5; & t \in [0,3) \\ 0.5; & t \in [3,10] \end{cases}, \quad p_2(t) = \begin{cases} 0.5t - 1.25; & t \in [0,3) \\ -0.5(t-3) - 0.75; & t \in [3,10] \end{cases}$$

$$(3.36) \qquad p_3(t) = \begin{cases} -0.25t^2 + 1.25t - 1.5; & t \in [0,3) \\ 0.25(t-3)^2 - 5(t-3) + 5; & t \in [3,10] \end{cases}$$

and the control

$$(3.37) \qquad u = \begin{cases} -2; & t \in [0,2) \\ 2; & t \in [2,3) \\ 0; & t \in [3,7) \\ -2; & t \in [7,8) \\ 2; & t \in [8,10] \end{cases}$$

The flat maximum for $x_1(t)$ is $x_1(t) = 11/3$ for $t \in (3,7)$. Figure 3 depicts the histories of $x_1(t)$ and $u(t)$.

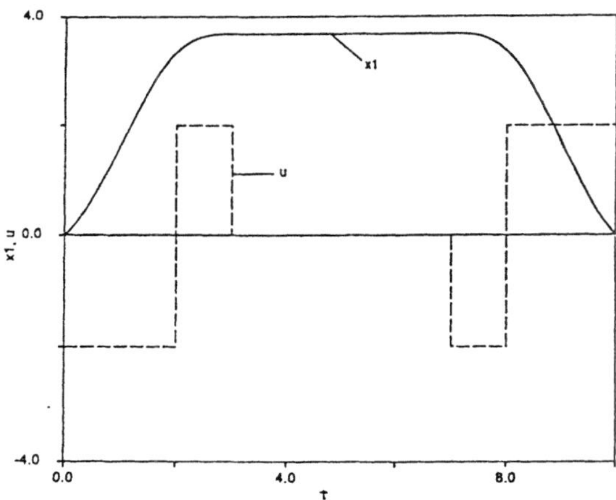

Fig. 3 Variations of $x_1(t)$ and $u(t)$ for Example 3.

Remarks:

(1) $x_1(t)$ has a unique maximum ($k = 1$). It is seen that the control has $(k+1)(n-1) = (1+1)(3-1) = 4$ switchings.

(2) The first two switchings occur at $t_1 = 2$ and $t_2 = 3$. It is easy to show that in general

$$(3.38) \qquad t_1 = \frac{1}{u_{max}}(2 + \sqrt{u_{max}+2}), \quad t_2 = \frac{2}{u_{max}}(1 + \sqrt{u_{max}+2})$$

and when u_{max} is very large

$$(3.39) \qquad \max_{0 \leq t \leq 10} x_1(t) \approx \frac{4}{u_{max}} + \sqrt{\frac{1}{u_{max}}}$$

From the system equation $\dot{x}_1 = x_2$ and the initial conditions $x_1(0) = 0$ and $x_2(0) = 1$, we must have $\max x_1(t) > 0$. But (3.39) indicates that $\max x_1(t) \to 0$ as $u_{max} \to \infty$. Thus this minimax problem has no solution if the control is unbounded.

4. Concluding Remarks

The minimax optimal control problem and some closely related problems are considered. The necessary conditions of optimality for these problems are given. Analytical properties of the problems, such as control continuity, singularity and control switchings, are discussed. Examples are presented to support the analysis.

As for other optimal control problems, the solution of a minimax-type problem usually can only be obtained through a numerical process. The necessary conditions not only can serve as the basis for the numerical algorithm, but often times, a close examination of the necessary conditions when applied to a particular problem can reveal distinct features of the problem before the numerical solution is obtained (e.g., Vinh *et al* (1988) and Lu *et al* (1988)).

References

[1] Johnson, C. D., *Optimal Control Problem with Chebyshev Minimax Performance Index*, Journal of Basic Engineering, **89**, (1967) 251–262.

[2] Powers, W. F., *A Chebyshev Minimax Technique Oriented to Aerospace Trajectory Optimization Problems*, AIAA Journal., **10**, (1972), 1291–1296.

[3] Miele, A., Mohanty, B. P., Venkataraman, P. and Kuo, Y. M., *Numerical Solution of Minimax Problems of Optimal Control, Part 1*, Journal of Optimization Theory and Applications, **38**, No. 1, (1982), 97–109.

[4] Miele, A. and Basapur, V. K., *Approximate Solution to Minimax Optimal Control Problems for Aeroassisted Orbital Transfer*, Acta Astronautica, **12**, No. 10, (1985), 809–818.

[5] Miele, A., Wang, T., Melvin, W. W. and Bowles, R. L., *Acceleration, Gamma, and Theta Guidance for Abort Landing in a Windshear*, Journal of Guidance, Control, and Dynamics, **12**, No. 6, (1989), 815–821.

[6] Warga, J., *Minimax Problems and Unilateral Curves in the Calculus of Variations*, SIAM Transactions on Control , **3**, No. 1, (1965), 91–105.

[7] Vinh, N. X. and Lu, P., *Necessary Conditions For Maximax Problems with Application to Aeroglide of Hypervelocity Vehicles*, Acta, Astronautica, **15**, No. 5–6, (1987), 413–420.

[8] Vinh, N. X. and Lu, P., *Chebyshev Minimax Problems for Skip Trajectories*, The Journal of the Astronautical Sciences, **36**, No. 1–2, (1988), 179–197.

[9] Lu, P. and Vinh, N. X., *Minimax Optimal Control for Atmospheric Fly-Through Trajectories*, Journal of Optimization Theory and Applications, **57**, No. 1, (1988), 41–58.

[10] Lu, P. and Vinh, N. X., *Optimal Control Problems with Maximum Functional*, Journal of Guidance, Control, and Dynamics, **14**, No. 6, November–December, (1991), 1215–1223.

[11] Pontryagin, L. S., Boltyanskii, V. G., Gramkrelidze, R. V. and Mishchenko, E F., *The Mathematical Theory of Optimal Processes*, Intersciences, New York, 1962

Authors' addresses

Dr. Ping Lu
Department of Aerospace Engineering and Engineering Mechanics
Iowa State University
Ames, IA 50011, USA

Dr. Nguyen X. Vinh
Department of Aerospace Engineering
The University of Michigan
Ann Arbor, MI 48109, USA

Numerical Methods

International Series of Numerical Mathematics, Vol. 111, ©1993 Birkhäuser Verlag Basel

Trajectory Optimization Using Sparse Sequential Quadratic Programming

John T. Betts

Abstract. One of the most effective numerical techniques for the solution of trajectory optimization and optimal control problems is the direct transcription method. This approach combines a nonlinear programming algorithm with a discretization of the trajectory dynamics. The resulting mathematical programming problem is characterized by matrices which are large and sparse. Constraints on the path of the trajectory are then treated as algebraic inequalities to be satisfied by the nonlinear program. This paper describes a nonlinear programming algorithm which exploits the matrix sparsity produced by the transcription formulation. Numerical experience is reported for trajectories with both state and control variable equality and inequality path constraints.

1   Introduction

It is well known that the solution of an optimal control or trajectory optimization problem can be posed as the solution of a two-point boundary value problem. This approach has been successfully utilized by many authors (eg. Ascher, et al. 1988, Betts 1990a, Dickmanns 1980, and Bulirsch, et al. 1991). Difficulties with indirect methods caused by the adjoint equations, are avoided in the direct transcription or collocation methods (cf. Betts and Huffman 1991a, 1992, Enright 1991, Hargraves and Paris 1987). For this method the dynamic equations are discretized, and the optimal control problem is transformed into a nonlinear program, which can be solved directly. The nonlinear programming problem is large and sparse and a technique for solving it is presented in Betts and Huffman (1992). This paper extends that method to efficiently handle inequality constraints and presents a nonlinear programming algorithm designed to exploit the properties of the problem which results from direct transcription of the trajectory optimization application.

2   Trajectory Optimization

2.1 The Optimal Control Problem   Let us find the n_u-dimensional control vector $\mathbf{u}(t)$ to minimize the performance index $\phi[\mathbf{y}(t_f), t_f]$ evaluated at the final time t_f. The dynamics of the system are defined by the state equations

$$(2.1) \qquad \dot{\mathbf{y}} = \mathbf{h}[\mathbf{y}(t), \mathbf{u}(t), t]$$

where $\mathbf{y}$ is the n_e dimension state vector. Initial conditions at time t_0 are defined by

$$(2.2) \qquad \boldsymbol{\psi}[\mathbf{y}(t_0), \mathbf{u}(t_0), t_0] \equiv \boldsymbol{\psi}_0 = 0,$$

and terminal conditions at the final time t_f are defined by

$$(2.3) \qquad \boldsymbol{\psi}[\mathbf{y}(t_f), \mathbf{u}(t_f), t_f] \equiv \boldsymbol{\psi}_f = 0.$$

In addition the solution must satisfy path constraints of the form

$$(2.4) \qquad \boldsymbol{\Psi}_L \leq \boldsymbol{\Psi}[\mathbf{y}(t), \mathbf{u}(t), t] \leq \boldsymbol{\Psi}_U,$$

where $\boldsymbol{\Psi}$ is a vector of size n_p, as well as simple bounds on the state variables

$$(2.5) \qquad \mathbf{y}_L \leq \mathbf{y}(t) \leq \mathbf{y}_U,$$

and control variables

$$(2.6) \qquad \mathbf{u}_L \leq \mathbf{u}(t) \leq \mathbf{u}_U.$$

Note that a path variable equality contraint can be imposed if the upper and lower bounds are equal, e.g. $(\Psi_L)_k = (\Psi_U)_k$ for some k. For simplicity in presentation we have chosen to eliminate discussion of trajectories with more than one phase, optimization variables independent of time, and alternate forms of the objective function. However the formulation is easily modified to accomodate these generalizations, and the reader is referred to Betts and Huffman (1991a) for details.

2.2 Transcription Formulation   The basic approach for solving the optimal control problem by transcription has been presented in detail elsewhere (cf. Betts and Huffman 1992, Enright 1991, Hargraves and Paris 1987) and will only be summarized here. All approaches divide the time interval into n_s segments $t_0 < t_1 < t_2 < \ldots < t_f = t_{n_s}$, where the points are referred to as mesh or grid points. Let us introduce the notation $\mathbf{y}_j \equiv \mathbf{y}(t_j)$ to indicate the value of the state variable at a grid point. In like fashion denote the control at a grid point by $\mathbf{u}_j \equiv \mathbf{u}(t_j)$. Results will be presented for three different discretization schemes namely trapezoidal, Hermite-Simpson, and

Runge-Kutta. Each scheme produces a distinct set of nonlinear programming (NLP) variables and constraints.

For the trapezoidal discretization, the NLP variables are

$$(2.7) \quad \mathbf{x} = [\mathbf{y}_0, \mathbf{u}_0, \mathbf{y}_1, \mathbf{u}_1, \ldots, \mathbf{y}_f, \mathbf{u}_f, t_f]^\top .$$

The state equations (2.1) are approximately satisfied by setting *defects*

$$(2.8) \quad \zeta_j = \mathbf{y}_j - \mathbf{y}_{j-1} - \frac{1}{2}\kappa_j \left[\mathbf{h}_j + \mathbf{h}_{j-1}\right]$$

to zero for $j = 1, \ldots, n_s$. The step size is denoted by $\kappa_j \equiv t_j - t_{j-1}$, and the right hand side of the differential equations (2.1) are given by $\mathbf{h}_j \equiv \mathbf{h}[\mathbf{y}(t_j), \mathbf{u}(t_j), t_j]$

For the Hermite-Simpson and fourth order Runge-Kutta discretization schemes, the NLP variables are augmented to include values of the control at the midpoints of the intervals. The defects for these schemes are given in Betts and Huffman (1991b). As a result of the transcription process the optimal control constraints (2.1)-(2.4) are replaced by the NLP constraints

$$(2.9) \quad \mathbf{c}_L \leq \mathbf{c}(\mathbf{x}) \leq \mathbf{c}_U,$$

with $\mathbf{c}(\mathbf{x}) = \left[\zeta_1, \zeta_2, \ldots, \zeta_f, \psi_0, \psi_f, \Psi_0, \Psi_1, \ldots, \Psi_f\right]^\top$ and $\mathbf{c}_L = [0, \ldots, 0, \Psi_L, \ldots, \Psi_L]^\top$ and a corresponding definition of $\mathbf{c}_U$. The first $n_e n_s$ equality constraints require that the defect vectors from each of the n_s segments be zero thereby approximately satisfying the differential equations (2.1). The boundary conditions are enforced directly by the equality constraints on ψ, and the nonlinear path constraints are imposed at the grid points. Note that nonlinear equality path constraints are accomodated by setting $\mathbf{c}_L = \mathbf{c}_U$ In a similar fashion the state and control variable bounds (2.5) and (2.6) become simple bounds on the NLP variables. The path constraints and variable bounds are always imposed at the grid points, and for the Hermite-Simpson, and Runge-Kutta discretization methods additional bounds are imposed on the control variables at the interval midpoints.

3 The Nonlinear Programming Problem

The nonlinear programming problem can be stated as follows: Find the N-vector $\mathbf{x}$ which minimizes the *objective function* $f(\mathbf{x})$ subject to the *constraints* $\mathbf{c}_L \leq \mathbf{c}(\mathbf{x}) \leq \mathbf{c}_U$, where $\mathbf{c}(\mathbf{x})$ is an m-vector of constraint functions, and the simple *bounds* $\mathbf{x}_L \leq \mathbf{x} \leq \mathbf{x}_U$.

Equality constraints are imposed by setting $c_L = c_U$ and variables can be fixed by setting $x_L = x_U$.

The solution point x^* must satisfy the Kuhn-Tucker necessary conditions for a local minimum which relate the Lagrange multipliers λ and ν such that

$$(3.1) \quad g = G^T\lambda + \nu$$

where $\nabla_x f(x) = g(x) = g$ is the N-dimensional gradient vector, and G is the $m \times N$ Jacobian matrix of constraint gradients.

4   A Sparse Nonlinear Programming Algorithm

4.1 QP Subproblem   The basic approach utilized by the algorithm is to solve a sequence of quadratic programming (QP) subproblems. Solution of the QP subproblem is used to define new estimates for the variables according to the formula

$$(4.1) \quad \bar{x} = x + \alpha p,$$

where the vector p is referred to as the *search direction*. The scalar α determines the step length and is typically set to one. The search direction p is found by minimizing the quadratic

$$(4.2) \quad g^T p + \frac{1}{2} p^T H p$$

subject to the linear constraints

$$(4.3) \quad b_\ell \leq \begin{bmatrix} Gp \\ p \end{bmatrix} \leq b_u$$

where H is a symmetric $N \times N$ positive definite approximation to the Hessian matrix. The upper bound vector is defined by

$$(4.4) \quad b_u = \begin{bmatrix} c_U - c \\ x_U - x \end{bmatrix}$$

with a similar definition for the lower bound vector b_ℓ. Gill et al. (1987) proposed a quadratic programming method which requires solving the Kuhn-Tucker or KT system

$$(4.5) \quad \begin{bmatrix} H & A^T \\ A & 0 \end{bmatrix} \begin{bmatrix} -p \\ \pi \end{bmatrix} \equiv K_0 \begin{bmatrix} -p \\ \pi \end{bmatrix} = \begin{bmatrix} g \\ 0 \end{bmatrix}.$$

once for an initial set of 'free' variables. Subsequent changes in the active set can be computed using factorizations of the KT matrix, and a small dense *Schur complement* of $\mathbf{K}_0$. This sparse symmetric indefinite KT system can be solved very efficiently using the *multifrontal* algorithm described in Ashcraft and Grimes (1988).

4.2 Merit Function   When a quadratic program is used to approximate a general nonlinearly constrained problem it may be necessary to adjust the steplength α in order to achieve "sufficient reduction" in a merit function that in some way combines the objective function and constraint violations. The merit function used is similar to that proposed by Gill et al. (1986b);

$$M(\mathbf{x},\lambda,\boldsymbol{\nu},\mathbf{s},\mathbf{t}) = f - \lambda^T(\mathbf{c}-\mathbf{s}) - \boldsymbol{\nu}^T(\mathbf{x}-\mathbf{t})$$
$$(4.6) \qquad +\frac{1}{2}(\mathbf{c}-\mathbf{s})^T\mathbf{Q}(\mathbf{c}-\mathbf{s}) + \frac{1}{2}(\mathbf{x}-\mathbf{t})^T\mathbf{R}(\mathbf{x}-\mathbf{t}).$$

The diagonal penalty matrices are defined by $Q_{ii} = \rho_i$ and $R_{ii} = \gamma_i$. For this merit function the slack variables $\mathbf{s}$ at the beginning of a step are defined by

$$(4.7) \qquad \mathbf{s}_i = \max\left[c_{Li}, \min\left(c_i - \frac{\lambda_i}{\rho_i}, c_{Ui}\right)\right].$$

with a corresponding definition for the bound slacks. The search direction in the real variables $\mathbf{x}$ as given by (4.1) is augmented to permit the multipliers and the slack variables to vary according to

$$(4.8) \qquad \begin{bmatrix} \overline{\mathbf{x}} \\ \overline{\lambda} \\ \overline{\boldsymbol{\nu}} \\ \overline{\mathbf{s}} \\ \overline{\mathbf{t}} \end{bmatrix} = \begin{bmatrix} \mathbf{x} \\ \lambda \\ \boldsymbol{\nu} \\ \mathbf{s} \\ \mathbf{t} \end{bmatrix} + \alpha \begin{bmatrix} \mathbf{p} \\ \xi \\ \eta \\ \mathbf{q} \\ \delta \end{bmatrix}$$

The multiplier search directions ξ and η are defined using the QP multipliers μ and ω according to $\xi \equiv \mu - \lambda$ and $\eta \equiv \omega - \boldsymbol{\nu}$. From the QP (4.2)-(4.4) the predicted slack variables are just

$$(4.9) \qquad \overline{\mathbf{s}} = \mathbf{Gp} + \mathbf{c} = \mathbf{s} + \mathbf{q}.$$

Using this expression define the slack vector step by

$$(4.10) \qquad \mathbf{q} = \mathbf{Gp} + (\mathbf{c} - \mathbf{s}).$$

Similarly the bound slack vector search direction is given by $\delta = \mathbf{p} + (\mathbf{x} - \mathbf{t})$. Note that when a full step is taken $\alpha = 1$, the updated estimate for the Lagrange multipliers

$\overline{\lambda}$ and $\overline{\nu}$ are just the QP estimates μ and ω. The slack variables s and t are just the linear estimates of the constraints and the terms $(\mathbf{c} - \mathbf{s})$ and $(\mathbf{x} - \mathbf{t})$ in the merit function are measures of the *deviation* from linearity.

4.3 Parameter Definitions   In Gill et al. (1986b) it is shown that the penalty weights ρ_i and γ_i are finite provided the Hessian matrix $\mathbf{H}$ used in the QP subproblem is positive definite. However in general the Hessian of the Lagrangian

$$(4.11) \quad \mathbf{H}_L = \nabla_x^2 f - \sum_{i=1}^{m} \lambda_i \nabla_x^2 c_i.$$

is not positive definite. In fact it is only necessary that the projected Hessian be positive semi-definite at the solution with the correct active set of constraints (cf Gill et al. 1981). Consequently we use the modified matrix

$$(4.12) \quad \mathbf{H} = \mathbf{H}_L + \tau(|\sigma| + 1)\mathbf{I}$$

The Levenberg parameter τ is chosen such that $0 \le \tau \le 1$ and is normalized using the Gerschgorin bound for the most negative eigenvalue of $\mathbf{H}_L$, i.e.

$$(4.13) \quad \sigma = \min_{1 \le i \le N} \left\{ h_{ii} - \sum_{i \ne j}^{N} |h_{ij}| \right\}$$

with h_{ij} is used to denote the nonzero elements of $\mathbf{H}_L$.

The proper choice for the Levenberg parameter τ can greatly effect the performance of the nonlinear programming algorithm. Quadratic convergence can only be obtained when $\tau = 0$ and the correct active set has been identified. On the other hand, if $\tau = 1$ in order to guarantee a positive definite Hessian, the search direction $\mathbf{p}$ is significantly biased toward a gradient direction and convergence is degraded. The strategy employed to adjust τ at a particular step in the nonlinear programming iteration is as follows:

1. (Inertia Test) If the inertia of the KT matrix $\mathbf{K}_0$ (cf eq. 4.5) is correct go to step 2, otherwise increase τ and repeat step 1.

2. (Trust Region Strategy)

    (a) Compute the ratio of actual reduction to predicted reduction

    $$(4.14) \quad \varrho_1 = \frac{M^{(k)} - M^{(k-1)}}{\tilde{M}^{(k)} - M^{(k-1)}}$$

(b) Compute the rate of change in the projected gradient norm

$$(4.15) \quad \varrho_2 = \frac{\|\vartheta^{(k)}\|}{\|\vartheta^{(k-1)}\|}$$

where $\vartheta = \mathbf{g} - \mathbf{G}^T\lambda - \nu$

(c) If $\varrho_1 \leq 0.25$ reduce the trust radius, i.e. set $\tau^{(k+1)} = \min(2\tau^{(k)}, 1)$,

(d) If $\varrho_1 \geq 0.75$ increase the trust radius, i.e. set $\tau^{(k+1)} = \tau^{(k)}\min(0.5, \varrho_2)$.

The inertia (i.e. the number of positive, negative, and zero eigenvalues) of the related KT matrix (4.5), is used to infer that the projected Hessian is positive definite. Basically the philosophy is to reduce the Levenberg parameter when the predicted reduction in the merit function agrees with the actual reduction, and increase it when the agreement is poor. The process is accelerated by making the change in τ proportional to ϱ_2.

Although the Levenberg parameter is used to ensure that the projected Hessian approximation is positive definite, it is still necessary to define the penalty weights $\mathbf{Q}$ and $\mathbf{R}$. In Gill et al. (1986b) it is shown that convergence of the method requires choosing the weights such that

$$(4.16) \quad M_0' \leq -\frac{1}{2}\mathbf{p}^T\mathbf{H}\mathbf{p}$$

where M_0' denotes the direction derivative of the merit function (4.6) with respect to the steplength α evaluated at $\alpha = 0$. To achieve this, choose the penalty parameters r_i as the *minimum norm* solution to (4.16). In essence then the penalty weights are chosen to be as small as possible consistent with the descent condition (4.16).

4.4 Algorithm Strategy   Three different approaches have been implemented. The first strategy is probably the most aggressive, while the last strategy is the more conservative. The strategies are:

"m" (<u>m</u>inimize): Beginning at $\mathbf{x}^0$ solve a sequence of quadratic programs until the solution $\mathbf{x}^*$ is found.

"fm" (<u>f</u>easible point then <u>m</u>inimize): Beginning at $\mathbf{x}^0$ solve a sequence of quadratic programs to locate a feasible point $\mathbf{x}^f$, and then beginning from $\mathbf{x}^f$ solve a sequence of quadratic programs until the solution $\mathbf{x}^*$ is found.

"fme" (<u>f</u>easible point then <u>m</u>inimize subject to <u>e</u>qualities): Beginning at $\mathbf{x}^0$ solve a sequence of quadratic programs to locate a feasible point $\mathbf{x}^f$, and then beginning from $\mathbf{x}^f$ solve a sequence of quandratic programs while maintaining feasible equalities until the solution $\mathbf{x}^*$ is found.

4.5 Finding a Feasible Point   The first step in either the "fm" or "fme" strategy is to determine a point which is feasible with respect to the constraints. A fourth strategy, "f", to just locate a feasible point is also available in the software. The approach employed is to take a series of steps of the form given by (4.1) with the search direction computed to solve a *least distance program.* This can be accomplished if we impose the requirement that the search direction have minimum norm, i.e. $\|\mathbf{p}\|$. The minimum norm solution can be obtained from (4.2) by setting $\mathbf{H} = \mathbf{I}$ and $\mathbf{g} = \mathbf{0}$. Since the solution of this subproblem is based on a linear model of the constraint functions it may be necessary to adjust the step length α in (4.1) to produce a reduction in the constraint error. Specifically a quadratic line search is used to adjust α so that $\Upsilon(\overline{\mathbf{x}}) \leq \Upsilon(\mathbf{x})$, where $\Upsilon(\mathbf{x})$ is the error in the violated constraints.

4.6 The Minimization Process   All three strategies "m","fm" and "fme" execute a series of steps to minimize the merit function (4.6). In the case of strategy "m" the iteration begins from the arbitrary and possibly infeasible point $\mathbf{x}^0$. On the other hand strategy "fm" begins the minimization of the merit function from a feasible point $\mathbf{x}^f$. Finally, the "fme" strategy not only begins the minimization at a feasible point, but *maintains* feasibility with respect to the equality constraints. Let us denote the equalities by $\mathbf{e}$, with Jacobian $\mathbf{E}$.

The iteration begins at the point $\mathbf{x}$, and proceeds as follows:

1. Evaluate gradient information $\mathbf{g}$ and $\mathbf{G}$ and then

   (a) Terminate if the Kuhn-Tucker conditions are satisfied;

   (b) otherwise define τ from (4.14)-(4.15), compute $\mathbf{H}_L$ from (4.11)

2. Construct the optimization search direction;

   (a) compute $\mathbf{H}$ from (4.12)

   (b) compute $\mathbf{p}$ by solving the QP subproblem (4.2)-(4.3)

   (c) if inertia of $\mathbf{H}$ is incorrect increase τ and return to step (a)

   (d) compute ξ and η from their definitions,

   (e) compute $\mathbf{q}$ and δ from (4.10),

   (f) compute penalty parameters to satisfy (4.16)

   (g) and initialize $\mathbf{p}^{(2)} = \mathbf{p}^{(3)} = 0$, $\alpha = 1$, $\tilde{\alpha} = 0$;

3. Compute the predicted point for

   (a) the variables from $\overline{\mathbf{x}} = \mathbf{x} + \alpha\mathbf{p} + \alpha^2\mathbf{p}^{(2)} + \alpha^3\mathbf{p}^{(3)}$,

   (b) the multipliers and slacks from (4.8),

    (c) then evaluate the constraints $\bar{c} = c(\bar{x})$ at the predicted point and then

    (d) if $\|\bar{e}\| \leq \epsilon$ or not strategy "fme", set $\hat{x} \Leftarrow \bar{x}$ and go to step 7;

4. Solve the underdetermined system $\mathbf{E}\mathbf{d} = \bar{e}$, for $\mathbf{d}$ and set $v = 1$;

5. Compute the corrected point $\hat{x} = \bar{x} - v\mathbf{d}$,

6. Evaluate the constraints $\hat{e}$ at the corrected point $\hat{x}$ then,

    (a) if $\|\hat{e}\| \leq \epsilon$ and $\tilde{\alpha} = 0$ compute $\mathbf{p}^{(2)} = \frac{1}{\alpha^2}[\hat{x} - x - \alpha\mathbf{p}]$ save the corrected poirt (set $\tilde{x} \Leftarrow \hat{x}$, and $\tilde{\alpha} \Leftarrow \alpha$) and then go to step 7,

    (b) else if $\|\hat{e}\| \leq \epsilon$ and $\tilde{\alpha} \neq 0$ compute the elements of $p_i^{(2)}$ and $p_i^{(3)}$ for $i = 1, \ldots, N$ from the system

$$
\begin{bmatrix} \alpha^2 & \alpha^3 \\ \tilde{\alpha}^2 & \tilde{\alpha}^3 \end{bmatrix} \begin{bmatrix} p_i^{(2)} \\ p_i^{(3)} \end{bmatrix} = \begin{bmatrix} \hat{x}_i - x_i - \alpha p_i \\ \tilde{x}_i - x_i - \tilde{\alpha} p_i \end{bmatrix}
$$

    save the corrected point (set $\tilde{x} \Leftarrow \hat{x}$, and $\tilde{\alpha} \Leftarrow \alpha$) and then go to step 7,

    (c) else if $\|\hat{e}\| \leq \|\bar{e}\|$ update the corrected point (set $\bar{x} \Leftarrow \hat{x}$ and $\bar{e} \Leftarrow \hat{e}$) and return to step 4,

    (d) else reduce the step length v to achieve constraint reduction and return to step 5;

7. Evaluate the merit function $M(\hat{x}, \bar{\lambda}, \bar{\nu}, \bar{s}, \bar{t}) = \widehat{M}$ and

    (a) if the merit function $\widehat{M}$ is 'sufficiently' less than M then $\hat{x}$ is an improved point – update all quantities, and return to step 1,

    (b) else change the step length α to minimize M and return to step 3

The steps outlined describe the fundamental elements of the optimization process, however a number of points deserve additional clarification. First note that the algorithm consists of an 'outer' loop (steps 1-7) to minimize the merit function, and an 'inner' loop (steps 3-6) to eliminate the error in the binding constraints. The 'outer' loop can be viewed as a univariate (line) search in the direction $\mathbf{p}$, with the step length α adjusted to minimize the merit function. The 'inner' loop can be viewed as a nonlinear root solving process designed to eliminate the error in the equality constraints for the specified value of α. Note that steps 4-6 are only executed for the "fme" strategy. The 'inner' constraint elimination process must be initiated with an estimate of the variables, and this estimate is given in step (3a). Notice that the first prediction is based on a linear model for the constraints since $\mathbf{p}^{(2)} = \mathbf{p}^{(3)} = 0$ in step 2. However, after the constraint error has been eliminated, the value of $\mathbf{p}^{(2)}$ is updated in step (6a) and the second prediction is based on a quadratic model of the constraints. After the second corrected point is obtained, subsequent predictions utilize the cubic model defined by solving the system in step (6b). Adjusting the value of the steplength α as required in step (7b) is accomplished using a safeguarded line-search procedure which constructs a quadratic and/or cubic model of the merit function.

Because the constraint elimination process requires the solution of an underdetermined system in step 4 there is some ambiguity in the algorithm. This ambiguity is eliminated by choosing the minimum norm direction which can be obtained by solving the augmented system

$$(4.17) \qquad \begin{bmatrix} \mathbf{I} & \mathbf{E}^{\mathsf{T}} \\ \mathbf{E} & \mathbf{0} \end{bmatrix} \begin{bmatrix} \mathbf{d} \\ -\varpi \end{bmatrix} = \begin{bmatrix} \mathbf{0} \\ \bar{\mathbf{e}} \end{bmatrix}.$$

Notice that the Jacobian $\mathbf{E}$ is evaluated at the reference point $\mathbf{x}$, and not reevaluated during the 'inner' loop iteration, even though the right hand side $\bar{\mathbf{e}}$ does change. Because the Jacobian is not reevaluated the 'inner' loop will have a linear convergence rate. Nevertheless this approach has been found attractive because: (a) the coefficient matrix can be factored only once per 'outer' optimization iteration, thereby significantly reducing the linear algebra expense, and (b) the corrections defined by $\mathbf{d}$ are orthogonal to the constraint tangent space at the reference point $\mathbf{x}$, and hence tend to produce a well-conditioned constraint iteration process.

In order to evaluate the Hessian matrix (4.11) an estimate of the Lagrange multipliers is needed. The values obtained by solving the QP with $\mathbf{H} = \mathbf{I}$ are used for the first iteration, and thereafter the values $\bar{\lambda}$ from (4.8) are used. Furthermore, for the very first iteration the multiplier search direction $\xi = 0$ so that the multipliers will be initialized to the QP estimates μ. The Levenberg parameter τ in (4.12) and the penalty weights are initialized to zero, and consequently the merit function is initially just the Lagrangian. Gradient and Hessian information can either be computed analytically or using finite difference estimates. All numerical results construct this information using sparse finite differencing as described in Betts and Huffman (1991a),(1992). Although central difference estimates must be used during optimization, forward difference estimates are used when finding a feasible point. Numerical experience also suggests that it is not necessary to compute a Hessian matrix for every iteration (step 1b), when the algorithm is progressing well. Instead we simply use the current (presumably best) estimate, until it is necessary to recompute the estimate to maintain good progress.

5   Computational Results

Computational results on a series of trajectory optimization problems are summarized in Betts and Huffman (1991b). Also described is a strategy for grid refinement–that is an approach for changing the location and number of grid points to control discretization error. As the grid is refined, it is necessary to solve a sequence of NLP problems. The test set consists of 74 problems with no grid refinement and 10 problems with refinement. Space does not permit a complete presentation of results for all problems. Instead the grid refinement performance will be presented for a minimum time to climb problem (Bryson et al.,1969). Also the method will be compared with results generated by the OTIS software (Hargraves and Paris, 1987).

Table 1 presents a summary of the grid refinement iterations for this problem. Observe that most of the NLP optimization iterations were performed on the first refinement iteration, with the number of Hessian calls dropping to one for the final refinement iterations. Also note that the number of right hand side evaluations increases significantly for iterations 2-6 because the refinement strategy uses trapezoidal quadrature for the first iteration and Hermite quadrature thereafter. The results obtained for this problem can also be compared with a standard dense optimization method. The OTIS computer program (Hargraves and Paris, 1987) uses the sequential quadratic programming code NPSOL (Gill et al. 1986a), and a slightly different Hermite transcription approach. Table 2 summarizes the results of this comparison. Clearly, as the number of grid points increase the computational cost is dominated by the linear algebra, which for NPSOL is proportional to N^3 and the overhead penalty becomes less significant. Although storage comparisons are difficult, experience suggests that the dense method is limited to approximately 350 variables and constraints for the Sun IPC workstation configuration. In contrast the sparse algorithm can easily accomodate problems with ten times as many variables and constraints within the same memory limitations. Furthermore, previously reported results (Betts and Huffman, 1992) have demonstrated that problems with nearly 13000 variables can be solved on larger computers. In general, high fidelity applications of direct transcription methods require a large number of grid points. Grid refinement and sparse optimization methods make these applications feasible.

Table 1: Iteration Summary

i_r	n_g	HC	FE	RHS	T
1	10	11	886	8860	27.38
2	25	5	1005	49245	71.94
3	47	1	319	29667	49.32
4	71	1	319	44979	84.51
5	92	1	282	51606	101.68
6	99	1	282	55554	116.32
		20	3093	239911	451.15

i_r Refinement Iteration No.

n_g No. gridpoints ($n_g = n_s + 1$)

Table 2: Dense Comparison

n_g	10	20	40
T_o	73.72	759.32	7794.65
T	67.83	165.31	295.47
R	1.08	4.59	26.38

HC	No. Hessian Calls
FE	No. Function Eval.
RHS	No. Right Hand Side Eval.
T_o	OTIS: Solution time (sec)
T	Solution time (sec)
R	Speedup $R = T_o/T$

6   Summary and Conclusions

This paper presents a method for solving path constrained trajectory optimization problems using a sparse nonlinear programming algorithm. A comparison of three different strategies for using a sparse quadratic programming algorithm suggest that an approach which first locates a feasible point and then stays "near" the constraints, produces a reasonable compromise between speed and robustness. A grid refinement strategy has been used to exploit the benefits of trapezoidal discretization, followed by refinement with a higher order Hermite-Simpson approach. When coupled with the sparse nonlinear programming algorithm, the technique has demonstrated both speed and reliability when solving optimal control problems with both equality and inequality constraints.

Acknowledgements

The author wishes to acknowledge the continuing support and interaction provided by Dr. William Huffman, in all aspects of this research. In addition the insightful contributions of Dr. Paul Frank should be acknowledged, especially in the development of an efficient, robust sparse quadratic programming algorithm.

References

Ashcraft, C.C., and Grimes, R.G. (1988). 'The Influence of Relaxed Supernode Partitions on the Multifrontal Method,' Boeing Computer Services Technical Report ETA-TR-60.

Ascher, U., Mattheij, R., and Russell, R.D. (1988)]. *Numerical Solution of Boundary Value Problems for Ordinary Differential Equations*, Prentice Hall, Englewood Cliffs, N.J.

Betts, J.T. (1990). 'Sparse Jacobian Updates in the Collocation Method for Optimal Control Problems,' *Journal of Guidance, Control, and Dynamics*, Vol. 13, No. 3, May-June.

Betts, J.T., and Huffman, W.P. (1991a). 'Trajectory Optimization on a Parallel Processor,' *Journal of Guidance, Control, and Dynamics*, Vol. 14, No. 2, March-April.

Betts, J.T., and Huffman, W.P. (1991b). 'Path Constrained Trajectory Optimization Using Sparse Sequential Quadratic Programming,' AIAA-91-2739-CP, pp 1236-1259, Proceedings of the AIAA Guidance, Navigation, and Control Conference, New Orleans, LA.

Betts, J.T., and Huffman, W.P. (1992). 'Application of Sparse Nonlinear Programming to Trajectory Optimization,' *Journal of Guidance, Control, and Dynamics*, Vol 15, No. 1, January-February.

Bryson, A.E., Desai, M.N., and Hoffman, W.C., (1969)]. 'Energy-State Approximation in Performance Optimization of Supersonic Aircraft,' *Journal of Aircraft*, Vol. 6, No. 6, Nov-Dec.

Bulirsch, R., Montrone, F., and Pesch, H.J. (1991). 'Abort Landing in the Presence of Windshear as a Minimax Optimal Control Problem, Part 2: Multiple Shooting and Homotopy,' *Journal of Optimization Theory and Applications*, Vol 70, No. 2, pp 223-254, August.

Dickmanns, E.D., (1980). 'Efficient Convergence and Mesh Refinement Strategies for Solving General Ordinary Two-Point Boundary Value Problems by Collocated Hermite Approximation,' 2nd IFAC Workshop on Optimisation, Oberpfaffenhofen, Sept. 15-17.

Enright, P.J. (1991). 'Optimal Finite-Thrust Spacecraft Trajectories Using Direct Transcription and Nonlinear Programming,' Ph.D. Thesis, University of Illinois.

Gill, P.E., Murray, W., and Wright, M.H., (1981). *Practical Optimization*, Academic Press.

Gill, P.E., Murray, W., Saunders, M.A., and Wright, M.H., (1986a). 'User's Guide for NPSOL (Version 4.0): a Fortran package for nonlinear programming,' Report SOL 86-2, Department of Operations Research, Stanford University.

Gill, P.E., Murray, W., Saunders, M.A., and Wright, M.H., (1986b). 'Some Theoretical Properties of an Augmented Lagrangian Merit Function', Report SOL 86-6, Department of Operations Research, Stanford University.

Gill, P.E., Murray, W., Saunders, M.A., and Wright, M.H., (1987). 'A Schur-Complement Method for Sparse Quadratic Programming,' Report SOL 87-12, Department of Operations Research, Stanford University.

Hargraves, C.R., and S.W. Paris, (1987). 'Direct Trajectory Optimization Using Nonlinear Programming and Collocation,' *J. of Guidance, Control, and Dynamics*, Vol. 10, No.4, July-Aug, p338.

Author's Address

Dr. John T. Betts, Senior Principal Scientist
Boeing Computer Services, P.O.Box 24346, MS 7L-21
Seattle, WA 98124-0346
jbetts@espresso.boeing.com

Numerical Solution of Optimal Control Problems by Direct Collocation

Oskar von Stryk

Abstract

By an appropriate discretization of control and state variables, a constrained optimal control problem is transformed into a finite dimensional nonlinear program which can be solved by standard SQP-methods [10]. Convergence properties of the discretization are derived. From a solution of this method known as direct collocation, these properties are used to obtain reliable estimates of adjoint variables. In the presence of active state constraints, these estimates can be significantly improved by including the switching structure of the state constraint into the optimization procedure. Two numerical examples are presented.

1   Statement of problems

Systems governed by ordinary differential equations arise in many applications as, e. g., in astronautics, aeronautics, robotics, and economics. The task of optimizing these systems leads to the optimal control problems investigated in this paper.

The aim is to find a control vector $u(t)$ and the final time t_f that minimize the functional

$$(1.1) \qquad J[u, t_f] = \Phi(x(t_f), t_f)$$

subject to a system of n nonlinear differential equations

$$(1.2) \qquad \dot{x}_i(t) = f_i(x(t), u(t), t), \quad i = 1, \ldots, n, \quad 0 \le t \le t_f,$$

boundary conditions

$$(1.3) \qquad r_i(x(0), x(t_f), t_f) = 0, \quad i = 1, \ldots, k \le 2n,$$

and m inequality constraints

$$(1.4) \qquad g_i(x(t), u(t), t) \ge 0, \quad i = 1, \ldots, m, \quad 0 \le t \le t_f.$$

Here, the l vector of control variables is denoted by $u(t) = (u_1(t), \ldots, u_l(t))^T$ and the n vector of state variables is denoted by $x(t) = (x_1(t), \ldots, x_n(t))^T$. The functions $\Phi : \mathrm{I\!R}^{n+1} \to \mathrm{I\!R}$, $f : \mathrm{I\!R}^{n+l+1} \to \mathrm{I\!R}^n$, $r : \mathrm{I\!R}^{2n+1} \to \mathrm{I\!R}^k$, and $g : \mathrm{I\!R}^{n+l+1} \to \mathrm{I\!R}^m$ are assumed to be continuously differentiable. The controls $u_i : [0, t_f] \to \mathrm{I\!R}$, $i = 1, \ldots, l$, are assumed to be bounded and measureable and t_f may be fixed or free.

2   Discretization

This section briefly recalls the discretization scheme as described in more detail in [18]. Some of the basic ideas of this discretization scheme have been formerly outlined by Kraft [14] and Hargraves and Paris [11].
A discretization of the time interval

$$(2.1) \qquad 0 = t_1 < t_2 < \ldots < t_N = t_f$$

is chosen. The parameters Y of the nonlinear program are the values of control and state variables at the grid points t_j, $j = 1, \ldots, N$, and the final time t_f

$$(2.2) \qquad Y = (u(t_1), \ldots, u(t_N), x(t_1), \ldots, x(t_N), t_N) \in \mathbf{R}^{N(l+n)+1}.$$

The controls are chosen as piecewise linear interpolating functions between $u(t_j)$ and $u(t_{j+1})$ for $t_j \le t < t_{j+1}$

$$(2.3) \qquad u_{\mathrm{app}}(t) = u(t_j) + \frac{t - t_j}{t_{j+1} - t_j}(u(t_{j+1}) - u(t_j)).$$

The states are chosen as continuously differentiable functions and piecewise defined as cubic polynomials between $x(t_j)$ and $x(t_{j+1})$ with $\dot{x}_{\mathrm{app}}(s) := f(x(s), u(s), s)$ at $s = t_j$, t_{j+1},

$$(2.4) \qquad x_{\mathrm{app}}(t) = \sum_{k=0}^{3} c_k^j \left(\frac{t - t_j}{h_j}\right)^k, \quad t_j \le t < t_{j+1}, \quad j = 1, \ldots, N-1,$$

$$(2.5) \qquad c_0^j = x(t_j),$$

$$(2.6) \qquad c_1^j = h_j f_j,$$

$$(2.7) \qquad c_2^j = -3x(t_j) - 2h_j f_j + 3x(t_{j+1}) - h_j f_{j+1},$$

$$(2.8) \qquad c_3^j = 2x(t_j) + h_j f_j - 2x(t_{j+1}) + h_j f_{j+1},$$

$$\text{where } f_j := f(x(t_j), u(t_j), t_j), \quad h_j := t_{j+1} - t_j.$$

The approximating functions of the states have to satisfy the differential equations (1.2) at the grid points t_j, $j = 1, \ldots, N$, and at the centers $t_{c,j} := t_{j+1/2} := (t_j + t_{j+1})/2$,

$j = 1, \ldots, N - 1$, of the discretization intervals. This scheme is also known as cubic collocation at Lobatto points. The chosen approximation (2.4) – (2.8) of $x(t)$ already fulfills these constraints at t_j. Therefore, the only remaining constraints in the nonlinear programming problem are

- the collocation constraints at $t_{c,j}$

$$(2.9) \qquad f(x_{\text{app}}(t_{c,j}), u_{\text{app}}(t_{c,j}), t_{c,j}) - \dot{x}_{\text{app}}(t_{c,j}) = 0, \; j = 1, \ldots, N - 1,$$

- the inequality constraints at the grid points t_j

$$(2.10) \qquad g(x_{\text{app}}(t_j), u_{\text{app}}(t_j), t_j) \geq 0, \; j = 1, \ldots, N,$$

- and the initial and end point constraints at t_1 and t_N

$$(2.11) \qquad r(x_{\text{app}}(t_1), x_{\text{app}}(t_N), t_N) = 0.$$

In the following, the index "app" for approximation will be suppressed.

By this scheme the number of four free parameters for each cubic polynomial is reduced to two and the number of three collocation constraints per subinterval is reduced to one. Compared with other collocation schemes we have a reduced number of constraints to be fulfilled and a reduced number of free parameters to be determined by the numerical procedure. This results in a better performance of an implementation of this method in terms of convergence, reliability, and efficiency compared with other schemes.

3   Convergence properties of the discretization

In the sequel, we assume that, for example, the controls u_i, $i = 1, \ldots, l$, appear nonlinearly in f, the optimal control is continuous and the final time t_f is fixed. Furthermore, we assume that the number of inequality constraints m is 1 and that the constraint $g = g_1$ is active within an interval $[t_{\text{entry}}, t_{\text{exit}}]$ along the optimal trajectory, where $0 < t_{\text{entry}} < t_{\text{exit}} < t_f$.

3.1   Necessary first order optimality conditions of the continuous problem

There exist an n-vector function of adjoint or costate variables $\lambda(t) = (\lambda_1(t), \ldots, \lambda_n(t))^T$ and a multiplier function $\eta(t)$. With the Hamiltonian

$$(3.1) \qquad H(x, u, t, \lambda, \eta) = \sum_{k=1}^{n} \lambda_k f_k(x, u, t) \; + \; \eta(t) g(x, u, t),$$

the necessary first order conditions of optimality result in a multi-point boundary value problem

$$(3.2) \qquad \dot{x}_i(t) \;=\; \frac{\partial H}{\partial \lambda_i} \;=\; f_i(x,u,t),$$

$$(3.3) \qquad \dot{\lambda}_i(t) \;=\; -\frac{\partial H}{\partial x_i} \;=\; -\sum_{k=1}^{n} \lambda_k(t)\frac{\partial f_k(x,u,t)}{\partial x_i} - \eta(t)\frac{\partial g(x,u,t)}{\partial x_i}, \quad i=1,\ldots,n,$$

$$(3.4) \qquad 0 \;=\; \frac{\partial H}{\partial u_j} \;=\; \sum_{k=1}^{n} \lambda_k(t)\frac{\partial f_k(x,u,t)}{\partial u_j} + \eta(t)\frac{\partial g(x,u,t)}{\partial u_j}, \quad j=1,\ldots,l,$$

$$(3.5) \qquad g(x,u,t) > 0 \ \text{and} \ \eta(t) = 0, \quad \text{or} \quad g(x,u,t) = 0 \ \text{and} \ \eta(t) \le 0.$$

The original boundary constraints (1.3) and additional constraints on $\lambda(t)$ at 0, t_{entry}, t_{exit}, and t_f also have to be fulfilled. In general, at junction points t_{entry}, t_{exit}, the adjoint variables may have discontinuities. For more details cf. Bryson, Ho [4] and Hestenes [12] and also Jacobson, Lele, Speyer [13], Maurer [15], and the results of Maurer cited in Bulirsch, Montrone, Pesch [5] for the necessary conditions of optimality in the constrained case.

In the sequel, we shall see that the necessary first order optimality conditions of the continuous problem are reflected in the necessary first order optimality conditions of the discretized problem.

3.2   Necessary first order optimality conditions of the discretized problem

For the sake of simplicity, we now assume that $n = 1$ and $l = 1$. In this section, we will use the notations

$$(3.6) \qquad u_i := u(t_i), \quad x_i := x(t_i), \quad i = 1,\ldots,N,$$

and

$$(3.7) \qquad f_i := f(x_i, u_i, t_i), \quad f_{i+1/2} := f(x(t_{i+1/2}), u(t_{i+1/2}), t_{i+1/2}).$$

The Lagrangian of the nonlinear program of the discretized problem from Sec. 2 can then be written as

$$
\begin{aligned}
L(Y,\mu,\sigma,\nu) \;=\;\; & \Phi(x_N, t_N) \\
& - \sum_{j=1}^{N-1} \mu_j\left(f(x(t_{c,j}), u(t_{c,j}), t_{c,j}) - \dot{x}(t_{c,j})\right) \\
& - \sum_{j=1}^{N} \sigma_j g(x_j, u_j, t_j) - \sum_{j=1}^{k} \nu_j r_j(x(t_1), x(t_N), t_N)
\end{aligned}
$$
$$(3.8)$$

with $\mu = (\mu_1, \ldots, \mu_{N-1})^T \in \mathbf{R}^{N-1}$, $\sigma = (\sigma_1, \ldots, \sigma_N)^T \in \mathbf{R}^N$ and $\nu = (\nu_1, \ldots, \nu_k)^T \in \mathbf{R}^k$. A solution of the nonlinear program fulfills the necessary first order optimality conditions of Karush, Kuhn, and Tucker, cf., e. g., [9]. Among others, these are

$$(3.9) \quad \frac{\partial L}{\partial u_i} = 0, \quad \frac{\partial L}{\partial x_i} = 0, \quad \frac{\partial L}{\partial \mu_j} = 0, \quad i = 1, \ldots, N, \ j = 1, \ldots, N - 1.$$

$$(3.10) \quad g(x_i, u_i, t_i) > 0 \text{ and } \sigma_i = 0, \quad \text{or} \quad g(x_i, u_i, t_i) = 0 \text{ and } \sigma_i \leq 0.$$

As the "fineness" of the grid, we define

$$(3.11) \quad h := \max\{h_j = t_{j+1} - t_j : j = 1, \ldots, N - 1\}.$$

In detail, we find for $i = 2, \ldots, N$,

$$
\begin{aligned}
0 = \frac{\partial L}{\partial u_i} = \ & -\mu_{i-1}\left(\frac{\partial f(x(t_{i-1/2}), u(t_{i-1/2}), t_{i-1/2})}{\partial u_i} - \frac{\partial \dot{x}(t_{i-1/2})}{\partial u_i}\right) \\
& -\mu_i\left(\frac{\partial f(x(t_{i+1/2}), u(t_{i+1/2}), t_{i+1/2})}{\partial u_i} - \frac{\partial \dot{x}(t_{i+1/2})}{\partial u_i}\right) \\
& -\sigma_i \frac{\partial g(x(t_i), u(t_i), t_i)}{\partial u_i}
\end{aligned}
$$
$$(3.12)$$

Using the basic relations (3.8) – (3.16) of [18] and the notation from (3.6), (3.7), we obtain after some calculations and by using the chain rule of differentiation

$$
\begin{aligned}
\frac{\partial L}{\partial u_i} = \ & -\frac{1}{2}\left(\frac{\partial f_{i-1/2}}{\partial u}\mu_{i-1} + \frac{\partial f_{i+1/2}}{\partial u}\mu_i\right) - \frac{1}{4}\frac{\partial f_i}{\partial u}(\mu_i + \mu_{i-1}) \\
& +\frac{1}{8}\frac{\partial f_i}{\partial u}\left(h_{i-1}\mu_{i-1}\frac{\partial f_{i-1/2}}{\partial x} - h_i\mu_i\frac{\partial f_{i+1/2}}{\partial x}\right) - \sigma_i\frac{\partial g(x(t_i), u(t_i), t_i)}{\partial u}.
\end{aligned}
$$
$$(3.13)$$

Letting $h \to 0$ and keeping $t = t_i$ fixed, we have

$$(3.14) \quad \frac{\partial L}{\partial u_i} = -\frac{1}{2}\left(\frac{\partial f_i}{\partial u}\mu_i + \frac{\partial f_i}{\partial u}\mu_i\right) - \frac{1}{4}\frac{\partial f_i}{\partial u}(\mu_i + \mu_i) - \sigma_i\frac{\partial g(x(t_i), u(t_i), t_i)}{\partial u}$$

and finally

$$(3.15) \quad \frac{3}{2}\mu_i\frac{\partial f(x(t_i), u(t_i), t_i)}{\partial u} + \sigma_i\frac{\partial g(x(t_i), u(t_i), t_i)}{\partial u} = 0.$$

This equation is equivalent to the condition (3.4).
On the other hand, for $i = 2, \ldots, N - 1$,

$$
\begin{aligned}
0 = \frac{\partial L}{\partial x_i} = \ & -\mu_{i-1}\left(\frac{\partial f(x(t_{i-1/2}), u(t_{i-1/2}), t_{i-1/2})}{\partial x_i} - \frac{\partial \dot{x}(t_{i-1/2})}{\partial x_i}\right) \\
& -\mu_i\left(\frac{\partial f(x(t_{i+1/2}), u(t_{i+1/2}), t_{i+1/2})}{\partial x_i} - \frac{\partial \dot{x}(t_{i+1/2})}{\partial x_i}\right) \\
& -\sigma_i\frac{\partial g(x(t_i), u(t_i), t_i)}{\partial x_i}.
\end{aligned}
$$
$$(3.16)$$

Using again the basic relations (23) – (31) of [18] and the notation from (3.6), (3.7), we obtain after some calculations and by using the chain rule of differentiation

$$
\begin{aligned}
\frac{\partial L}{\partial x_i} \;=\; & -\frac{3}{2}\left(\frac{\mu_i}{h_i} - \frac{\mu_{i-1}}{h_{i-1}}\right) \\
& -\frac{1}{4}\frac{\partial f_i}{\partial x}(\mu_{i-1} + \mu_i) - \frac{1}{2}\left(\mu_{i-1}\frac{\partial f_{i-1/2}}{\partial x} + \mu_i\frac{\partial f_{i+1/2}}{\partial x}\right) \\
& +\frac{1}{8}\frac{\partial f_i}{\partial x}(h_{i-1}\mu_{i-1} - h_i\mu_i) - \sigma_i\frac{\partial g(x(t_i), u(t_i), t_i)}{\partial x}.
\end{aligned}
\tag{3.17}
$$

For convenience, we now suppose an equidistant grid, i.e.

$$
(3.18)\quad h \;=\; h_i \;=\; t_{i+1} - t_i \;=\; \frac{t_f - t_0}{N-1}, \quad i = 1, \dots, N-1.
$$

Now letting $h \to 0$ and keeping $t = t_i$ fixed, we have (cf. [18])

$$
(3.19)\quad \frac{3}{2}\dot\mu_i + \frac{3}{2}\mu_i\frac{\partial f(x(t_i), u(t_i), t_i)}{\partial x} + \sigma_i\frac{\partial g(x(t_i), u(t_i), t_i)}{\partial x} \;=\; 0.
$$

This equation is equivalent to the adjoint differential equation (3.3).
Similar results hold for a non-equidistant grid under additional conditions and for $n > 1$. They can also be extended to more general problems.

4   Estimates of adjoint variables

It has been shown in the previous section that the necessary conditions of optimality of the discretized problem reflect the necessary conditions of the original continuous problem. More precisely, it has been shown that Eq. (3.17) and Eq. (3.13), resp., are discretized versions of the adjoint differential equation (3.3) and the condition (3.4), respectively.

Therefore, we obtain an estimate of $\lambda(t)$ from the multipliers of the discretized problem by

$$
(4.1)\quad \lambda(t_{i+1/2}) \;=\; -\frac{3}{2}\rho_i\,\mu_i, \quad i = 1, \dots, N-1,
$$

where ρ_i is a scaling factor depending on the discretization. In addition, an estimate of $\eta(t_i)$ can be obtained from σ_i.

Another approach for estimating adjoint variables in combination with a direct collocation method has been used by Enright and Conway [8]. They used the multipliers ν_j from Eq. (3.8) of the boundary conditions in the discretized problem in order to estimate $\lambda(t_f)$. This estimate is then used as an initial value for the backward integration of the adjoint differential equations (3.3). It is a well-known matter of fact that this backward integration is crucial for highly nonlinear problems. Also, state constraints were not considered.

A further approach for estimating adjoint variables is based on an interpretation of the adjoint variables as sensitivities connected to the gradient of the cost function

$$(4.2) \quad \lambda(t) = \frac{\partial \Phi}{\partial x}(t) \quad \text{at} \quad u = u_{\text{optimal}}$$

where x satisfies the differential equations (1.2). This relation can be found, e. g., in Breakwell [2] or in Bryson, Ho [4]. In a discretized version as, e. g.,

$$(4.3) \quad \lambda(t) = \frac{\tilde{\Phi}(x(t) + \delta) - \tilde{\Phi}(x(t) - \delta)}{2\delta} \quad \text{at} \quad u = \tilde{u}_{\text{optimal}},$$

it can be used in combination with a direct shooting method as, e. g., [1], with a suitable steplength δ for the difference quotient. Here, the superscript $\tilde{\ }$ denotes that the variable or value has been obtained numerically, e. g., by a direct shooting method. For more details, cf., e. g., Eq. (3.14) in [1].

The guess of adjoint variables by direct methods is usually affected by several sources of inaccuracies and troubles. First, the suboptimal control $\tilde{u}_{\text{optimal}}$ calculated by a direct method is often inaccurate and can differ significantly from the optimal control. Second, the accuracy of the calculated objective $\tilde{\Phi}$ is often not better than one percent. In addition, the case of nearly active or inactive state variable inequality constraints has not yet been included in a reliable manner in previous attempts.

In contrast to the former approaches, the quality of the estimated adjoints does neither depend crucially on a highly accurate computation of the cost function or the calculated suboptimal control nor the appearance of active state constraints following our approach. As it is shown from the examples and the reported numerical results in this paper and in [6], [17], and [18], the new way of estimating adjoint variables herein proposed is very reliable and accurate even for complicated and highly nonlinear problems and problems including state constraints. Furthermore, convergence properties of the discretization scheme have been derived.

5 Examples and numerical results

The results reported in this section have been obtained by using the implementation DIRCOL (cf. [17]) of the direct collocation method mentioned in the previous sections. The use of grid refinement techniques yields a sequence of related nonlinear programs with increasing dimensions. In each macro iteration step, one nonlinear program has to be solved by the Sequential Quadratic Programming method NPSOL due to Gill, Murray, Saunders, and Wright [10]. The reported estimates of adjoint variables are direct outputs of DIRCOL.

5.1 Optimal ascent of the lower stage of a Sänger-type vehicle

This problem describes the lifting of an airbreathing lower stage of a two-stage-to-orbit Sänger-type launch vehicle. We focus on the Ramjet-powered second part of the trajectory. The four state variables are the velocity v, the flight path angle γ, the altitude h, and the mass m. The three control variables are the lift coefficient c_L, the thrust angle ϵ and the throttle setting δ, $\delta \in [0,1]$. The equations of motion are

$$(5.1) \quad \dot{v} \;=\; \frac{T(v,h;\delta)}{m}\cos\epsilon - \frac{D(v,h;c_L)}{m} - g(h)\sin\gamma,$$

$$(5.2) \quad \dot{\gamma} \;=\; \frac{1}{v}\left(\frac{T(v,h;\delta)}{m}\sin\epsilon + \frac{L(v,h;c_L)}{m} - \left(g(h) - \frac{v^2}{r_0+h}\right)\cos\gamma\right),$$

$$(5.3) \quad \dot{h} \;=\; v\sin\gamma,$$

$$(5.4) \quad \dot{m} \;=\; b(v,h)\,\delta, \quad b(v,h) = \text{ maximum mass flow.}$$

The considered time interval is $[0,t_f]$ and t_f is free. The following formulae are used for the thrust, the lift and the drag forces

$$
\begin{aligned}
T(v,h;\delta) &= T_m(v,h)\,\delta, \quad T_m(v,h) = \text{maximum thrust,}\\
L(v,h;c_L) &= q(v,h)\,S\,c_L,\\
D(v,h;c_L) &= q(v,h)\,S\,c_D(m_a(v,h),c_L),\\
\text{where}\quad q(v,h) &= \frac{v^2}{2}\rho_0\exp(-\beta h), \quad g(h) = g_0\left(\frac{r_0}{r_0+h}\right)^2.
\end{aligned}
$$

The lift and drag model has a quadratic polar

$$
\begin{aligned}
c_D(m_a(v,h),c_L) &= c_{D0}(m_a(v,h)) + k(m_a(v,h))\,c_L^2, \quad a(h) = \text{ speed of sound,}\\
m_a(v,h) &= \frac{v}{a(h)}, \quad k(m_a) = \text{ a characteristic function of the vehicle.}
\end{aligned}
$$

The quantities S, g_0, r_0, and ρ_0 are constants. For more details of the problem and for a three dimensional formulation cf. Chudej [7]. The boundary conditions are

$$(5.5) \quad
\begin{aligned}
h(0) &= 20\text{ km}, & h(t_f) &= 30\text{ km},\\
v(0) &= 925\text{ m/s}, & v(t_f) &= 1700\text{ m/s},\\
\gamma(0) &= 0.05, & \gamma(t_f) &= 0.04,\\
m(0) &= 332400\text{ kg}.
\end{aligned}
$$

The objective is to maximize the final mass, i. e.,

$$(5.6) \quad J[c_L, \epsilon, \delta, t_f] = -m(t_f) \quad \rightarrow \quad \text{min!}$$

Here, the direct collocation method was applied on a rather bad initial estimate of the optimal trajectory. For the states, the boundary values have been interpolated linearly and the controls have been set to zero. The direct collocation method DIRCOL converges in two macro iteration steps to a solution with 21 grid points. From this solution, the optimal states and the adjoint variables have been estimated. Based on this estimate, the multiple shooting method was applied to solve the boundary value problem arising from the optimality conditions (see [7]). The final solutions are $m(t_f) = 321243.$ kg and $t_f = 179.75$ s. For these values, the solution of the direct collocation method was accurate to four digits.

In Figs. 1 to 6, the solution of the direct collocation method is shown by a dashed line and the highly accurate solution of the multiple shooting method is shown by a solid line. In the figures, there is no visible difference between the suboptimal and the optimal state variables.

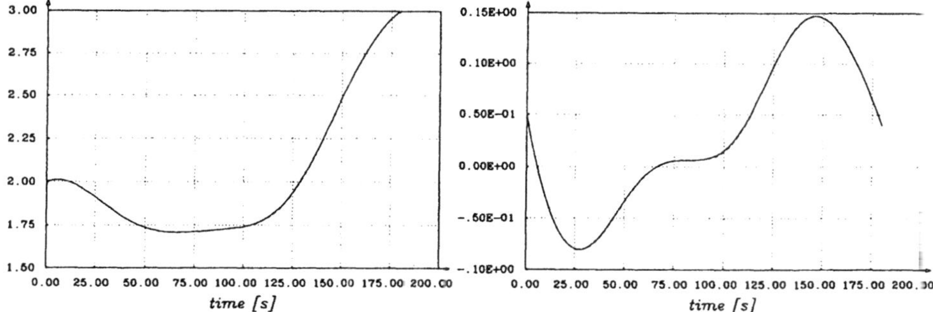

Fig. 1: The altitude h[10km].

Fig. 2: The flight path angle γ[1].

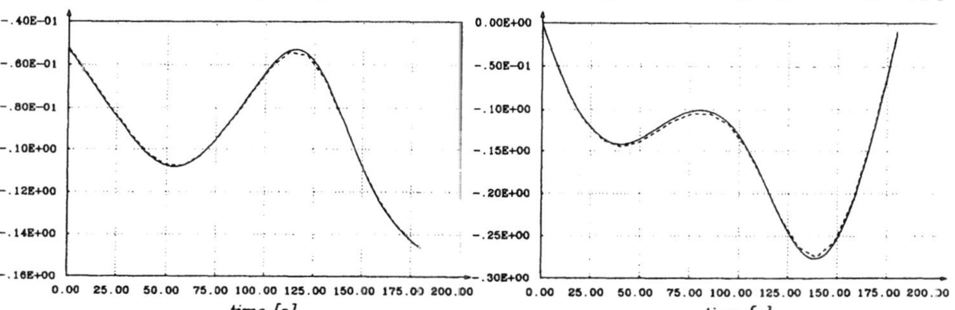

Fig. 3: The adjoint variable λ_h.

Fig. 4: The adjoint variable λ_γ.

Also, the estimated adjoint variables and the suboptimal controls of the direct collocation method show a pretty good conformity with the highly accurate ones. The approximation quality can furthermore be improved by increasing the number of grid points to more than 21. The optimal throttle setting δ equals one within the whole time interval as it is found by both methods.

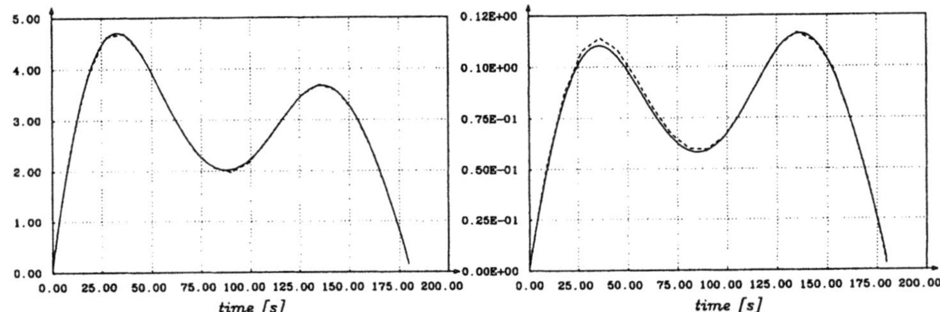

Fig. 5: The lift coefficient c_L.   Fig. 6: The thrust angle $\epsilon[1]$.

5.2 A problem with a second order state variable inequality constraint

This well-known problem is due to Bryson, Denham, and Dreyfus [3]. After a transformation, the differential equations and boundary conditions are

$$(5.7) \quad \begin{array}{llllll} \dot{x} &=& v, & x(0) &=& 0, \quad x(1) = 0, \\ \dot{v} &=& u, & v(0) &=& 1, \quad v(1) = -1, \\ \dot{w} &=& u^2/2, & w(0) &=& 0, \quad w(1) \text{ is free.} \end{array}$$

The objective is

$$(5.8) \quad J[u] = w(1) \quad \rightarrow \quad \text{min!}$$

The state constraint to be taken into account is of order 2 here

$$(5.9) \quad g(x) = l - x(t) \geq 0, \quad \frac{\partial}{\partial u}\left(\frac{\mathrm{d}}{\mathrm{d}t}g\right) \equiv 0, \quad \frac{\partial}{\partial u}\left(\frac{\mathrm{d}^2}{\mathrm{d}t^2}g\right) \not\equiv 0.$$

Explicit formulae of the solution depending on the value of l can be given, cf. [3], [4]. For $l = 1/9$, there exists an interior boundary arc $[t_{\text{entry}}, t_{\text{exit}}] = [t_{\text{I}}, t_{\text{II}}] = [3l, 1 - 3l]$

where the state constraint is active. The minimum objective value is $w(1) = 4/(9l) = 4$. With the Hamiltonian $H = \lambda_x v + \lambda_v u + \lambda_w u^2/2 + \eta(l-x)$ the minimum principle yields for the adjoint variables

$$(5.10) \quad \lambda_x(t) = \begin{cases} 2/(9l^2), & 0 \le t < t_\mathrm{I}, \\ 0, & t_\mathrm{I} \le t < t_\mathrm{II}, \\ -2/(9l^2), & t_\mathrm{II} \le t \le 1, \end{cases}$$

$$\lambda_v(t) = \begin{cases} 2\left(1 - t/(3l)\right)/(3l), & 0 \le t < t_\mathrm{I}, \\ 0, & t_\mathrm{I} \le t < t_\mathrm{II}, \\ 2\left(1 - (1-t)/(3l)\right)/(3l), & t_\mathrm{II} \le t \le 1, \end{cases}$$

and $\lambda_w \equiv 1$. The adjoint variable λ_x suffers discontinuities when entering or leaving the state constraint. A first solution is obtained by using DIRCOL with an equidistant grid of $N = 11$ grid points resulting in a minimum objective value of $w(1) = 3.99338$

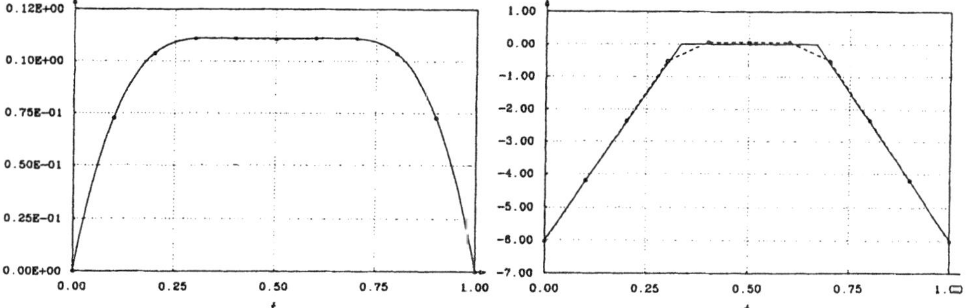

Fig. 7: The state variable x.

Fig. 8: The control variable u.

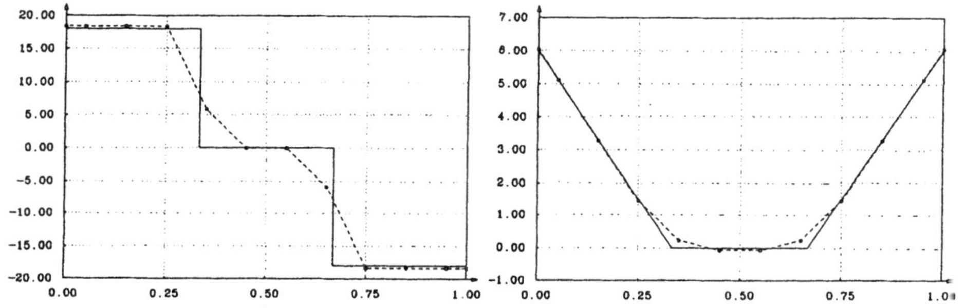

Fig. 9: The adjoint variable λ_x.

Fig. 10: The adjoint variable λ_v.

In Figs. 7 to 10 these first suboptimal solutions are shown by dashed lines and the exact solutions are shown by solid lines. In addition, the grid points of the discretization are marked.

The solution is now refined by using a "three-stage" collocation approach that includes the switching structure of the state constraint, i. e. the switching points t_I and t_{II} are included as two additional parameters with two additional equality conditions in the optimization procedure

$$(5.11) \quad g(x) \begin{cases} \geq 0, & 0 \leq t < t_I, \\ = 0, & t_I \leq t < t_{II}, \\ \geq 0, & t_{II} \leq t \leq 1, \end{cases} \quad x(t_I - 0) = l, \quad x(t_{II} + 0) = l.$$

The method DIRCOL is now applied to the reformulated problem with a separate grid of 4 grid points in each of the three stages $[0, t_I]$, $[t_I, t_{II}]$, and $[t_{II}, 1]$. This results in a minimum objective value of $w(1) = 3.99992$ and a more accurately satisfied state constraint. In Figs. 11 to 14 the refined solutions are shown. In addition, two dotted vertical lines show the entry and exit points of the state constraint that are computed with an error of one percent. The quality of the estimated adjoint variables and also of the control variable has been significantly improved while the dimension of the resulting nonlinear program has not been increased.

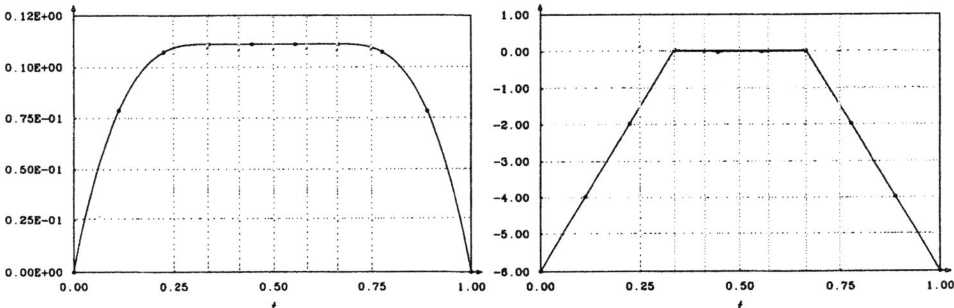

Fig. 11: The state variable x. $\hspace{3cm}$ Fig. 12: The control variable u.

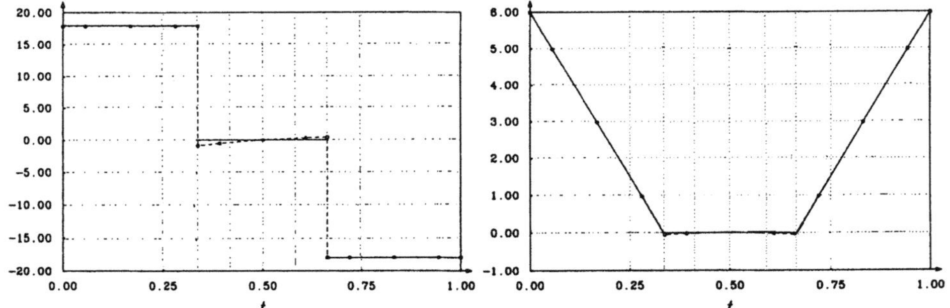

Fig. 13: The adjoint variable λ_x. $\hspace{2.5cm}$ Fig. 14: The adjoint variable λ_v.

6 Conclusions

A way of estimating adjoint variables of optimal control problems by a direct collocation method has been described. The method seems to be superior to previous approaches for estimating adjoint variables in terms of reliability and the ability to include discontinuities in adjoints at the junction points of state constraint subarcs. Furthermore, the estimates of the adjoint variables and the suboptimal controls have been improved by including the switching structure of active state constraints in the optimization procedure.

Acknowledgement. The author is indebted to Prof. Dr. P. E. Gill who supplied him with a version of the SQP-method NPSOL for this scientific research. The author is also indebted to the numerical analysis and optimal control group of Prof. Dr. R. Bulirsch for helpful discussions.

References

[1] Bock, H.G.; K.J. Plitt. *A multiple shooting algorithm for direct solution of optimal control problems.* Proc. of the IFAC 9th World Congress, Budapest, Hungary, July 2-6 (1984) 242-247.

[2] Breakwell, J.V. *The optimization of trajectories.* SIAM J. Appl. Math. **7** (1959) 215-247.

[3] Bryson, A.E.; W.F. Denham; S.E. Dreyfus. *Optimal programming problems with inequality constraints. I: Necessary conditions for extremal solutions.* AIAA J. **1**, 11 (1963) 2544-2550.

[4] Bryson, A.E.; Y.-C. Ho. *Applied Optimal Control.* Rev. Printing. (Hemisphere Publishing Corporation, New York, 1975).

[5] Bulirsch, R.; F. Montrone; H.J. Pesch. *Abort landing in the presence of a windshear as a minimax optimal control problem, Part 1: Necessary conditions.* J. Optim. Theory Appl. **70**, 1 (1991) 1-23.

[6] Bulirsch, R.; E. Nerz; H.J. Pesch; O. von Stryk. *Combining direct and indirect methods in optimal control: Range maximization of a hang glider.* In: R. Bulirsch, A. Miele, J. Stoer, K.H. Well (eds.): *Optimal Control and Variational Calculus,* Oberwolfach, 1991, International Series of Numerical Mathematics (Birkhäuser, Basel) this issue.

[7] Chudej, K. *Optimal ascent of a hypersonic space vehicle.* In: R. Bulirsch, A. Miele, J. Stoer, K.H. Well (eds.): *Optimal Control and Variational Calculus,* Oberwolfach, 1991, International Series of Numerical Mathematics (Birkhäuser, Basel) this issue.

[8] Enright, P.J.; B.A. Conway. *Discrete approximations to optimal trajectories using direct transcription and nonlinear programming.* AIAA Paper 90-2963-CP (1990).

[9] Fletcher, R. *Practical methods of optimization.* (2nd ed., J. Wiley & Sons, Chichester/New York/Brisbane/Toronto/Singapore, 1987).

[10] Gill, P.E.; W. Murray; M.A. Saunders; M.H. Wright. *User's guide for NPSOL (Version 4.0).* Report SOL 86-2. Department of Operations Research, Stanford University, California, USA (1986).

[11] Hargraves, C.R.; S.W. Paris. *Direct trajectory optimization using nonlinear programming and collocation.* AIAA J. Guidance **10**, 4 (1987) 338-342.

[12] Hestenes, M.R. *Calculus of variations and optimal control theory.* (J. Wiley & Sons, New York, London, Sydney, 1966).

[13] Jacobson, D.H.; M.M. Lele; J.L. Speyer. *New necessary conditions of optimality for control problems with state-variable inequality constraints.* Journal of Mathematical Analysis and Applications **35** (1971) 255-284.

[14] Kraft, D. *On converting optimal control problems into nonlinear programming codes.* In: K. Schittkowski (ed.): *Computational Mathematical Programming*, NATO ASI Series **15** (Springer, 1985) 261-280.

[15] Maurer, H. *Optimale Steuerprozesse mit Zustandsbeschränkungen.* Habilitationsschrift, University of Würzburg, Würzburg, Germany (1976).

[16] Stoer, J.; R. Bulirsch. *Introduction to Numerical Analysis.* 2nd rev. ed. (Springer Verlag, 1983).

[17] von Stryk, O. *Ein direktes Verfahren zur Bahnoptimierung von Luft- und Raumfahrzeugen unter Berücksichtigung von Beschränkungen.*
(a) Diploma Thesis, Department of Mathematics, Munich University of Technology (May 1989).
(b) Lecture at the Institute for Flight Systems Dynamics, German Aerospace Research Corporation (DLR), Oberpfaffenhofen, Germany (June 16th, 1989).
(c) ZAMM **71**, 6 (1991) T705-T706.

[18] von Stryk, O.; R. Bulirsch. *Direct and indirect methods for trajectory optimization.* Annals of Operations Research **37** (1992) 357-373.

Author's address

Dipl. Math. Oskar von Stryk, Mathematisches Institut,
Technische Universität München, P.O.Box 20 24 20, D-W-8000 München 2, Germany
stryk@mathematik.tu-muenchen.de

Reduced SQP Methods for Nonlinear Heat Conduction Control Problems

F.-S. Kupfer, E.W. Sachs

Abstract

In this paper a problem is considered which arises in the control of heat conduction processes governed by nonlinear diffusion equations. We present the discretized form of such a problem and apply a reduced SQP method for the numerical solution of the optimization problem. This method makes use of the sparsity and offers the advantage to approximate second order information by a quasi Newton update which is practicable with regard to storage. The convergence result is a local 1-step q-superlinear convergence rate which is an improvement over results published previously. The algorithm is tested on the parabolic control problem and the results document a sizeable reduction in the number of iterations.

1 Introduction

This paper deals with the numerical solution of boundary control problems which occur in nonlinear diffusion processes. The application we have in mind lies in the heating of kilns in ceramic industry, see e.g. [1], [2], [3]. We are interested in the efficient numerical solution of discretized versions of these problems using modern optimization techniques.

A common problem in the ceramic industry is to control the heating process such that the temperature inside the product to be heated follows a certain firing curve, say p. Since the control law u should minimize the expenditure of energy, the objective is to

$$(1.1) \qquad \text{minimize } \int_0^T [(y(1,t) - p(t))^2 + \alpha u^2(t)] dt$$

for a given constant $\alpha \geq 0$. The temperature y solves the following nonlinear heat equation

$$
\begin{aligned}
C(y(x,t))\tfrac{\partial y}{\partial t}(x,t) - \nabla(\lambda(y(x,t))\nabla y(x,t)) &= q(x,t) &&\text{on } \Omega \times [0,T], \\
\lambda(y(x,t))\nabla y(x,t) &= b(x,t) &&\text{on } \partial\Omega \times [0,T], \\
y(x,0) &= y_0(x) &&\text{on } \Omega,
\end{aligned}
$$

(1.2)

where the spatial domain Ω is $[0,1]$. The specific heat capacity C and the heat conduction λ are real–valued functions which both depend on the temperature y.

The control function u is interpreted as the temperature of the ambiant air which controls the heat flux at the accessible surface of the product and therefore enters only the boundary input:

$$
\begin{aligned}
b(0,t) &= g[y(0,t) - u(t)], \\
b(1,t) &= 0;
\end{aligned}
$$

(1.3)

g is a real number which is usually normalized to be 1. The second equation in (1.3) means that there is no heat flux in the interior of the probe, where we use the convention that $x = 1$ and $x = 0$ are located in the inside and at the boundary of the probe, respectively.

The optimal control problem (1.1)–(1.3) has been considered in [3], [4] and [5] and we use it in order to test the efficiency of the method which we propose in this paper. Sometimes modelling includes some constraints, for example the maximum temperature gradient being bounded where the physical and chemical product qualities are changing or the duration of a certain temperature value to be reached, so that the product is baked up to the heart. Since in practice these constraints are often incorporated in the design of the firing curve p, state constraints are omitted in the problem formulation.

Under proper assumptions on λ and C it is shown in [3] that there exists a solution to (1.1)–(1.3) for $\alpha > 0$. In the same reference, a weak variational formulation is used in order to obtain a discrete version of the control problem. Specifically, the time derivative is approximated by difference ratios and a standard finite element approach is used for the space discretization. Let t^j denote the grid points in the time interval with time step $\tau = T/M$ and let N be the number of finite elements in the space interval where a grid consists of the points x_j. Then the Parabolic Boundary Control Problem (PBCP) can be written as a finite dimensional minimization problem as follows:

(1.4)   Minimize $f(y,u)$ subject to $h(y,u) = 0$,

where

$$f : \mathbb{R}^{M(N+2)} \to \mathbb{R} \qquad h : \mathbb{R}^{M(N+2)} \to \mathbb{R}^{M(N+1)}.$$

The objective function is given by

$$(1.5) \qquad f(y,u) = \tau \sum_{j=2}^{M+1} [(y_{N+1}^j - p^j)^2 + \alpha(u^j)^2].$$

The components $h^j : \mathbb{R}^{M(N+2)} \to \mathbb{R}^{N+1}$ are defined as

$$(1.6) \qquad h^1(y,u) = A\Gamma(y^2) + D\beta(y^2) + g(y_1^2 - u^2)e^1 - (q^1 + A\Gamma(y^1))$$

with $(\tau A)y^1 = ((y_0, b_i))$, and for $j = 2, ..., M$

$$(1.7) \qquad h^j(y,u) = A(\Gamma(y^{j+1}) - \Gamma(y^j)) + D\beta(y^{j+1}) + g(y_1^{j+1} - u^{j+1})e^1 - q^j,$$

where e^1 denotes the first unit vector in $\mathbb{R}^{N+1}$ and

$$
\begin{aligned}
A &= (1/\tau)((b_i, b_j)), \quad D = ((b_{ix}, b_{jx})), \\
\beta(y) &= \int_0^y \lambda(s)\,ds, \quad \Gamma(y) = \int_0^y C(s)\,ds, \quad y \in \mathbb{R}, \\
q_i^j &= \tfrac{1}{\tau} \int_{t_j}^{t_{j+1}} (q(\cdot, t), b_i)\,dt, \\
p^{j+1} &= \tfrac{1}{\tau} \int_{t_j}^{t_{j+1}} p(t)\,dt, \\
\\
y &= ((y^2)^T, ..., (y^{M+1})^T)^T \in \mathbb{R}^{M(N+1)}, \\
u &= (u^2, ..., u^{M+1})^T \in \mathbb{R}^M.
\end{aligned}
$$

Here $(\cdot, \cdot)$ denotes the L^2-inner product, b_i is the usual linear spline function with $b_i(x_j) = \delta_{ij}$ and β, Γ in (1.6), (1.7) are interpreted in the componentwise sense. The derivation of this optimization problem is described in detail in [4].

Sequential Quadratic Programming (SQP) Quasi Newton methods offer a number of positive aspects for a numerical solution of the discrete problem. Linear approximations to the constraints, the use of secant update formulas and superlinear convergence are general advantages of the SQP approach. Also, if properly implemented, these methods allow to maintain the sparsity inherent in the control problem. However, the use of SQP is severely limited by the size of the quasi Newton approximations which are matrices in $\mathbb{R}^{M(N+2) \times M(N+2)}$. Therefore, we are interested in numerical methods which are implementable with regard to storage facilities while retaining all the advantages of SQP. Ideally, we want to design an algorithm with the following properties: Infeasible iterates, maintaining the problem structure, positive definite secant updates, economizing on storage, superlinear convergence. The first four items motivated us to look at reduced SQP methods which will be introduced in the next section. However, the convergence theory for reduced secant algorithms shows that these methods produce iterates x^k which are usually only

$$\text{two-step superlinearly convergent,} \qquad \lim_{k \to \infty} ||x^{k+2} - x^*||/||x^k - x^*|| = 0.$$

In order to retain the advantage of the SQP method which is

$$\text{one-step superlinearly convergent,} \qquad \lim_{k \to \infty} ||x^{k+1} - x^*||/||x^k - x^*|| = 0,$$

we propose a modification of the original reduced iteration. In the third section we formulate a reduced secant algorithm and its modification for the discrete control problem and in the fourth section we present computational results which document the improved convergence rate numerically.

2   Reduced SQP Methods

In order to describe reduced SQP methods in a more compact form let us replace the set of variables (y, u) by x and consider the optimization problem

$$(2.1) \qquad \text{Minimize } f(x) \qquad \text{subject to } h(x) = 0,$$

where

$$f : \mathbb{R}^s \to \mathbb{R}, \qquad h : \mathbb{R}^s \to \mathbb{R}^q.$$

We make the usual assumptions on the smoothness of the problem data:

Assumption 1 *f and h are twice differentiable on a ball D which contains the solution x^* of (2.1). Furthermore, $f''(\cdot)$ and $h''(\cdot)$ are Lipschitz-continuous on D.*

We suppose that $h'(x)$ has full rank for $x \in D$. Then, to explain the principle of reduced SQP we let $T(x)$ be a basis of the null space of $h'(x)$ and we introduce the notion of a right-inverse $R(x)$:

Assumption 2 *For each $x \in D$ let $T(x) \in \mathbb{R}^{s \times (s-q)}$ be a matrix of full rank with*

$$\mathcal{N}(h'(x)) = \mathcal{R}(T(x)).$$

Furthermore let $R(x) \in \mathbb{R}^{s \times q}$ satisfy $h'(x)R(x) = I_q$.

We will discuss appropriate choices for R and T in the third section when we consider the application to the optimal control problem.

The basic idea of SQP methods is to approximate the original optimization problem (2.1) at a given iterate $x \in D$ by a quadratic programming problem:

$$(2.2) \qquad \text{Minimize } \nabla f(x)^T d + \frac{1}{2} d^T L_{xx}(x, l) d \quad \text{subject to } h'(x)d + h(x) = 0,$$

where $L = f(x) - l^T h(x), l \in \mathbb{R}^q$, is the Lagrangian. This quadratic problem is uniquely solvable if the Hessian of the Lagrangian restricted to the null space of the Jacobian, $T(x)^T L_{xx}(x, l)T(x)$, which is often called the reduced Hessian, is positive definite. The latter can be ensured if the second order sufficient optimality condition is satisfied at the solution:

Assumption 3 *There exists some $m > 0$ such that*

$$\xi^T L_{xx}(x^*, l^*)\xi \geq m\|\xi\|^2 \quad \text{for all } \xi : h'(x^*)\xi = 0.$$

If Assumption 3 is satisfied, the SQP step can be calculated from (2.2) and admits the following closed form expression (compare e.g. [6]):

$$(2.3) \quad d = -T(x)[T(x)^T L_{xx}(x, l)T(x)]^{-1} T(x)^T [\nabla f(x) - L_{xx}(x, l)R(x)h(x)] - R(x)h(x).$$

We want to use a quasi Newton update in place of exact second derivatives in (2.3). Instead of approximating L_{xx} it is more sensible to approximate the reduced Hessian $T^T L_{xx} T$ by a matrix, say B, in $\mathbb{R}^{(s-q) \times (s-q)}$. In this case, secant formulae like DFP and BFGS can be used successfully, since the requirement of positive definiteness is in line with the second order sufficiency condition (Assumption 3). Another advantage is the relatively small dimension of the approximating matrix. For example, in the application to the discrete control problem (1.4)–(1.7) the size of the matrix B is M as opposed to $M(N + 2)$ for a full Hessian approximation. Consequently, if one ignores the term $L_{xx}(x, l)R(x)h(x)$ in (2.3) and uses a secant approximation B for $T^T L_{xx} T$, then one is led to a practicable algorithm where the step is given by

$$(2.4) \qquad p = -T(x)B^{-1}T(x)^T \nabla f(x) - R(x)h(x).$$

We call p from (2.4) a reduced SQP step. It consists of a tangent step and a restoration step and can be interpreted as a particular SQP step in the sense that only a two-sided projection of the Hessian is approximated in the corresponding quadratic programming problem (2.2). The matrix B is modified at each iteration usually by a projected version of the DFP or BFGS update. Various formulae can be used in the definition of the update and we decided to present one of them in the following algorithm.

Algorithm 2.1 (Reduced BFGS Method)

Given $x \in \mathbb{R}^s$ and $B \in \mathbb{R}^{(s-q)\times(s-q)}$, B positive definite.

(1)   Solve $Bw = -T(x)^T \nabla f(x)$.

(2)   Set $x_+ = x + T(x)w - R(x)h(x)$.

(3)   Compute $v = T(x_+)^T \nabla f(x_+) - T(x - R(x)h(x))^T \nabla f(x - R(x)h(x))$.

(4)   Set

$$B_+ = B + \frac{vv^T}{v^T w} - \frac{(Bw)(Bw)^T}{w^T Bw},$$

if it is well defined, else set $B_+ = B$.

Various reduced SQP algorithms and numerous convergence results can be found in the literature [6], [7], [8], [9], [10], [11], [12], [13], [14]. Recently, there has also been progress in the infinite-dimensional theory [15] and application of reduced methods, like for example in parameter identification [16], optimal control [17], [4], [5], and optimal design problems [18].

Under certain assumptions it can be proved that Algorithm 2.1 is two-step super-linearly convergent and examples [19], [20] have shown that this is the best one can expect, since only a two-sided projection of the Hessian is correctly approximated. In [5] a modification of the algorithm is presented which leads to an improvement of the convergence behavior: The 2-step rate can be replaced by a 1-step rate. The modification requires no additional cost over the original iteration (2.4) and maintains all its advantages. To give a motivation of the new variant we reconsider step (2.3) once again and take a closer look at its null space coordinate. From the definition of T it is easy to see that the following identity is true

$$T(x)^T[\nabla f(x) - L_{xx}(x,l)R(x)h(x)] = T(x)^T[L_x(x,l) - L_{xx}(x,l)R(x)h(x)]$$

for each multiplier $l \in \mathbb{R}^q$. The idea is to replace the second order part on the right-hand side by suitable first order information. Therefore, note that the term in square brackets can be interpreted as a linear approximation to the following gradient of the Lagrangian:

$$L_x(x - R(x)h(x), l) \approx L_x(x, l) - L_{xx}(x, l)R(x)h(x).$$

Now, if we use the gradient $L_x(x - R(x)h(x), l)$ in the computation of the null space coordinate of the step (2.3) and replace again the reduced Hessian by an approximation B, then we are led to the following modified reduced SQP iteration:

$$(2.5) \qquad x_+ = x - R(x)h(x) - T(x)B^{-1}T(x)^T L_x(x - R(x)h(x), l).$$

An important difference between the original method (2.4) and the modified iteration (2.5) is that the tangent step and the restoration step in (2.5) are no longer independent of each other, because the new information acquired after the restoration enters into the computation of the tangent step.

The formula (2.5) has the disadvantage that a Lagrange multiplier is needed for the argument of L_x. However, in the application to the optimal control problem a clever choice of l ensures that the modified step is no longer explicitly dependent on the multiplier. More precisely, it is shown in [5] that under certain assumptions one can replace $T(x)^T L_x(x - R(x)h(x), l)$ by the reduced gradient $T(x - R(x)h(x))^T \nabla f(x - R(x)h(x))$ if the Lagrange multiplier $l = R(x - R(x)h(x))^T \nabla f(x - R(x)h(x))$ is used. Then the new iterate is

$$(2.6) \qquad x_+ = x - R(x)h(x) - T(x)B^{-1}T(x - R(x)h(x))^T \nabla f(x - R(x)h(x)).$$

Note that the reduced gradient $T(x - R(x)h(x))^T \nabla f(x - R(x)h(x))$ which enters into the computation of the modified step is also needed in Algorithm 2.1, namely for the computation of the vector v in Step (3). But this gradient can already be computed before solving the system with the reduced Hessian approximation in Step (1). Furthermore, since B in (2.6) is interpreted as an approximation to $T^T L_{xx} T$ one can adopt the update procedure of Algorithm 2.1. Consequently, when used with this version of the BFGS update, the modified method does not require more work per iteration than reduced SQP in the original form which can be seen in the following algorithmic formulation:

Algorithm 2.2 (Modified Reduced BFGS Method)

All steps are identical to Algorithm 2.1 except Step (1) which is replaced by

$$(1') \quad \text{Solve } Bw = -T(x - R(x)h(x))^T \nabla f(x - R(x)h(x)).$$

Before we state a convergence rate theorem for Algorithm 2.2, we make a few comments about the choice of T and R. In the context of mathematical programming the implementation of SQP methods is often based on null space techniques, where the basis $T(x)$ is usually obtained from a QR-factorization of $h'(x)$. However, for optimal control problems, where the set of variables can be naturally partitioned into state and control variables, the Jacobian can be written as

$$h'(x) = (h_y(y, u), h_u(y, u)),$$

where $h_y(y, u)$ is often invertible as an approximation of a solution operator for a linear differential equation. Then one can define T and R in the following way

$$(2.7) \quad T(y, u) = \begin{pmatrix} -h_y(y, u)^{-1} h_u(y, u) \\ I \end{pmatrix} \quad \text{and} \quad R(y, u) = \begin{pmatrix} h_y(y, u)^{-1} \\ O \end{pmatrix}.$$

In the following we refer to the choice (2.7) for R and T as the *separability framework*. In this framework Algorithm 2.2 exhibits the following local convergence rate:

Theorem 1 ([5]) *Consider problem (2.1) with variables $x = (y, u)$. Let Assumptions 1 and 3 be satisfied and suppose that $h_y(y, u)$ is invertible on D. Furthermore, choose T and R according to (2.7). Then, for (y_0, u_0) and B_0 sufficiently close to (y^*, u^*) and $T(y^*, u^*)^T L_{xx}(y^*, u^*, l^*) T(y^*, u^*)$, respectively, the sequence $\{(y_k, u_k)\}$ generated by Algorithm 2.2 converges to the solution at a 1-step q-superlinear rate.*

In the next section we apply the reduced BFGS method and its modification to the discrete control problem presented in the introduction.

3   Reduced BFGS Algorithms for Optimal Control

In this and the next section we consider the implementation of reduced SQP for the discrete optimal control problem (1.4)–(1.7), where we choose R and T according to (2.7). In order to describe the Jacobian of h we let $Q_j \in \mathbb{R}^{(N+1) \times M}$, $j = 2, \ldots, M+1$, denote the matrices with the elements

$$(3.1) \quad (Q_j)_{ik} = \begin{cases} -g & , \text{ if } i = 1 \text{ and } k = j - 1 \\ 0 & \text{ else} \end{cases}$$

For $w \in \mathbb{R}^{N+1}$ we define the following tridiagonal matrices in $\mathbb{R}^{(N+1) \times (N+1)}$

$$(3.2) \quad \begin{aligned} G(w) &= AC_d(w) + D\lambda_d(w) + \operatorname{diag}(g, 0, \ldots, 0), \\ E(w) &= -AC_d(w) \end{aligned}$$

with the diagonal matrices

$$C_d(w) = \mathrm{diag}(C(w_1), \dots, C(w_{N+1})) \quad \text{and} \quad \lambda_d(w) = \mathrm{diag}(\lambda(w_1), \dots, \lambda(w_{N+1})).$$

Then we can decompose the Jacobian of h into

$$(3.3) \quad h_y(y, u) = \begin{pmatrix} G(y^2) & 0 & 0 & 0 & 0 \\ E(y^2) & G(y^3) & 0 & 0 & 0 \\ 0 & \ddots & \ddots & 0 & 0 \\ 0 & 0 & \ddots & G(y^M) & 0 \\ 0 & 0 & 0 & E(y^M) & G(y^{M+1}) \end{pmatrix} \in \mathbb{R}^{M(N+1)\times M(N+1)}$$

and

$$(3.4) \quad h_u(y, u) = (Q_2^T, Q_3^T, \dots, Q_{M+1}^T)^T \in \mathbb{R}^{M(N+1)\times M}.$$

Note that $h_y(y, u) = h_y(y)$ only depends on the state variables, and $h_u(y, u) = h_u$ is a constant matrix.

The invertibility of (3.3) is satisfied under usual assumptions which guarantee the existence and the uniqueness of a solution of the discrete state equation. The precise conditions are not important for the following discussion and for details we refer the interested reader to Lemma 3.1 in [4].

We consider the computation of the reduced SQP step when T and R are chosen from (2.7). The coordinates of the tangent step are given by the control space component Δu of the total displacement $(\Delta y, \Delta u)$ and are obtained from solving a linear equation with the positive definite matrix B. That is,

$$\Delta u = -B^{-1}T(y, u)^T \nabla f(y, u).$$

Then, with h_y and h_u defined by (3.3), (3.4), both the tangent step, $T(y, u)\Delta u$, and the restoration step, $-R(y, u)h(y, u)$, can be computed from successively solving M tridiagonal systems, where each is of the dimension $N+1$. This shows that the natural choice (2.7) allows to maintain the sparsity in the Jacobian in the course of the iteration.

In order to be able to formulate the reduced BFGS algorithms completely for the parabolic control problem (1.4)–(1.7) it remains to investigate the calculation of the projected gradient $T(y, u)^T \nabla f(y, u)$. In the following we let $e^1 = (1, 0, \dots, 0)^T$ and $e^{N+1} = (0, \dots, 0, 1)^T$ denote unit vectors in $\mathbb{R}^{N+1}$.

Lemma 3 *Assume that $G(w)$ from (3.2) is nonsingular for $w \in \mathbb{R}^{N+1}$. Then*

$$T(y,u)^T \nabla f(y,u) = g(\pi_1^2, \dots, \pi_1^{M+1})^T + 2\alpha\tau u,$$

where $\pi \in \mathbb{R}^{M(N+1)}$ solves

$$G(y^{M+1})^T \pi^{M+1} = 2\tau(y_{N+1}^{M+1} - p^{M+1})\, e^{N+1},$$
$$G(y^j)^T \pi^j = 2\tau(y_{N+1}^j - p^j)\, e^{N+1} - E(y^j)^T \pi^{j+1}, \quad j = M, \dots, 2.$$

The proof follows by a direct inspection and the lemma shows that the computation of the reduced gradient corresponds to the solution of an adjoint differential equation.

Now we can state the Algorithm 2.1 for the Parabolic Boundary Control Problem (PBCP) in its entirety:

Algorithm 3.1 (Reduced BFGS Method for PBCP)

        Given $y \in \mathbb{R}^{M(N+1)}$, $u \in \mathbb{R}^M$, $B \in \mathbb{R}^{M\times M}$, B positive definite.

        Step 1   (Computation of $(y,u) - R(y,u)h(y,u)$)

            Solve $G(y^2)\eta^2 = -h^1(y,u)$ and

$$G(y^{j+1})\eta^{j+1} = -h^j(y,u) - E(y^j)\eta^j, \quad j = 2, \dots, M.$$

            Set $\bar{y} = y + \eta$.

        Step 2   (Computation of $c(\bar{y},u) = T(\bar{y},u)^T \nabla f(\bar{y},u)$)

            Compute $\pi \in \mathbb{R}^{M(N+1)}$ from

$$G(\bar{y}^{M+1})^T \pi^{M+1} = 2\tau(\bar{y}_{N+1}^{M+1} - p^{M+1})\, e^{N+1},$$
$$G(\bar{y}^j)^T \pi^j = 2\tau(\bar{y}_{N+1}^j - p^j)\, e^{N+1} - E(\bar{y}^j)^T \pi^{j+1}, \quad j = M, \dots, 2.$$

            Set $c(\bar{y},u) = g(\pi_1^2, \dots, \pi_1^{M+1})^T + 2\alpha\tau u$.

        Step 3   (Computation of (y_+, u_+))

            Solve $B\Delta u = -T(y,u)^T \nabla f(y,u)$.

            Solve $G(y^2)\xi^2 = g(\Delta u)^2 e^1$ and

$$G(y^{j+1})\xi^{j+1} = g(\Delta u)^{j+1} e^1 - E(y^j)\xi^j, \quad j = 2, \dots, M.$$

            Set $y_+ = \bar{y} + \xi$, $u_+ = u + \Delta u$.

Step 4 (Computation of $c(y_+, u_+) = T(y_+, u_+)^T \nabla f(y_+, u_+)$)

Compute the adjoint state $\pi_+ \in \mathbb{R}^{M(N+1)}$ from

$$G(y_+^{M+1})^T \pi^{M+1} = 2\tau((y_+)_{N+1}^{M+1} - p^{M+1}) e^{N+1},$$

$$G(y_+^j)^T \tau^j = 2\tau((y_+)_{N+1}^j - p^j) e^{N+1} - E(y_+^j)^T \pi^{j+1}, \quad j = M, \dots, 2.$$

Set $c(y_+, u_+) = g((\pi_+)_1^2, \dots, (\pi_+)_1^{M+1})^T + 2\alpha\tau u_+.$

Step 5 (Computation of B_+)

Set $v = c(y_+, u_+) - c(\bar{y}, u).$

Set $B_+ = B + \mathrm{BFGS}(B, \Delta u, v)$, if it is defined.

Else set $B_+ = B.$

A similar reduced BFGS algorithm for the heat conduction control problem can also be found in [4], but there the update is used with a different definition for the vector v. We also note that Orozco and Ghattas [18] apply a reduced Quasi Newton method in the separability framework to the shape optimization of airfoils.

It is the choice (2.7) for R, T which allows to maintain the sparsity in the Jacobian. Another advantage of the separability framework becomes apparent if one considers the modified step proposed in (2.5). It was already mentioned in the previous section that if l in (2.5) is chosen as the solution of $h_y(\bar{y}, u)^T l = \nabla_y f(\bar{y}, u)$, where $(\bar{y}, u) = (y, u) - R(y, u)h(y, u)$, then the modified method is not explicitly depending on a multiplier and reduces to a quite elegant variant (2.6) of the original method. In the setting of Algorithm 3.1 this means that one only has to use the new gradient information available from Step 2 for the computation of Δu in the subsequent Step 3.

Algorithm 3.2 (Modified Reduced BFGS Method for PBCP)

All steps are identical to Algorithm 3.1 except Step 3 which is replaced by

Step 3' (Computation of (y_+, u_+))

Solve $B\Delta u = -c(\bar{y}, u).$

Solve $G(y^2)\xi^2 = g\,(\Delta u)^2 e^1$ and

$$G(y^{j+1})\xi^{j+1} = g\,(\Delta u)^{j+1} e^1 - E(y^j)\xi^j, \quad j = 2, \dots, M.$$

Set $y_+ = \bar{y} + \xi, \quad u_+ = u + \Delta u.$

4   Numerical Results

In this section we present computational experiments with the Algorithms 3.1 and 3.2. For the purpose of comparative testing we disregard the globalization strategy and concentrate on the local behavior of the methods. The computations were done in double precision FORTRAN on a SUN Sparcstation 1 and LINPACK was used for linear algebra manipulations. The first test example documents the efficiency of the modified reduced BFGS algorithm.

Example 4.1.

We consider the nonlinear boundary value problem (1.2), (1.3) for $q = 0$, $g = 1$, and with C, λ, and y_0 given by linear functions:

$$\begin{aligned}
C(y) &= 0.3 + 0.05\,y, & y &\in \mathbb{R}, \\
\lambda(y) &= 1.0 + 0.1\,y, & y &\in \mathbb{R}, \\
y_0(x) &= 0.8 - 0.1\,x, & x &\in [0,1].
\end{aligned}$$

The values p^j for the reference profile are computed as follows: We give a control function u^* as represented in Fig. 1 and apply Newton's method to solve the discrete state equation (1.6), (1.7) using u^* as input. The components of the computed state which are associated with the boundary $x = 1$ then serve as data for the firing curve p so that u^* is optimal for (PBCP) with $\alpha = 0$.

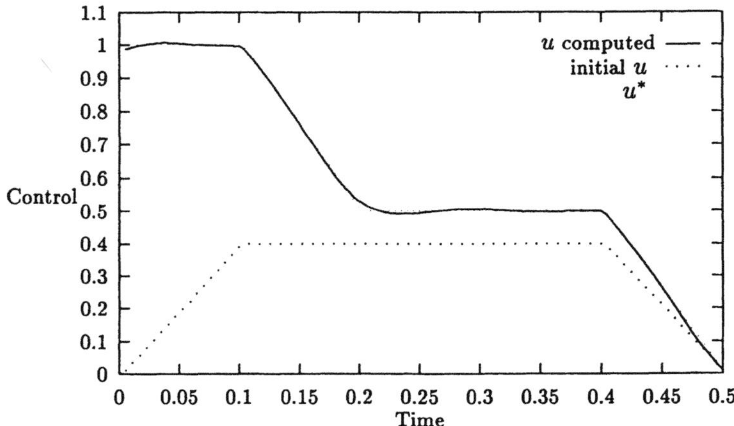

Figure 1: Control Computed by Algorithm 3.2 (Ex. 4.1)

The starting components of the state variable are set identical to -0.5 and the start control is depicted in Fig 1. From these rather rough initial values Algorithm 3.2 produces an accurate approximation of the expected solution after 17 iterations. The solid line in Fig. 1 represents the computed control, and the computed temperature distribution is shown in Fig. 2.

The situation is less satisfactory with the original version of the reduced BFGS method: To reach the same accuracy the Algorithm 3.1 needs nearly as twice as many iterations as Algorithm 3.2.

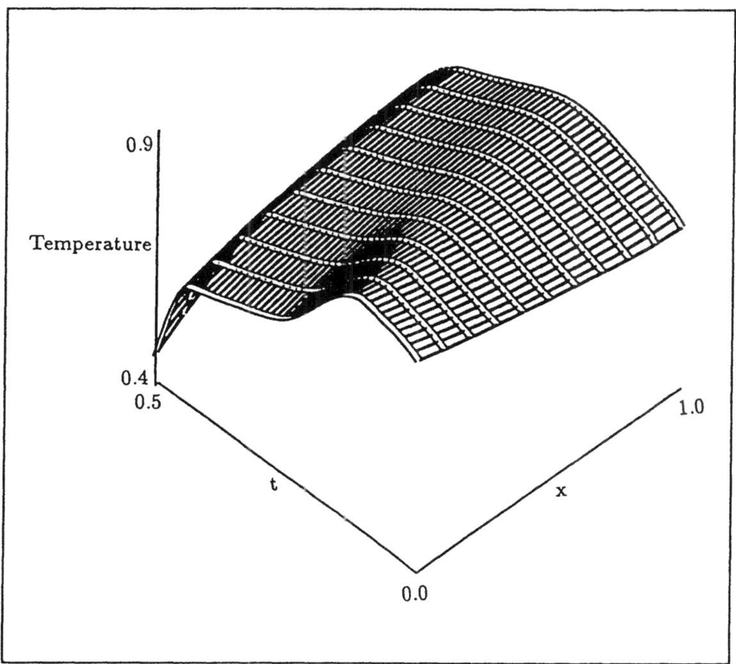

Figure 2: Temperature Distribution Computed by Algorithm 3.2 (Ex. 4.1)

Example 4.2.

In this example the firing curve p comes from interpolation of an attainable state for the infinite–dimensional unregularized control problem. The parabolic boundary control problem used in this process is taken from [21] with slight variations in the data. For a precise description of the test set we refer to [5].

In Table 1 we use the notation

$$\sigma_k = (\|u_{M,k} - u_{M,k-1}\|_U^2 + \|y_{MN,k} - y_{MN,k-1}\|_Y^2)^{1/2},$$

where the subscript k denotes the iteration number and $y_{MN,k}, u_{M,k}$ denote the functions in the finite-dimensional discrete state and control space with the coefficient vectors y_k and u_k, respectively. Under standard assumptions for linear convergence results it can be shown [5] that the iterates generated by a reduced BFGS method converge l-step superlinearly, $l = 1, 2$, if and only if the corresponding sequences for the steps σ_k/σ_{k-1} respectively σ_k/σ_{k-2} tend to zero. We monitor these ratios in Table 1 and, obviously, the convergence rate predicted by the theory can be observed numerically. The first and second column clearly show that the reduced BFGS method (Algorithm 3.1) is two-step but apparently not one-step superlinearly convergent. The convergence rate is vastly improved by the modified method (Algorithm 3.2) as can be seen from the last column of the table.

	Algorithm 3.1		Algorithm 3.2
k	σ_k/σ_{k-1}	σ_k/σ_{k-2}	σ_k/σ_{k-1}
2	0.74444		0.74444
3	1.42995	1.064509	0.29765
4	0.38438	0.549639	0.25959
5	2.00620	0.771133	0.09761
6	0.65838	1.320844	0.09004
7	1.07583	0.708309	0.00737
8	0.07357	0.079145	0.00071
9	1.80346	0.132673	0.00064
10	0.01303	0.023502	
11	2.15606	0.028097	
12	0.00276	0.005946	
13	0.27819	0.000767	
14	0.00005	0.000015	
15	0.02228	0.000001	

Table 1: Rates of Convergence for Reduced BFGS and Modification (Ex. 4.2)

Acknowledgement

This research was supported by the Deutsche Forschungsgemeinschaft.

References

[1] E. Weiland and J. P. Babary. A solution to the IHCP applied to ceramic product firing. In *Proc. 5th Int. Conf. on Numerical Methods in Thermal Problems, Montreal, 1987, 5*, pages 1358–1367. Chichester, 1988.

[2] D. Barreteau, M. Hemati, J. P. Babary, and E. Weiland. On modelling and control of intermittent kilns in the ceramic industry. In *12th I.M.A.C.S World Congress on Scientific Computation*, pages 339–341, 1988.

[3] J. Burger and M. Pogu. Functional and numerical solution of a control problem originating from heat transfer. *J. Optim. Theory Appl.*, 68:49 – 73, 1991.

[4] F.-S. Kupfer and E. W. Sachs. Numerical solution of a nonlinear parabolic control problem by a reduced SQP method. *Computational Optimization and Applications*, 1:113–135, 1992.

[5] F.-S. Kupfer. *Reduced Successive Quadratic Programming in Hilbert space with applications to optimal control*. doctoral thesis, Universität Trier, 1992.

[6] D. Gabay. Reduced quasi–Newton methods with feasibility improvement for nonlinearly constrained optimization. *Math. Programming Study*, 16:18–44, 1982.

[7] T. F. Coleman and A. R. Conn. On the local convergence of a quasi–Newton method for the nonlinear programming problem. *SIAM J. Numer. Anal.*, 21:755–769, 1984.

[8] J. Nocedal and M. L. Overton. Projected Hessian updating algorithms for nonlinearly constrained optimization. *SIAM J. Numer. Anal.*, 22:821–850, 1985.

[9] J. C. Gilbert. Une méthode à métrique variable réduite en optimisation avec contraintes d' égalité non linéaires. Technical Report RR–482, INRIA, 1986.

[10] C. B. Gurwitz. *Sequential quadratic programming methods based on approximating a projected Hessian matrix*. PhD thesis, Computer Science Department, New York University, New York, 1986.

[11] C. B. Gurwitz and M. L. Overton. Sequential quadratic programming methods based on approximating a projected Hessian matrix. *SIAM J. Sci. Stat. Comput.*, 10:631–653, 1989.

[12] J. Zhang and D. Zhu. A trust region type dogleg method for nonlinear optimization. *Optimization*, 21:543–557, 1990.

[13] R. H. Byrd. On the convergence of constrained optimization methods with accurate Hessian information on a subspace. *SIAM J. Numer. Anal.*, 27:141–153, 1990.

[14] R. H. Byrd and J. Nocedal. An analysis of reduced Hessian methods for constrained optimization. *Math. Programming*, 49:285–323, 1991.

[15] F.-S. Kupfer. An infinite dimensional convergence theory for reduced SQP methods in Hilbert space. Technical report, Universität Trier, Fachbereich IV – Mathematik, 1990.

[16] K. Kunisch and E. W. Sachs. Reduced SQP methods for parameter identification problems. *SIAM J. Numer. Anal.* to appear.

[17] F.-S. Kupfer and E. W. Sachs. A prospective look at SQP methods for semilinear parabolic control problems. In K.-H. Hoffmann and W. Krabs, editors, *Optimal Control of Partial Differential Equations, Irsee 1990*, volume 149, pages 143–157. Springer Lect. Notes in Control and Inform. Sciences, 1991.

[18] C. E. Orozco and O. N. Ghattas. Massively parallel aerodynamic shape optimization. In *Proceedings of the Symposium on High Performance Computing for Flight Vehicles*. to appear.

[19] R. H. Byrd. An example of irregular convergence in some constrained optimization methods that use the projected Hessian. *Math. Programming*, 32:232–237, 1985.

[20] Y. Yuan. An only 2-step q-superlinear convergence example for some algorithms that use reduced Hessian approxmations. *Math. Programming*, 32:224–231, 1985.

[21] J. Burger and C. Machbub. Comparison of numerical solutions of a one-dimensional nonlinear heat equation. *Communications in Applied Numerical Methods*, 7:1 – 14, 1991.

Authors' addresses

Prof. Dr. E.W. Sachs and Prof. Dr. F.-S. Kupfer
Universität Trier
FB IV – Mathematik
D - W– 5500 Trier
Germany
sachs@uni-trier.dbp.de.

Analysis and Synthesis of Nonlinear Systems

Decomposition and Feedback Control of Nonlinear Dynamic Systems

Felix L. Chernousko

Abstract

A nonlinear dynamic system governed by Lagrange equations and subject to bounded control forces is considered. Under certain conditions the feedback control is proposed which brings the system to the prescribed terminal state in finite time. This control is obtained through the decomposition of the system into subsystems with one degree of freedom each and applying the approach of differential games. The obtained feedback control is time-suboptimal and robust with respect to small disturbances and parameter variations. The proposed control can be used in robotics for control of manipulators.

1 Introduction

Designing a feedback control for nonlinear systems is a challenging problem which has important applications and attracts attention of many researchers. One possible approach to this problem is given by the feedback linearization technique. Leitmann [1] and Corless and Leitmann [2] developed the method based on Lyapunov functions that makes it possible to obtain controls that guarantee the desired behaviour of the system in the presence of uncertain disturbances. The mentioned approaches give feedback controls which bring the nonlinear system to the prescribed state in infinite time.

In our earlier papers [3], [4] we proposed the feedback control method which brings the nonlinear Lagrangian system to the prescribed terminal state in finite time. In these papers certain conditions were imposed on the system which were to be verified in the 2n-dimensional phase space of generalized coordinates and velocities.

In this paper we also consider the nonlinear Lagrangian system subject to bounded control forces. However, here the imposed conditions are much less restrictive and are to be verified in the n-dimensional coordinate space. We obtain the feedback control in the explicit form using the ideas of decomposition, differential games and optimal control. Under the obtained control, the system reaches the prescribed terminal state in finite time which is estimated.

2   Statement of the problem

We consider a nonlinear dynamic system governed by the Lagrange equations

$$(2.1) \qquad \frac{d}{dt}\frac{\partial T}{\partial \dot{q}_i} - \frac{\partial T}{\partial q_i} = Q_i + R_i + S_i, \ i = 1, \ldots, n.$$

Here t is the time, $q = (q_1, \ldots, q_n)$ is an n-vector of generalized coordinates, T is the kinetic energy, $Q = (Q_1, \ldots, Q_n)$ are control forces, $R = (R_1, \ldots, R_n)$ are dissipative forces, $S = (S_1, \ldots, S_n)$ include all other external forces. We assume that all motions of the system (2.1) lie in some domain $D \subset R^n$, so that always $q \in D$.

We impose the following conditions on the kinetic energy and forces. The kinetic energy is given by

$$(2.2) \qquad T = \frac{1}{2}(A(q)\,\dot{p},\ q) = \frac{1}{2}\sum_{i,j} a_{ij}(q)\,\dot{q}_i\,\dot{q}_j$$

where $A(q)$ is a symmetric positive definite n×n-matrix whose elements $a_{ij}(q)$ are continuously differentiable in D. All sums in (2.2) and below are taken with indices i, j running from 1 till n. We assume that there exist such numbers m, M that

$$(2.3) \qquad m\,|z|^2 \le (A(q)\,z\,,\ z) \le M\,|z|^2,\ 0 < m < M,\ q \in D$$

for any n-vector $z \in R^n$ and for all $q \in D$. We assume also that

$$(2.4) \qquad |\partial a_{ij}(q)/\partial q_k| \le C,\ q \in D;\ i, j, k = 1, \ldots, n$$

where $C > 0$ is a constant.

The controls Q_i in (2.1) are bounded

$$(2.5) \qquad |Q_i| \le Q_i^0,\ i = 1, \ldots, n.$$

Here Q_i^0 are given constants.

The dissipative forces $R(q, \dot{q}, t)$ satisfy the condition

$$(2.6) \qquad (R,\ \dot{q}) \le 0$$

for all $q \in D$, all $\dot{q}$ and all $t \geq t_0$ where t_0 is the initial time instant. We assume also that there exists a number $\nu_0 > 0$ such that if $|\dot{q}| \leq \nu \leq \nu_0$ for all $i = 1, \ldots, n$, then

$$(2.7) \quad |R_i| \leq R_i^0(\nu), \; i = 1, \ldots, n.$$

Here $R_i^0(\nu)$ are some continuous functions increasing monotonically with $\nu \in [0, \nu_0]$, and $R_i^0(0) = 0$.

The forces $S_i(q, \dot{q}, t)$ are assumed to be bounded

$$(2.8) \quad |S_i| \leq S_i^0, \; i = 1, \ldots, n$$

for all $q \in D$, all $\dot{q}$ and $t \geq t_0$. These forces may include unknown (uncertain) disturbances; only the bounds $S_i^0 > 0$ are to be known.

The problem is to find a feedback control $Q_i(q, \dot{q})$, $i = 1, \ldots, n$, satisfying the conditions (2.5) and bringing the system (2.1) from the orbitrary initial state

$$(2.9) \quad q_i(t_0) = \dot{q}_i, \; \dot{q}_i(t_0) = \dot{q}_i$$

to the prescribed terminal state

$$(2.10) \quad q_i(t_*) = q^*, \; \dot{q}_i(t_*) = 0.$$

3 Feedback control

We introduce the following two sets in the 2n-dimensional space $q, \dot{q}$

$$(3.1) \quad \begin{aligned} \Omega_1 &= \{(q, \dot{q}) \; : \; q \in D; \quad \exists i, \quad |\dot{q}_i| > \varepsilon\} \\ \Omega_2 &= \{(q, \dot{q}) \; : \; q \in D; \quad \forall i, \quad |\dot{q}_i| < \varepsilon\}. \end{aligned}$$

These sets depend on the number $\varepsilon > 0$ which will be chosen later. We define the feedback control in Ω_1 as

$$(3.2) \quad Q_i = -Q_i^0 \operatorname{sign} \dot{q}_i, \; i = 1, \ldots, n \quad (\text{in } \Omega_1)$$

It follows from (2.1), (2.6), (2.8), (3.2) that

$$(3.3) \quad dT/dt = ((Q + R + S), \dot{q}) \; \leq \; -\sum_i (Q_i^0 - S_i^0) \, |\dot{q}_i|.$$

We assume that $Q_i^0 > S_i^0$, $i = 1, \ldots, n$, and denote

$$(3.4) \quad r_0 = \left[\sum_i \left(Q_i^0 - S_i^0 \right)^2 \right]^{1/2} > 0.$$

Then the inequality (3.3) can be transformed into

$$(3.5) \quad dT/dt \le -r_0 \, |\dot{q}| \le -r_0 \, (2\,T/M)^{1/2}.$$

Here the right inequalitiy (2.3) is used. Integrating (3.5) we obtain

$$(3.6) \quad T^{1/2} - T_0^{1/2} \le -r_0 \, (2\,M)^{-1/2}(t - t_0)$$

where T_0 is the initial value of the kinetic energy (2.2) at $t = t_0$. It follows from (3.6) that the system reaches the boundary between the sets Ω_1 and Ω_2 at some instant t_1. To evaluate t_1, we note that the kinetic energy T_1 at t_1 can be estimated by means of (2.3) and (3.1)

$$(3.7) \quad T_1 \ge m \, |\dot{q}|^2/2 \ge m \, \varepsilon^2/2.$$

Using (3.6) and (3.7) we obtain

$$(3.8) \quad t_1 - t_0 \le (2\,M)^{1/2} \, r_0^{-1} \left[T_0^{1/2} - (m/2)^{1/2} \, \varepsilon \right].$$

The estimate similar to (3.8) is obtained also for the coordinates at t_1

$$(3.9) \quad |q_i(t_1) - q_i^0| \le (M/m^{1/2}) \, r_0^{-1}(T_0 - m \, \varepsilon^2/2).$$

Thus, the system enters the set Ω_2 from (3.1) at the instant $t = t_1$. Now we shall obtain feedback control in Ω_2 which will bring the system to the terminal state (2.10) without its leaving the set Ω_2.

Let us substitute the kinetic energy (2.2) into equations (2.1)

$$(3.10) \quad A \, \ddot{q} + \sum_{j,k} \Gamma_{jk} \, \dot{q}_j \, \dot{q}_k = Q + R + S.$$

Here the vectors Γ_{jk} have the components

$$(3.11) \quad \begin{aligned} \Gamma_{ij} &= (\gamma_{1jk}, \ldots, \gamma_{njk}) \\[2mm] \gamma_{ijk} &= \frac{\partial a_{ij}}{\partial q_k} - 0.5 \, \frac{\partial a_{jk}}{\partial q_i}, \quad i, j, k = 1, \ldots, n. \end{aligned}$$

The following estimates stems from (3.11) and (2.4)

$$(3.12) \quad |\gamma_{ijk}| \le 1.5\, C; \ |\Gamma_{jk}| = \left(\sum_i \gamma_{ijk}^2 \right)^{1/2} \le 1.5\, C\, n^{1/2}, \ i,j,k = 1,\dots,n.$$

Solving (3.10) with respect to $\ddot{q}$, we obtain

$$(3.13) \quad \ddot{q} = U + V$$

where the following notations are introduced

$$(3.14) \quad U = A^{-1}\, Q, \ V = A^{-1}\left(R + S - \sum_{j,k} \Gamma_{jk}\, \dot{q}_j\, \dot{q}_k \right).$$

We consider U as a new control vector subject to the following constraints

$$(3.15) \quad |U_i| \le U^0 = r\, M^{-1}\, n^{-1/2}, \ r = \min_i Q_i^0, \ i = 1,\dots,n.$$

If the constraints (3.15) are satisfied, we have $|U| \le r\, M^{-1}$, and it follows from (3.14) and (2.3) that

$$|Q| = |A\, U| \le M\, |U| \le r.$$

Therefore, if U satisfies the inequalities (3.15), then $Q = A\, U$ satisfies the imposed constraints (2.5).

The vector V from (3.14) can be estimated by means of the inequalities (2.3) and (3.12)

$$(3.16) \quad |V| \le m^{-1}\left(|R| + |S| + 1.5\, C\, n^{1/2} \sum_{j,k} |\dot{q}_j|\, |\dot{q}_k| \right).$$

Taking into account the constraints (2.7) and (2.8), properties of the functions R_i^0 from (2.7), and conditions $|\dot{q}_i| \le \varepsilon$ which are satisfied in Ω_2, see (3.1), we obtain from (3.16)

$$(3.17) \quad \begin{aligned} &|V| \le V^0 = m^{-1}\left[R^0(\varepsilon) + S^0 + 1.5\, C\, n^{5/2}\, \varepsilon^2 \right] \\ &R^0 = \left[\sum_i (R_i^0)^2 \right]^{1/2}, \ S^0 = \left[\sum_i (S_i^0)^2 \right]^{1/2}, \ \varepsilon \le \nu_0. \end{aligned}$$

Here $R^0(\varepsilon)$ is a continuous monotone function for $\varepsilon \le \nu_0$.

The system (3.13) can now be rewritten as

$$(3.18) \quad \ddot{q}_i = U_i + V_i, \ |U_i| \le U^0, \ |V_i| \le V^0$$

where U^0 and V^0 are defined by (3.15) and (3.17). We suppose that

$$(3.19) \quad \rho = V^0 \,/\, U^0 \le 1$$

and consider the ith equation of the system (3.18). We shall obtain the control U_i for this equation assuming that V_i can be an arbitrary function of time bounded by the constraint (3.18). Thus, we use the approach of differential games for the subsystems (3.18). Denote

$$(3.20) \quad q_i - q_i^* = x, \ \dot{q} = \dot{x} = y, \ U_i = u, \ V_i = v$$

and rewrite the ith equation (3.18)

$$(3.21) \quad \dot{x} = y, \ \dot{y} = u + v, \ |u| \le U^0, \ |v| \le V^0, \ \rho = V^0 \,/\, U^0 < 1.$$

The terminal conditions (2.10) can be rewritten as

$$(3.22) \quad x(t_*) = 0, \ y(t_*) = 0.$$

The system will not leave the set Ω_2 if

$$(3.23) \quad |y(t)| \le \varepsilon, \ t \ge t_1.$$

Now our initial control problem is reduced to the following control problem for the subsystems (3.21). We are to find the feedback control $u(x, y)$ satisfying the conditions (3.21), (3.23) and bringing the system (3.21) to the terminal state (3.22) in finite time under any v bounded by the constraint (3.21). We shall obtain the desired control by modifying the well-known solution of the differential game for the system (3.21).

Let u and v be controls of two players; the player u minimizes while v maximizes the time required to reach the terminal state (3.22). The optimal strategy u in this game, as it was shown Krasovsky [5], coincides with the time-optimal control for the system

$$(3.24) \quad \dot{x} = y, \ \dot{y} = (1 - \rho)\, u, \ |u| \le U^0$$

obtained from (3.21) with v replaced by the optimal control $v = -\rho\, u$ of the second player. We define our feedback control as follows

$$
(3.25) \qquad
\begin{aligned}
u(x, y) &= U^0 \operatorname{sign}[\psi(x) - y] & &\text{if} \ \ y \ne \psi(x) \\[4pt]
u(x, y) &= U^0 \operatorname{sign} x = -U^0 \operatorname{sign} y & &\text{if} \ \ y = \psi(x).
\end{aligned}
$$

If we insert $\psi = \psi_0$ in (3.25) where

$$(3.26) \quad \psi(x) \equiv -[2\,U^0(1-\rho)\,|x|]^{1/2}\operatorname{sign} x$$

then the formulas (3.25) describe the time-optimal feedback control for the system (3.24). This control, however, does not satisfy the state constraint (3.23). To take it into account, we modify the switching curve $y = \psi(x)$ and define it as follows

$$(3.27) \quad \begin{aligned} \psi(x) &= \psi_0(x), & |x| \le x^* \\[1mm] \psi(x) &= -\delta \operatorname{sign} x, & |x| > x^*. \end{aligned}$$

Here $\psi_0(x)$ is given by (3.26), and

$$(3.28) \quad \delta \in (0,\varepsilon),\ x^* = \delta^2\,[2\,U^0(1-\rho)]^{-1}.$$

The switching curve $y = \psi(x)$ defined by (3.26) - (3.28) is a smooth curve symmetric with respect to the point $x = y = 0$. It lies inside the band $|y| \le \varepsilon$ and consists of two arcs of parabolas (3.26) and two rays $y = \pm\delta$. The switching curve is represented in Fig. 1 by a thick line. It can be shown that any trajectory of the system (3.21) (for any possible v) which starts at $t = t_1$ in the band $|y| \le \varepsilon$, never leaves this band and reaches the origin of coordinates in finite time. Some trajectories are shown in Fig. 1 by thin lines. Each trajectories at first reaches the switching curve and then follows it in sliding regime. The time of motion can be easily estimated.

The above assertions are true if the imposed conditions $\varepsilon \le \nu_0$ and $\rho < 1$ are satisfied, see (3.17) and (3.19). We can rewrite these conditions using the formulas (3.15) and (3.17) as follows

$$(3.29) \quad \varepsilon \le \nu_0,\ R^0(\varepsilon) + S^0 + 1.5\,C\,n^{5/2}\,\varepsilon^2 \le m\,M^{-1}\,r\,n^{-1/2}.$$

As $R^0(\varepsilon)$ is a continuous monotone function. The number $\varepsilon > 0$ satisfying (3.29) exists if

$$(3.30) \quad S^0 < m\,M^{-1}\,r\,n^{-1/2}.$$

Note that if the system (2.1) is subject only to control and dissipative forces ($S^0 = 0$) then this condition is always satisfied.

Now we can summarize our results. Let the conditions (3.30) be fulfilled where r and S^0 are defined by formulas (3.15) and (3.17), respectively. Then we can choose $\varepsilon > 0$

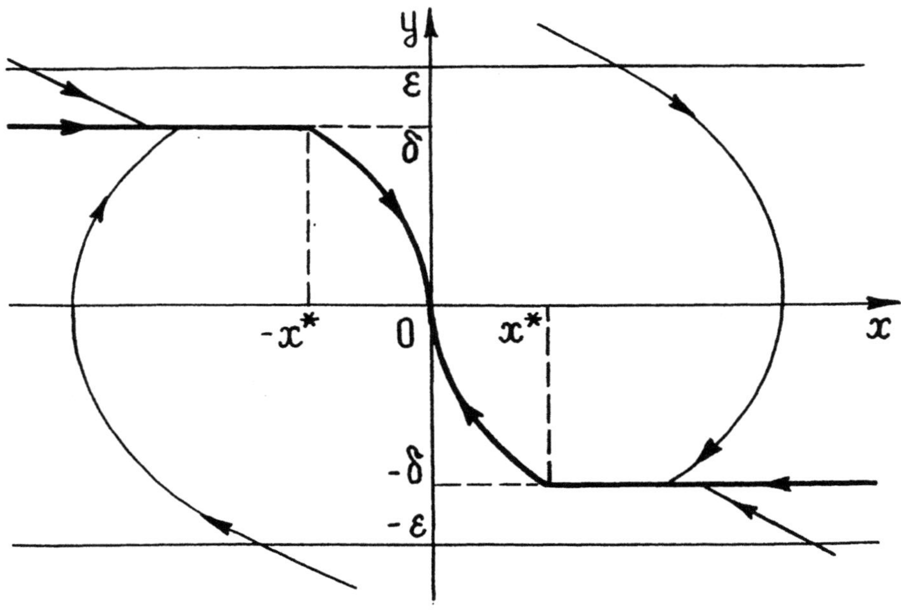

Figure 1:

satisfying the inequalities (3.29). The desired feedback control $Q(q, \dot{q})$ bringing the system (2.1) from any initial stae (2.9) to the prescribed terminal state (2.10) is given by equation (3.2) in the set Ω_1 and by following equations

$$
\begin{aligned}
Q(q, \dot{q}) &= A(q)\, U(q, \dot{q}) \\
U_i(q_i, \dot{q}_i) &= u(q_i - q_i^*, \dot{q}_i), \quad i = 1, \ldots, n
\end{aligned}
\tag{3.31}
$$

in the set Ω_2. The formulas (3.31) stem from (3.14) and (3.20). Here the function $u(x, y)$ is defined by the formulas (3.25) - (3.27) where the parameters U^0, ρ and x^* are given by (3.15), (3.17), (3.19) and (3.28). The parameter δ can be chosen arbitrarily from the interval $\delta \in (0, \varepsilon)$. The time of motion depends on δ and decreases when δ grows. Using the estimates (3.8) and (3.9) and evaluating the time of motion inside the set Ω_2, we obtain the following estimate for the total time of motion t_*

$$
\begin{aligned}
t_* - t_0 \;\leq\; & \delta^{-1} \max_i |q_i^0 - q_i^*| + 2\, M^{1/2}\, r_0^{-1}\, [T_0^{1/2} - (m/2)^{1/2}\, \varepsilon] \\
& + (M/m)^{1/2}\, r_0^{-1}\, \delta^{-1}\, (T_0 - m\, \varepsilon^2/2) \\
& + (2\, \varepsilon^2 + 4\, \varepsilon\, \delta + 3\, \delta^2)\, \delta^{-1}\, [2\, U^0\, (1 - \rho)]^{-1}.
\end{aligned}
$$

Here the maximum is taken with respect to $i = 1, \ldots, n$.

4   Conclusion

The system (2.1) may be subject to some unknown (uncertain) disturbances included in the forces S_i and parameter variations which can be reduced to some additional forces. If all these disturbances are small enough so that the condition (3.30) is satisfied, then the obtained feedback control can cope with them. In this sense, our feedback control is robust.

The obtained control can be called time-suboptimal because its switching curve near the terminal state coincides with the time-optimal switching curve.

The most obvious application of the obtained control are in robotics where the mechanical (or electro-mechanical) system has usually one control actuator per each degree of freedom. These applications are discussed in papers [3], [4] for another feedback control law. Computer simulation shows quite satisfactory behavior of the manipulation robot models under the feedback control described in this paper.

References

[1] G. Leitmann, *Deterministic control of uncertain systems.*
Acta Astronautica 7 (1980), 1457-1461.

[2] M. Corless and G. Leitmann, *Adaptive control of systems containing uncertain functions and unknown functions with uncertain bounds.*
I. Optimization Theory Appl. 41 (1983), 155-168.

[3] F.L. Chernousko, *Decomposition and suboptimal control in dynamic systems.*
Applied Mathematics and Mechanics (PMM), 54, No. 6 (1990), 727-734.

[4] F.L. Chernousko, *Decomposition and suboptimal control synthesis in dynamic systems.*
Soviet Journal of Computer and Sytem Sciences 29, No- 4 (1990), 64-82.

[5] N.N. Krasovsky, *Game Problems of Meeting of Motions*
(Russian), Nauka, Moscow, 1970

Author's address

Felix L. Chernousko, Institute for Problems in Mechanics, Russian Academy of Sciences, pr. Vernadskogo 101, 117526, Moscow, Russia.

A Discrete Stabilizing Study Strategy for a Student Related Problem under Uncertainty

G. Leitmann, C. S. Lee

Abstract. Raggett et al. (1981) first proposed an optimal control problem of a lazy and forgetful student who wishes to pass his/her final examination with a minimum expenditure of effort. Later on, Bondi (1982), Parlar (1984), Cheng and Teo (1987) as well as Lee and Leitmann (1990) discussed, modified and extended the problem. Lee and Leitmann (1991) also considered a related problem in which the system parameters are uncertain but bounded, and the results of Corless and Leitmann (1988, 1989) and Barmish et al. (1983) were employed to obtain study strategies for a student who wishes to achieve a specified desired level of knowledge and then to maintain his/her knowledge level within a calculable neighborhood of that desired level in a continuous assessment study environment. In this paper, we consider a discrete analog of the problem discussed in Lee and Leitmann (1991) and apply the results of Corless and Manela (1986) to obtain a corresponding discrete study strategy for the student.

1. Introduction

Raggett et al. (1981) first proposed the problem of a *lazy* and *forgetful* student who wishes to pass his/her final examination with a minimum expenditure of effort. They assumed that a student's knowledge level at any time t reflects his/her performance in an examination if he/she were given one at that time. Furthermore, while his/her intake of knowledge per unit of time (week) is proportional to the square root of the amount of work he/she puts in, he/she is also constantly forgetting a proportion c (> 0) of what he/she already knows. Later, Bondi (1982), Parlar (1984), Cheng and Teo (1987) as well as Lee and Leitmann (1990) discussed, modified and extended the above problem. Parlar (1984) assumed that the rate of change of the student's knowledge level is a linear function of the work rate. He also introduced the model of a *lazy, forgetful* but *ambitious* student who always attempts to acquire maximum knowledge with minimum effort. In their discussion of the latter model, Cheng and Teo (1987) adopted an objective functional which is quadratic in the work rate. They pointed out that the objective functional may be interpreted as a function of the work rate and that a quadratic cost function may reflect the study behavior of a lazy student better than a linear one. Lee and Leitmann (1990) also assumed that the rate of change of the student's knowledge level is a linear function of the work rate, however, the coefficient of the work rate W is not a constant but a linear function of the current knowledge level N. This is based on the observation that, under normal circumstances, a typical student at a higher knowledge level can absorb more than another who is at a lower knowledge level, the work rate being the same for both.

Later, Lee and Leitmann (1991) considered also a related problem in which the system parameters are uncertain but bounded and the student wishes to achieve a specified desired level of knowledge (which is assumed to reflect his/her examination performance) and then to maintain his/her knowledge level within a calculable neighborhood of that desired level in a continuous assessment study environment.

In this paper, we present a discrete model which has the same features as that introduced in Lee and Leitmann (1991) and employ the results of Corless and Manela (1986) to obtain a corresponding discrete time study strategy for the student.

2. Problem Formulation

Suppose that T, the total study period of a student, is divided into η equal subintervals of duration τ $(=\frac{T}{\eta})$; we let $N(k)$ represent his/her knowledge level (%) at sampling instant $t(= k\tau)$. His/her knowledge level at the next sampling instant may be described by the following model:

(2.1) $$N(k+1) = N(k) + [b_1 + b_2 N(k)]\, W(k) - cN(k)$$

(2.2) $$0 \leq W(k) \leq \overline{W}, \quad (\overline{W}\, given)$$

(2.3) $$N(0) = N_o \geq 0$$
$$k \in (0,1,\ldots,(\eta-1)).$$

where $-cN(k)$ accounts for forgetfulness and $[b_1 + b_2 N(k)]\, W(k)$ for learning, including the increased effectiveness of learning at a higher level of knowledge. N_o denotes the student's initial knowledge level and $W(k)$ represents the amount of work he/she puts into his/her studies during the subinterval $[k\tau,(k+1)\tau)$. It is bounded and its upper limit is given by $\overline{W}$; b_1, b_2, and c are parameters which involve uncertainties with known bounds and they are assumed to have the following forms:

$$b_1(k) = b_1^* + \Delta b_1(k)$$
$$(2.4) \qquad b_2(b) = b_2^* + \Delta b_2(k)$$
$$c(k) = c^* + \Delta c(k)$$

where b_1^*, b_2^* and c^* are known positive constants and $\Delta b_1(k)$, $\Delta b_2(k)$ and $\Delta c(k)$ are uncertainties satisfying

$$|\Delta b_1(k)| \leq \sigma_1 < b_1^*$$
$$(2.5) \qquad |\Delta b_2(k)| \leq \sigma_2 < b_2^*$$
$$|\Delta c(k)| \leq \sigma_3 \ .$$

Here, σ_1, σ_2 and σ_3 are known positive constants, and we assume that Δb_1, Δb_2, Δc and W are constant over each subinterval.

Consider the transformation

$$x(k) = N(k) - N^*$$
$$(2.6)$$
$$u(k) = W(k) - W^*$$

where

$$(2.7) \qquad W^* = \frac{N^* c^*}{b_1^* + b_2^* N^*}$$

is the constant work rate (days) corresponding to the desired steady state N^* in the absence of uncertainty. From (2.1), (2.4), (2.6) and (2.7), we obtain

$$(2.8) \qquad x(k+1) = f(k, x(k)) + B(k, x(k))u(k) + g_1(q(k), x(k), u(k))$$

where

$$q(k) = (\Delta b_1(k), \Delta b_2(k), \Delta c(k)),$$

$$(2.9) \qquad f(k, x(k)) = x(k)(1+b_2^* W^* - c^*)$$

$$(2.10) \qquad B(k, x(k)) = b_1^* + b_2^*(N^* + x(k))$$

$$(2.11) \qquad g_1(q(k), x(k), u(k)) = [\Delta b_1(k)W^* + \Delta b_2(k)N^* W^* - \Delta c(k)N^*]$$
$$+ [\Delta b_2(k)W^* - \Delta c(k)]x(k) + [\Delta b_1(k) + \Delta b_2(k)x(k) + \Delta b_2(k)N^*]u(k) .$$

From (2.2) and (2.6), the constraint on the work rate u becomes

$$(2.12) \qquad -W^* \le u(k) \le \overline{W} - W^*$$

3. Stabilizing Controllers and Simulation Results

In this section, we apply the results of Corless and Manela (1986) to obtain a discrete time
study strategy for a student who wishes to achieve a specified desired knowledge level in
the presence of bounded uncertainties in the system parameters and then to maintain
his/her knowledge level within a calculable neighborhood of that desired level in a continuous assessment study environment. The definition of a globally uniformly asymptotically stable (g.u.a.s.) system and a statement of the stability theorem used are given in the
reference. Interested readers may refer to the cited reference for a proof of the theorem.

For the one-dimensional uncertain discrete time system (2.8), Assumptions 1-3 of the
reference are satisfied with $P \equiv 1$, and

$$\tilde{f}(k, x) \equiv 0, \ \tilde{c} = 0, \ c_o = 0, \ c_1 = |\, 1 + b_2^* W^* - c^* \,|$$

$$(3.1) \qquad \begin{aligned} \rho_o &= \sigma_1 W^* + (\sigma_2 W^* + \sigma_3)N^* \\ \rho_1 &= \sigma_2 W^* + \sigma^3 \\ \rho_2 &= \sigma_1 + \sigma_2 N^* \end{aligned}$$

In the absence of control constraint (2.12), the stabilizing control proposed by Corless and
Manela (1986) is given by

$$u(k) = p(k, x(k)) = \begin{cases} 0 & \text{if } \rho_2 \geq 1 \\ -B(k, x(k))^{-1}f(k, x(k)) & \text{if } \rho_2 < 1 \end{cases}$$

If control constraint (2.12) is imposed *a posteriori*, that is, if one employs

$$\begin{cases} u(k) = -W^* & \text{if } p(k, x(k)) < W^* \\ u(k) = \overline{W} - W^* & \text{if } p(k, x(k)) > \overline{W} - W^*, \end{cases}$$

then the proposed control (study strategy) need not guarantee the achievement of a calcu-able neighborhood of the specified knowledge level N^* within a calculable time period. However, in the simulations we impose the constraint (2.12); furthermore, we desire reaching N^* in a prescribed time period.

To be consistent with the simulations performed in Lee and Leitmann (1989a,b), we again selected $b_1^* = 1(\%/\text{day})$. $b_2^* = 0.001(1/\text{day})$, $c^* = 0.055$, $\sigma_1 = 0.1(\%/\text{day})$, $\sigma_2 = 0.0001(1/\text{day})$, $\sigma_3 = 0.0055$, $T = 210$ days = 30 weeks, $\tau = 1(\text{week})$, $\eta = 30$, $\overline{W} = 7(\text{days})$. Two Values of N^* we considered are $N^* = 50\%$ and $N^* = 85\%$; the corresponding values of W^* (days) are $W^* = 2.619$ and $W^* = 4.309$, respectively. We also selected two realizations for the uncertainties, namely

$$(\Delta b_1, \Delta b_2, \Delta c) = (-0.1, -0.0001, 0.0055) \quad \text{and}$$

$$(\Delta b_1, \Delta b_2, \Delta c) =$$

$$\left(-0.1\sin\frac{2k\pi}{3}, 0.0001\cos(\pi - \frac{2k\pi}{3}5), -0.0055\cos\frac{2k\pi}{3}\right)$$

Figures 1a and 1b display the plots of knowledge level of a student who works at the steady state rates $W^* = 2.619$ and $W^* = 4.309$, respectively, in the presence of the first set of uncertainties. For this set of uncertainties, Figures 2a and 3a depict the student's knowledge levels as a result of utilizing the Corless-Manela stabilizing controllers (stabil-izing study strategies) shown in Figures 2b and 3b, respectively. Under the disturbance of the second set of uncertainties, Figures 4a and 5a depict the student's knowledge levels resulting from the utilization of the Corless-Manela stabilizing controllers shown in Figures 4b and 5b, respectively. The neighborhood of the desired knowledge level, $N^*(\%)$, which is reached in a calculable time period (not necessarily equal or less than the desired one) is given by $d(\%)$ provided no control constraint is imposed. For $N^* = 50\%$, $d = 0.57\%$, for $N^* = 85\%$, $d = 0.96\%$, and for $N^* = 100\%$, $d = 1.13\%$. Owing to the pres-ence of the control constraint (2.2) and the specified terminal time T, if the initial knowledge level of the student is very low, the stabilizing study strategy of Corless-

Manela may not be able to drive the system state into a sufficiently small neighborhood of a desired knowledge level. To demonstrate this, we use $N(0) = 5\%$, and $N^* = 100\%$ in Figure 6; thus, even though the student works at maximum rate (7 days) throughout the whole semester, his/her knowledge level reaches only 92.74% when the semester ends!

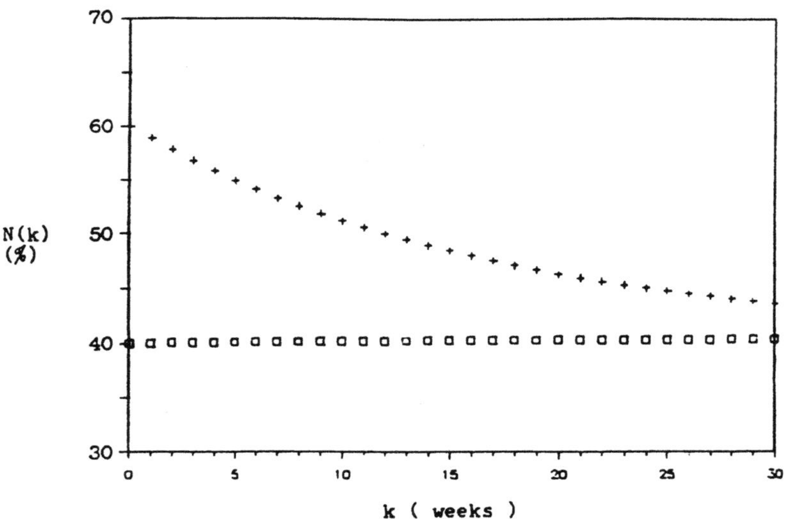

Fig. 1a Time histories of knowledge levels with $N^* = 50\%$, $N(0) = 40\%$ and 60%, respectively, utilizing $W^* = 2.619$(days) in the presence of the *first* set of uncertainties.

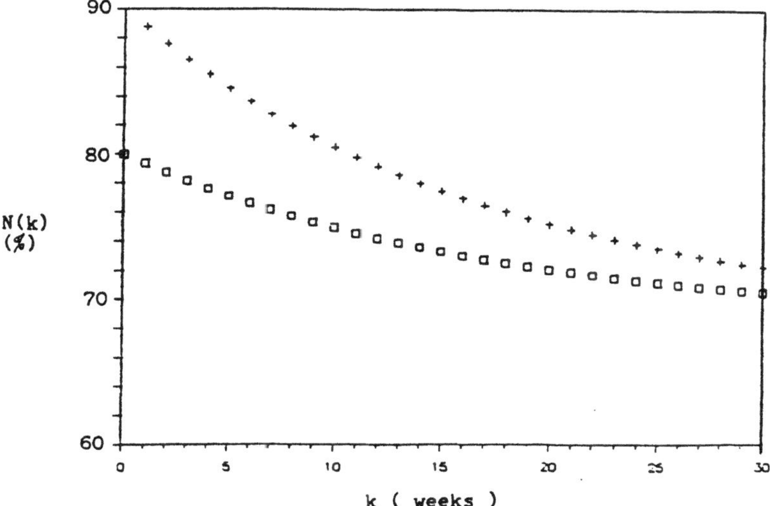

Fig. 1b Time histories of knowledge levels with $N^* = 85\%$, $N(0) = 80\%$ and 90%, respectively, utilizing $W^* = 4.309$ days in the presence of the *first* set of uncertainties.

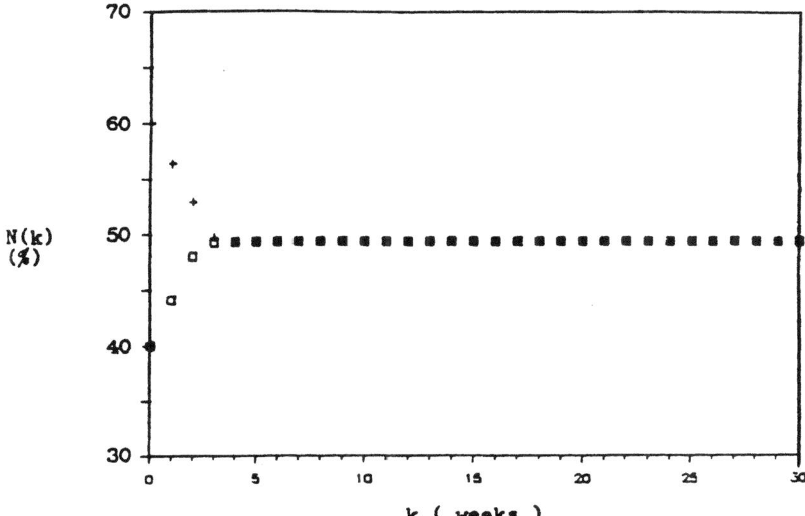

Fig. 2a Time histories of knowledge levels with $N^* = 50\%$, $N(0) = 40\%$ and 60%, respectively, utilizing the stabilizing study strategies shown in Fig. 2b in the presence of the *first* set of uncertainties.

                                        G. Leitmann, C. S. Lee

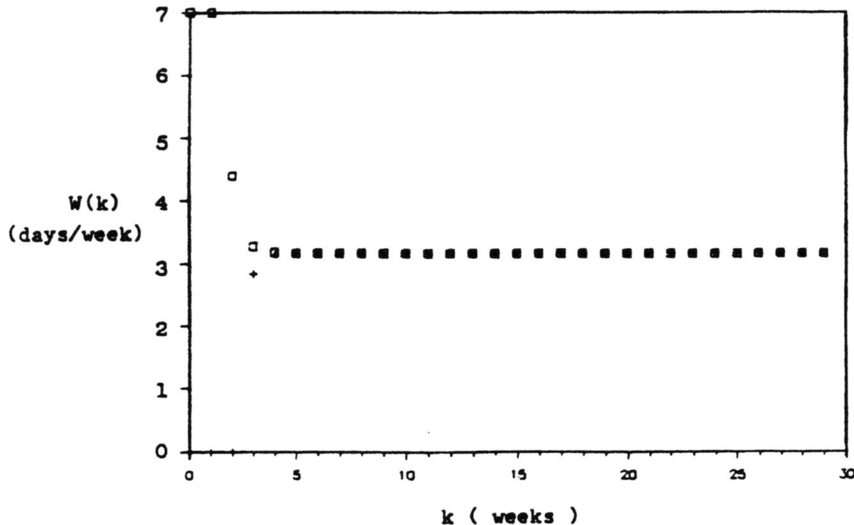

Fig. 2b Stabilizing study strategies according to Corless and Manela for $N^* = 50\%$.

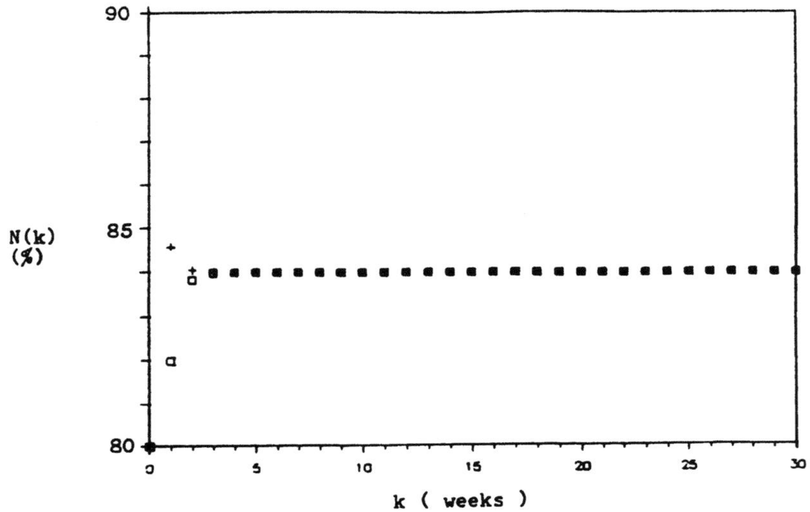

Fig. 3a Time histories of knowledge levels with $N^* = 85\%$, $N(0) = 80\%$ and 90 %, respectively, utilizing the stabilizing study strategies shown in Fig. 3b in the presence of the *first* set of uncertainties.

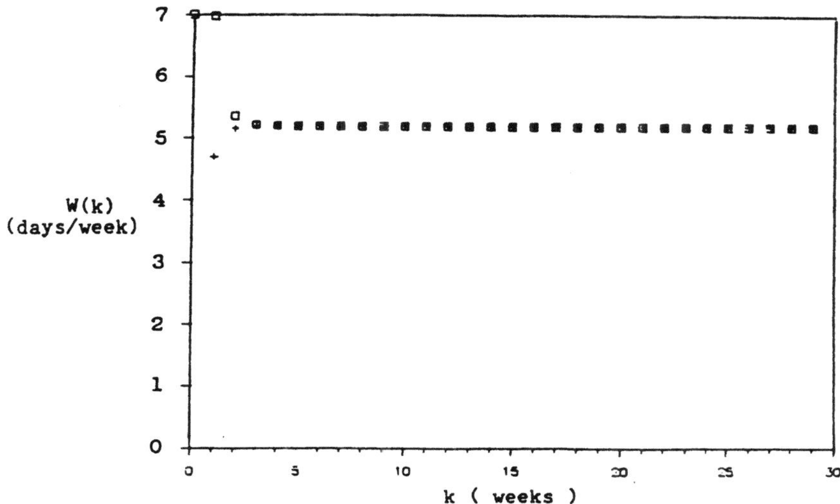

Fig. 3b Stabilizing study strategies according to Corless and Manela for $N^* = 85\%$.

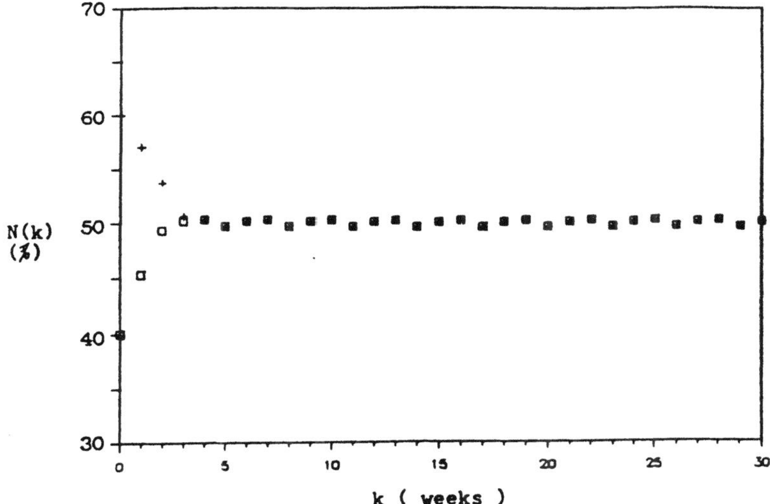

Fig. 4a Time histories of knowledge levels with $N^* = 50\%$, $N(0) = 40\%$ and 60%, respectively, utilizing the stabilizing study strategies shown in Fig. 4b in the presence of the *second* set of uncertainties.

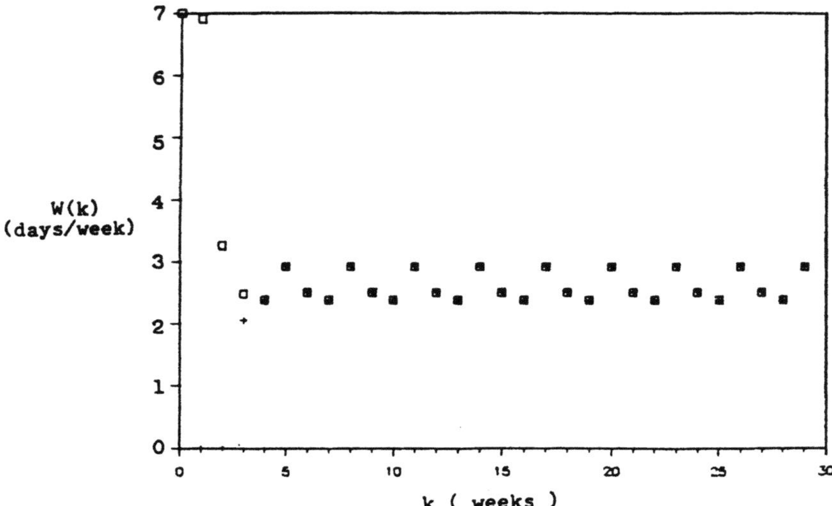

Fig. 4b Stabilizing study strategies according to Corless and Manela for $N^* = 50\%$.

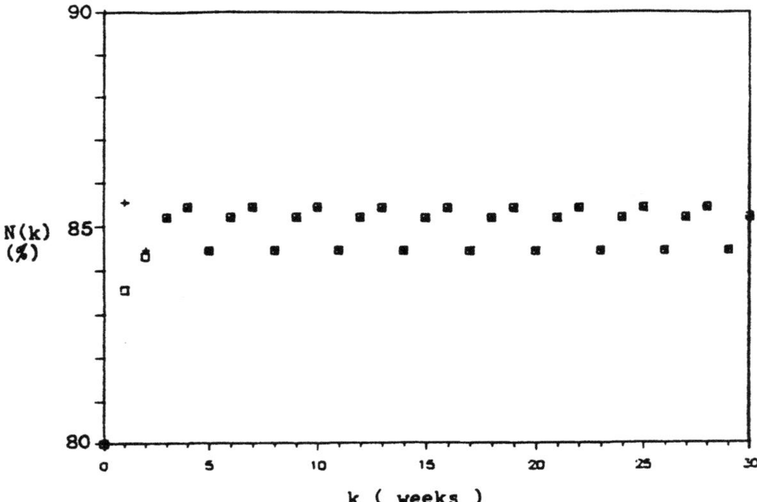

Fig. 5a Time histories of knowledge levels with $N^* = 85\%$, $N(0) = 80\%$ and 90%, respectively, utilizing the stabilizing study strategies shown in Fig. 5b in the presence of the *second* set of uncertainties.

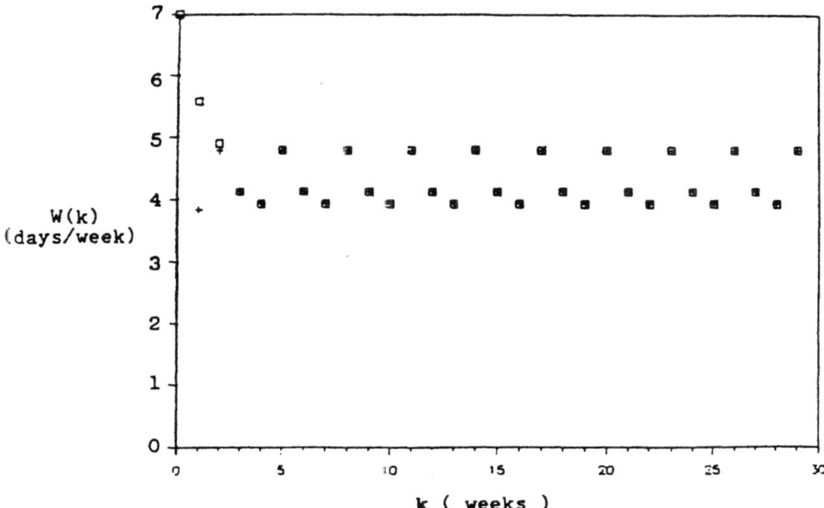

Fig. 5b Stabilizing study strategies according to Corless and Manela for $N^* = 85\%$.

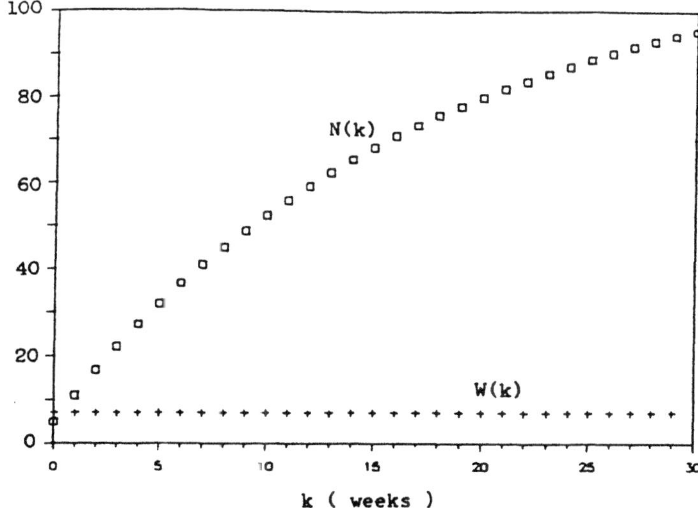

Fig. 6 Time histories of knowledge level and the corresponding stabilizing study strategy according to Corless and Manela for $N^* = 100\%$ with $N(0) = 5\%$ in the presence of the *first* set of uncertainties.

4. Conclusion

We have utilized a discrete stabilizing memoryless state feedback controller proposed by Corless and Manela (1986) to drive the response of a student learning model to and then maintain it in a calculable neighborhood of some specified desired knowledge level in the presence of uncertainties with known bounds. We have also observed that if a student's initial knowledge level is "very low" and his/her specified desired knowledge level is "very high", then even though he/she works at maximum rate throughout the whole semester, he/she may still not be able to achieve the desired level when semester ends.

References

1.   Barmish, B.R. et al. (1983), A new class of stabilizing controllers for uncertain dynamical systems. SIAM Journal of Control and Optimization 21, 246-255.

2.   Bondi, H. (1982), Note on A student related optimal control problem, by Raggett, Hempson and Jukes. Bulletin of The Institute of Mathematics and its Applications 18, 10-11.

3.   Cheng, T.C.E. and Teo, K.L. (1987), Further extensions of a student related optimal control problem. Mathematical and Computer Modelling (formerly Mathematical Modelling) 9, 499-506.

4.   Corless, M. and Leitmann, G. (1988), Deterministic control of uncertain systems. Proceedings of the Conference on Modeling and Adaptive Control (Sopron, Hungary), Lecture Notes in Control and Information Sciences 105, Springer Verlag, Berlin.

5.   Corless, M. and Leitmann, G. (1989), Deterministic control of uncertain systems: A Lyapunov theory approach. Deterministic Nonlinear Control of Uncertain Systems: Variable Structure and Lyapunov Control, ed. A.S.I. Zinober; IEE Press, London.

6.   Corless, M. and Manela, J. (1986), Control of uncertain discrete time systems, Proceedings of the American Control Conference, Seattle, Washington, 515-520.

7.   Lee, C.S. and Leitmann, G. (1990), On a student related optimal control problem, J. Optimization Theory and Applications 65, 129-138.

8. Lee, C.S. and Leitmann, G. (1991), Some stabilizing study strategies for a student related problem under uncertainty. Dynamics and Stability of Systems 6, 63-78.

9. Ogata, K. (1987), Discrete-Time Control Systems, Prentice Hall International, Inc. Englewood Cliffs, NJ, U.S.A.

10. Parlar, M. (1984), Some extensions of a student related optimal control problem. Bulletin of The Institute of Mathematics and its Applications 20, 180-181.

11. Raggett, G.F. et al. (1981), A student related optimal control problem. Bulletin of The Institute of Mathematics and its Applications 17, 133-136.

Authors' addresses

Prof. George Leitmann
Department of Mechanical Engineering
University of California
Berkeley, CA 94720
U.S.A.

Prof. Cho Seng Lee
Department of Mathematics
University of Malaya
59100
Kuala Lumpur, Malaysia

Stability Conditions in Terms of Eigenvalues of a Nonlinear Optimal Controlled System

Houria Bourdache-Siguerdidjane

Abstract. The stability conditions, in terms of nonlinear eigenvalues, may be deduced from the exact solution of a nonlinear controlled system. Those values are associated to eigenvectors which satisfy an algebraic nonlinear equation.

1. Introduction

It is shown in B. Siguerdidjane (1991, 1992) that the explicit solution of the problem of satellite angular momentum regulation, using jets only, may be directly written down in terms of characteristic values and characteristic vectors. Those values and vectors satisfy an algebraic nonlinear equation which may be derived by some differential manipulations.

The problem of stabilizing the angular velocity or the attitude of a rigid space vehicle has been extensively studied in the literature. The center manifold and Lyapunov approaches are used (see e.g. Bonnard 1982, Crouch 1984, Byrnes and Isidori 1991) to check whether or not the system is stable. In Bloch, Mardsden (1990) and Aeyels (1992), the Energy-Casimir method is used to study the system under only one control torque. So, in this paper we will discuss the stability conditions in terms of eigenvalues based on the exact solution.

Consider a rigid body in an inertial reference frame. Let ω_1, ω_2, ω_3, as usual denote the angular velocity components and let I_1, I_2, I_3, be the moments of inertia of the body about the principal axes which are the body axes.

The motion of the body under external forces is described by the Euler equations

$$I\dot{\omega} = S(\omega)I\omega + u \tag{1}$$

with

$$I = diag(I_1, I_2, I_3), \ \omega = (\omega_1, \omega_2, \omega_3)^T, \ \omega \in R^3.$$

and where

$$u = (u_1, u_2, u_3)^T, \ u \in R^3$$

is the control torque vector generated by the gas jets, the rotation matrix is

$$S(\omega) = \begin{pmatrix} 0 & \omega_3 & -\omega_2 \\ -\omega_3 & 0 & \omega_1 \\ \omega_2 & -\omega_1 & 0 \end{pmatrix}$$

The case where the feedback law $u = \beta\,\omega$, β is a negative constant or a time-varying function may be obtained by computing the nonlinear optimal feedback strategy (B. Siguerdidjane, 1987). The analytical solutions of the closed loop system has thus been obtained in (B. Siguerdidjane, 1991, 1992). With this control, the geometric and algebraic spectra are identical to those of the free motion .

Let us consider the following homogeneous polynomial system

$$\dot{x} = f(x) \tag{2}$$

where $f(x)$ is a homogeneous polynomial vector field of degree p, and each component $f_i(x)$ is homogeneous of degree p. x denotes the state vector of components $(x_1, x_2, ..., x_n)$, $x \in R^n$.

It is shown in Hahn(1967) and Samardzjia (1983) that, for an odd degree of homogeneity p, if system (2), with p odd, has a nonempty set, then the necessary conditions of asymptotic stability in the large is that for any eigenvector, the associated eigenvalue is negative. This property is observed from the solution of (2) which is given only for a real eigenvector.

However, the problem under study has complex eigenvectors as it will be described in the forthcoming sections. In addition, the degree of homogeneity is even. Let us then look for the stability conditions based on the exact solution.

2. Some background results

Set $x_1 = I_1\omega_1$, $x_2 = I_2\omega_2$, $x_3 = I_3\omega_3$. System (1) becomes

$$\begin{aligned}
\dot{x_1} &= k_x\, x_2\, x_3 + u_1 \\
\dot{x_2} &= k_y\, x_3\, x_1 + u_2 \\
\dot{x_3} &= k_z\, x_1\, x_2 + u_3
\end{aligned} \tag{3}$$

where $k_x = (I_y - I_z) / I_y I_z$, $k_y = (I_z - I_x) / I_z I_x$ and $k_z = (I_x - I_y) / I_x I_y$. Suppose that $I_1 > I_2 > I_3$ and $k_1.k_2.k_3 \neq 0$, we then have $k_1 > 0$, $k_2 < 0$ and $k_3 > 0$.

It is found in (Dwyer (1982), B. Siguerdidjane, 1987) that the optimal feedback law which minimizes the cost of fuel and energy is linear in the state such that $u = \beta\, x$. β is a negative constant $\beta = -(\lambda_1/\lambda_3)^{1/2}$ or a function of time

$$\beta(t) = \frac{-(\lambda_1/\lambda_3)^{1/2}\, tanh(\lambda_1/\lambda_3)^{1/2}\,(t_f - t) - s/2\lambda_3}{1 + s/2\lambda_3\,(\lambda_3/\lambda_1)^{1/2}\, tanh(\lambda_1/\lambda_3)^{1/2}\,(t_f - t)}$$

where $s > 0$ and λ_1, λ_3 are the weight coefficients in the cost function minimized

$$J = x^T S x + \int_0^{t_f} (\lambda_1(x_1^2 + x_2^2 + x_3^2) + \lambda_3(u_1^2 + u_2^2 + u_3^2)) \, dt,$$

λ_1, $\lambda_3 > 0$; $S = sI$, I is the identity matrix.

Moreover, it is shown in (B. Siguerdidjane 1991, 1992) that the explicit solution of the controlled system (1) may be directly written down in terms of characteristic values and characteristic vectors as follows

$$x_1(t) = i \, h(t) \, k \, v_1 \, \xi_0 \, cn \, (\, \xi_0 \, \lambda \, g(t))$$

$$x_2(t) = - \, h(t) \, k \, v_2 \, \xi_0 \, sn \, (\, \xi_0 \, \lambda \, g(t)) \qquad (4)$$

$$x_3(t) = i \, h(t) \, v_3 \, \xi_0 \, dn \, (\, \xi_0 \, \lambda \, g(t))$$

where λ denotes the eigenvalue and $v^T = (v_1, v_2, v_3)$ denotes the eigenvector. ξ_0 is an arbitrary constant. One may keep in mind the fact that the characteristic value must be postmultiplied by the same constant according to consequence 1 in the Appendix. $i, -1$ and i are introduced to satisfy the derivatives formulas hereafter. k is the so-called modulus of the elliptic function ($k < 1$). Jacobi elliptic functions are listed in tables of higher functions in Jahnke et al.(1945).

For β constant, it is found that

$$h(t) = e^{\beta t}$$

and

$$g(t) = e^{\beta t} / \beta.$$

For β time-varying function, it is found that

$$h(t) = C_1 / \cosh \alpha\tau + b \sinh \alpha\tau, \qquad (5)$$

where $\alpha = (\lambda_1/\lambda_3)^{1/2}$ and τ denotes the time to go $(t_f - t)$.

and

$$g(\tau) = \frac{C_1}{\alpha\sqrt{b^2 - 1}} \, Log \left(\frac{(b^2 - 1) + \sqrt{b^2 - 1} \, (\sinh \alpha\tau + b \cosh \alpha\tau)}{\sqrt{b^2 - 1} \, (\cosh \alpha\tau + b \sinh \alpha\tau)} \right) + C_2 \qquad (6)$$

where $\delta = \alpha\tau$, $b = s/\lambda_3 2\alpha$ and C_2 is an arbitrary constant.

3. Characteristic nonlinear equation

By using the derivative formula of Jacobi functions,

$$
\begin{aligned}
cn'(s) &= -sn(s)\, dn(s) \\
sn'(s) &= cn(s)\, dn(s) \\
dn'(s) &= -k^2\, sn(s)\, cn(s)
\end{aligned}
\tag{7}
$$

and differentiating equations (4), one may easily obtain

$$
\begin{aligned}
\lambda\, v_1 &= k_x\, v_2\, v_3 \\
\lambda\, v_2 &= k_y\, v_1\, v_3 \\
\lambda\, v_3 &= k_z\, v_1\, v_2
\end{aligned}
\tag{8}
$$

or in matrix notation

$$
f(v) - \lambda\, v = 0
$$

which is the characteristic nonlinear algebraic equation, $v = (v_1, v_2, v_3)^{\mathrm{T}}$.

One may notice that the system under study has the form

$$
\dot{x}(t) = f(x(t)) + a\, x
\tag{9}
$$

However, it is shown that the characteristic equation of system (7) is identical to the one of system (2). The second method to obtain this characteristic equation is recalled in the Appendix.

Let E and Δ be the geometric and algebraic spectra respectively, we obtain

$$
E = \left\{ V_1 = \begin{pmatrix} 1 \\ 0 \\ 0 \end{pmatrix}, V_2 = \begin{pmatrix} 0 \\ 1 \\ 0 \end{pmatrix}, V_3 = \begin{pmatrix} 0 \\ 0 \\ 1 \end{pmatrix}, V_4 = \begin{pmatrix} 1 \\ a \\ b \end{pmatrix}, V_5 = \begin{pmatrix} 1 \\ -a \\ b \end{pmatrix}, V_6 = \begin{pmatrix} 1 \\ a \\ -b \end{pmatrix}, V_7 = \begin{pmatrix} 1 \\ -a \\ -b \end{pmatrix} \right\}
\tag{10}
$$

where

$$
a = \sqrt{\frac{k_2}{k_1}} ; \quad b = \sqrt{\frac{k_3}{k_1}} ,
$$

$$
\Lambda = \left\{ \lambda_1 = \lambda_2 = \lambda_3 = 0; \ \lambda_4 = \lambda_7 = \sqrt{k_2 k_3}, \lambda_5 = \lambda_6 = -\sqrt{k_2 k_3} \right\}
\tag{11}
$$

4. Stability Conditions

Now we turn to the study of the behaviour of $x^T(t) = (x_1(t), x_2(t), x_3(t))$ expressed in terms of the initial state $x_0^T = (x_{01}, x_{02}, x_{03})$. In practice, it is required that the system be stable i.e. $x(t)$ must remain bounded for all subsequent times.

Case 1: Since β is negative, $e^{\beta t} \to 0$ as $t \to \infty$. In addition, $sn(0) = 0$, $cn(0) = dn(0) = 1$. So, roughly speaking, equations (4) are asymptotically stable.

The asymptotical stability may also be shown by using the Lyapunov function $L(x) = \dfrac{1}{2}(x_1^2 + x_2^2 + x_3^2)$ or by using algebraic geometric methods (Baillieul, 1983).

The constant ξ_0 and the modulus k may be evaluated by using the initial conditions. Suppose that the initial time $t = t_0$. So let $x_{10} = x_1(t_0)$, $x_{20} = x_2(t_0)$ and $x_{30} = x_3(t_0)$. From equations (4) and making use of the relationships between the functions sn, dn and cn we find that

$$\xi_0^2 = ((x_{20}^2 / v_2^2) - (x_{30}^2 / v_3^2)) / h_0^2$$

$$k^2 \xi_0^2 = ((x_{20}^2 / v_2^2) - (x_{10}^2 / v_1^2)) / h_0^2 \tag{12}$$

and

The angular momentum thus is

$$M^2(t) = ((v_1^2 + v_2^2 + v_3^2) k^2 sn^2 (\xi_0 \lambda e^{\beta t} / \beta) - k^2 v_1^2 - v_3^2) \xi_0^2 e^{2\beta t}, \tag{13}$$

whence

$$M^2(t) = (-k^2 v_1^2 - v_3^2) \xi_0^2 e^{2\beta t}. \tag{14}$$

It is obvious that $M^2(t) \to 0$ as $t \to \infty$.

Case 2: t_f is fixed, $h(t) \to C_1$ as $t \to t_f$ while equation $g(\tau)$ tends to a finite value so equations (4) are also stable in this case.

Remark. One may notice that case 1 also coresponds to the steady state regulation.

Even so, the functions sn, cn and $dn \to \infty$ as the arguments $\to iK'$ where iK' is the imaginary quarter period of Jacobi elliptic functions.

β depends on the weight coefficients in the cost function which is minimized to get the optimal control. So, in the first case, $x(t)$ is stable if we choose β such as the relationship $\lambda = iK'\beta e^{-\beta t} / \xi_0$ for a given t, never holds. In the second case, the stability is guaranteed if $\xi_0 \lambda g(t) \neq iK'$ for a given t, $t_0 \leq t \leq t_f$.

Appendix

Characteristic nonlinear equation (second method)

Given a continuous time nonlinear system of the form

$$\dot{x}(t) = f(x(t)) + a x \tag{A.1}$$

where $f(x)$ is a homogeneous polynomial vector field of degree p, $f(x)= (f_1(x), f_2(x) \dots,$ $f_n(x))$ and each component $f_i(x)$ is homogeneous of degree p. x denotes the state vector of components $(x_1, x_2, \dots, x_n)$, $x \in R^n$, $a \in R$.

So, $$\dot{x}_i = f_i(x) + a x_i \tag{A.2}$$

Recall that a polynomial field P is called homogeneous of degree p if, for any $\alpha \in R$, $P(\alpha x) = \alpha^p P(x)$, for all $x \in R^n$.

We introduce a new vector $z = (z_1, \dots, z_n)^T \in C^n$

such that for any non vanishing x_j

$$x = x_j z \quad , \quad j = 1, \dots, n, \tag{A.3}$$

clearly, when $i=j$, $z_j = 1$.

Next, differentiate (A.3) with respect to time

$$\dot{x} = \dot{x}_j z + x_j \dot{z}, \tag{A.4}$$

substitute $\dot{x}$ from (A.1) into (A.4) and use (A.2):

$$x_j \dot{z} = f(x) - f_j(x) z + a x - a x_j z. \tag{A.5}$$

using again (A.1) and the fact that f is homogeneous of degree p shows that

$$\dot{z} = x_j^{p-1} (f(z) - f_j(z) z) \tag{A.6}$$

or in terms of components

$$\dot{z}_i = x_j^{p-1} (f_i(z) - f_j(z) z_i), \quad i = 1, \dots, n. \tag{A.7}$$

Hence, the non trivial singular solutions are given by

$$f(z) - f_j(z) z = 0 \tag{A.8}$$

Let us denote by $v = (v_1, \dots, v_n)^T$ one solution of equation (A.8) for a given j, $v_i \in C$.

Finally, the nonlinear characteristic equation is

$$f(v) - f_j(v) v = 0. \tag{A.9}$$

Note that it reduces to the characteristic nonlinear equation of the system $\dot{x}(t) = f(x(t))$ presented in Samardzija (1983).

Postmultiplying the left hand side of (A.9) by v^T, we obtain

$$f_j(v) = \frac{v^T f(v)}{\|v\|^2} \tag{A.10}$$

Let $\lambda = f_j(v)$, the vector v and the value λ are the characteristic vector and its associated characteristic value respectively.

Let R_j denote, for a given j, the representation (A.9). It is clear that if the state vector x has no zero component, n representations R_j may then be obtained.

Definition 1: Two vectors V_1 and V_2, belonging to the set E of all solutions of the algebraic nonlinear equation $f(V) - f_j(V)\, V = 0$, $j=1,\dots, n$, are said to be equivalent if and only if there exists a non zero element c of the complex field C such that $V_1 = c\, V_2$.

Consequence 1: When an eigenvector is multiplied by a nonzero element $c \in C$, the associated eigenvalue is therefore multiplied by c^{p-1}.

References

Aeyels D. (1992). On stabilization by means of the Energy-Casimir method. Syst. Contr. Letters 18, pp. 325-328.

Baillieul J., (1980). The geometry of homogeneous polynomial dynamical systems. Nonlinear analysis. Theory, Methods & Applications, 4, 5, pp. 879-900.

Bonnard, B.(1982). Contrôle de l'attitude d'un satellite rigide. RAIRO, Automatique/ Syst.Analys. and Contr.,16, pp. 85-93.

Bloch A. M., J. E. Mardsden (1990). Stabilization of rigid body dynamics by Energy-Casimir method. Syst. Contr. Letters 14, pp. 341-346.

Bourdache-Siguerdidjane H. (1991). Feedback controlled satellite angular momentum: on the analytic solution of the angular velocity'. IEEE Contr. and Dec. Conf., Brighton, pp. 144-146.

Bourdache-Siguerdidjane H. (1992). Further results on the optimal regulation of satellite angular momentum. Optim. Contr. Appl. & Methods J., 13.

Bourdache-Siguerdidjane H. (1987). On applications of a new method of computing optimal nonlinear feedback controls. Optim. Contr. Appl. & Methods J., 8, pp. 397-409.

Byrnes C., A. Isidori (1991). On the attitude stabilization of rigid spacecraft. Automatica, Vol. 27, 1, pp.87-95.

Crouch P.E. (1984). Spacecraft attitude control and stabilization: application of geometric theory to rigid body models. IEEE Trans. Aut. Contr., AC-29,4, pp. 321-331.

Dwyer T. (1982). The control of angular momentum for asymmetric rigid bodies. IEEE Trans. Automatic Control, AC 27, 3, pp. 686-688.

Hahn W. (1967). Stability of motion, Spinger Verlag.

Jahnke E., F. Emde (1945). Tables of functions with formulas and curves, eds Dover.

Samardzija N. (1983). Stability properties of Autonomous homogeneous polynomial differential systems. J. Diff. Equat. 48, pp. 66-70.

Author's address

Dr. Houria Bourdache-Siguerdidjane
Ecole Supérieure d'Electricité
Plateau de Moulon
91192, Gif-sur-Yvette, France

SIGUER@FRESE51

Program-positional Optimization for Dynamic Systems

R Gabasov, F.M. Kirillova, N.V. Balashevich

Abstract

For the problem $I(u) = c'\, x(t^*) \rightarrow \max$, $\dot{x} = A\,x + b\,u + w(t)$, $x(0) = x_0$, $H\,x(t^*) = g$, $|u(t)| \leq 1$, $t \in [0, t^*]$ where $w(t)$, $t \in T$ is an unknown perturbation, an algorithm of constructing the optimal program-positional controller is justified.

1   Introduction

There are two principles of control in cybernetic programs (open loop) and positional (closed loop) ones [1]. Program controls are simpler for calculating and realizing, positional ones (feedback) are better in the presence of perturbations. Finite methods for calculating optimal program controls are described in [2]. In [3] a new method for constructing optimal positional controls is given. The purpose of the article is to suggest an algorithm for constructing an optimal controller producing controls of the mixed type: at fixed moments it constructs the feedback controls in accordance with the state of the system at these moments, in intervals between the moments the controller uses the program controls corresponding to the state of the system at the initial point of the interval. In systems closed by program-positional feedback unlike systems closed by positional feedbacks sliding motions do not appear under any perturbations. The investigated type of controls is natural for situations when in the course of the control process one can correct the control rule only at separate moments. In this case two possibilities are distinquished:

1. measurements of the system state are conducted continuously,

2. the state is also measured only at separarte moments.

The case when information about the states is not complete and exact will be the subject of another investigation.

2   Problem statement

In the class of piecewise continuous controls $u(t)$, $t \in T = [0, t^*]$, $t^* = N\vartheta$, we consider the terminal problem

$$(2.1) \qquad I(u) = c' x(t^*) \quad \rightarrow \quad \max,$$

$$(2.2) \qquad \dot{x} = A x + b u, \ x(0) = x_0,$$

$$(2.3) \qquad H x(t^*) = g,$$

$$|u(t)| \leq 1, \ t \in T,$$

$$(u \in R, \ x \in R^n, \ g \in R^m, \ \mathrm{rank}\, H = m \leq n).$$

An optimal program control $u^0(t)$, $t \in T$, and a trajectory $x^0(t)$, $t \in T$, are defined as usual. A parameter ϑ characterizes the points $\vartheta_k = k\vartheta$, $k = \overline{0, N-1}$, where correction of the control rule is possible.

To introduce optimal program-positional controls we embed the problem (2.1) - (2.3) in the family

$$(2.4) \qquad \begin{aligned} & I_\tau(u) = c' x(t^*) \quad \rightarrow \quad \max, \ \dot{x} = A x + b u, \ x(\vartheta_k) = z, \\ & H x(t^*) = g, \ |u(t)| \leq 1, \ t \in T_k = [\vartheta_k, t^*], \end{aligned}$$

depending on values $\vartheta_k = k\vartheta$, $k = \overline{0, N-1}$, and a n-vector z.

Denote by $u^0(t|\vartheta_k, z)$, $x^0(t|\vartheta_k, z)$, $t \in T_k$, the solution to the problem (2.4) . Let Ω_k be a totality of positions $\omega = (\vartheta_k, z)$ for which the problem (2.4) has a solution, $k = \overline{0, N-1}$.

A piecewise continuous function $u^0(t, \vartheta_k, z)$, $t \in T_k$, $(\vartheta_k, z) \in \Omega_k$, $k = \overline{0, N-1}$, is called the optimal program-positional control if it satisfies the following conditions:

1. $|u^0(t, \vartheta_k, z)| \leq 1$, $t \in T_k$, $(\vartheta_k, z) \in \Omega_k$, $k = \overline{0, N-1}$,

2. the trajectory $x(t)$, $t \in T_k$, of the system $\dot{x} = A\,x + b\,u^0(t, \vartheta_s, x(\vartheta_s))$, $x(\vartheta_k) = z$, $\vartheta_s \leq t \leq \vartheta_{s+1}$, $s = \overline{k, N-1}$, coincides with $x^0(t|\vartheta_k, z)$, $t \in T_k$, $k = \overline{0, N-1}$, for every $(\vartheta_k, z) \in \Omega_k$.

Suppose that under real conditions the dynamic system closed by the semi-feedback $u^0(t, \vartheta_k, x(\vartheta_k))$ is described not by the equation (2.2) but because of unknown piecewise continuous perturbations $w(t)$, $t \in T$, it moves according to the equation

$$
(2.5) \qquad
\begin{aligned}
\dot{x} &= A\,x + b\,u^0(t, \vartheta_k, x(\vartheta_k)) \qquad x(0) = x_0 \\[2mm]
\vartheta_k &\leq t \leq \vartheta_{k+1}, \qquad\qquad k = \overline{0, N-1}.
\end{aligned}
$$

Consider the system (2.5) with perturbation $w^*(t)$, $t \in T$. Denote by $x^*(t)$, $t \in T$, and $u^*(t) = u^0(t, \vartheta_k, x^*(\vartheta_k))$, $\vartheta_k \leq t < \vartheta_{k+1}$, $k = \overline{0, N-1}$, corresponding realizations of the trajectory and the control.

A device which controls the dynamic system such that the control $u^*(t)$ is produced, $t \in T$, in the real-time mode, will be called *an optimal program-positional controller* (briefly, an optimal controller).

The aim of the further exposition is to describe an algorithm for designing the optimal controller. This will be done in section 6 and now we give the necessary auxiliary results.

3 Defining equations of optimal controller

According to [2] the optimal control $u_\tau^0(t)$, $t \in T_\tau = [\tau, t^*]$, of the problem

$$
(3.1) \qquad
\begin{aligned}
I_\tau(u) &= c'\,x(t^*) \quad \longrightarrow \quad \max, \quad \dot{x} = A\,\dot{x} + b\,u, \quad x(\tau) = x^*(\tau), \\[2mm]
H\,x(t^*) &= g, \ |u(t)| \leq 1, \ t \in T_\tau,
\end{aligned}
$$

equals

$$
(3.2) \qquad u_\tau^0(t) = \operatorname{sign} \Delta_\tau^0(t), \ t \in T_\tau,
$$

where $\Delta_\tau^0(t) = \Delta^0(t|\tau, x(\tau)) = \psi'(t)\,b$ is the co-control,

$$
\dot{\psi} = -A'\,\psi, \ \psi(t^*) = c - H'\,y(\tau),
$$

$y(\tau) = y(\tau, x^*(\tau))$ is the optimal potential vector of the problem (3.1).

Thus the optimal control (3.2) is determined by

$$(3.3) \quad t_i(\tau), \ i \in P = \{1, 2, \ldots, p\}, \ y(\tau)$$

consisting of the zeroes $t_1(\tau) < t_2(\tau) < \ldots < t_p(\tau)$ of the co-control Δ^0_τ, $t \in T_\tau$, and the potential vector.

The elements (3.3) satisfy the system of equations

$$
(3.4) \quad
\begin{aligned}
&f(\tau; t_i(\tau), i \in P; x^*(\tau)) = 0, \\[2mm]
&q_j(t_i(\tau), i \in P; y(\tau)) = 0, \ j \in P,
\end{aligned}
$$

where

$$f(\tau; t_i, i \in P; x) = \sum_{i=0}^{p} k_i \int_{t_i}^{t_{i+1}} H(t)\, dt + H\, F(t^* - t)\, x - g,$$

$$t_0 = \tau, \ t_{p+1} = t^*, \ H(t) = H\, F(t^* - t)\, b, \ \dot{F} = A\, F, \ F(0) = E,$$

$$k_i = \mathrm{sign}\Delta^0_\tau(t_i + 0);$$

$$q_j(t_i, i \in P; y) = (c' - y'\, H)\, F(t^* - t_j)\, b, \ j \in P.$$

The system (3.4) is called *the defining equations of the optimal controller*.
Let $K^- = \{i \in \{0, \ldots, p\} : k_i = -1\}$, $K^+ = \{0, \ldots, p\} \backslash K^-$. The totality $S(\tau) = \{p(\tau), K^-(\tau), K^+(\tau)\}$ will be called the structure of the defining equations.

Further it is supposed that the perturbation $w^*(t)$ acts in the interval $T^0 = [0, t^0]$, $t^0 < t^*$, and the function $x^*(\tau)$, $\tau \in T^0$, and the moment t^0 are such that

$$(3.5) \quad \mathrm{rank}\{H(t), \ t \in T_*(\tau)\} = m$$

where $T_*(\tau)$ is the totality of break points of the control $u^0_\tau(t)$, $t \in T_\tau$.

When increasing $\tau \in T^0$ the structure of the equations may change because of one of the following reasons [3]:

1. At the moment $\bar{\tau} \in T^0$ the co-control $\Delta_{\bar{\tau}}^0(t)$, $t \in T_{\bar{\tau}}$, becomes equal to zero at its stationary point $\bar{t} \in]\bar{\tau}, t^*[$: $\partial \Delta_{\bar{\tau}}^0(t)/\partial t|_{t=\bar{t}} = 0$. One can detect the moment $\bar{\tau}$ by observing the behaviour of stationary points of the co-control which satisfy equations similar to (3.4). At the moment $\bar{\tau}$ we replace $p = p(\tau)$, $\tau < \bar{\tau}$ by $\bar{p} = p(\bar{\tau} + 0) = p + 2$, renumber the points $t_i(\bar{\tau})$, $i \in \bar{P} = \{1, 2, \ldots, \bar{p}\}$, and coordinate the values $\bar{k}_i = \pm 1$, $i \in \{0, \ldots, \bar{p}\}$.

2. At the moment $\bar{\tau} \in T^0$ two adjacent zeroes of the co-control can coalesce $t_s(\bar{\tau}) = t_{s+1}(\bar{\tau})$, $s \in P \setminus \{p\}$. In this case we define $\bar{p} = p - 2$, renumber the zeroes and coordinate k_i, $i \in \bar{P} \cup \{0\}$.

3. Zeroes placed on T_τ can go out through the ends of the interval. The moment $\bar{\tau}$ of their leaving is detected by $t_1(\tau)$, $t_p(\tau)$. In this case $\bar{p}(\tau) = p(\bar{\tau} + 0) = p - 1$.

4. An additional zero of the co-control can come in T_τ through the ends of the interval. The moment of appearing can be detected by observing the values $\Delta_\tau^0(t)$, $t \in T_\tau$, at the points $t = \tau$, $t = t^*$. In this case $\bar{p} = p(\bar{\tau} + 0) = p - 1$.

Without loss in generality one can regard that simultaneously at least two of the presented reasons of change of the structure of the equations (3.4) are not realized.
The cases when the function $x^*(t)$, $t \in T^0$, is such that

1. $\Delta_\tau^0(\bar{t}) = \dot{\Delta}_\tau^0(\bar{t}) = \ddot{\Delta}_\tau^0(\bar{t}) = 0$ for certain $\bar{t} \in [\tau, t^*]$,

2. there exists such a moment $\bar{\tau} < t^0$ and a function $\bar{t}(\tau) \in]\tau, t^*[$, $\tau \in T^+(\bar{\tau})$, that $\Delta_\tau^0(\bar{t}(\tau)) = \dot{\Delta}_\tau^0(\bar{t}(\tau)) \equiv 0$, $\tau \in T^+(\bar{\tau})$, or $\Delta_\tau^0(t^*) \equiv 0$, $\tau \in T^+(\bar{\tau})$, are also exclusive. Here $T^+(\tau)$ is the right side neighbourhood of the point τ.

4 Computational method for solving defining equations in real time with continuous state measurements

As it was intended we investigate the situation when correction of control is possible only at the moments $\vartheta_k, k = \overline{0, N-1}$. However measurements of the system state can be made either constantly or only at discrete instants of time. In this item we consider the first case. This means that values $x^*(t)$, $t \in T^0$, are given continuously and the defining equations are solved continuously in time.

We split the interval T^0 into parts with constant structure of (3.4), describe a method of solving the equations (3.4) for them and point out rules of transition between adjacent parts [3].

Under the condition (3.5) and the conditions

$$\partial \Delta_\tau^0(t)/\partial t|_{t=t_i(\tau)} \neq 0, \ i \in P;$$

the Jacobi matrix $G(t_i(\tau), \ i \in P; \ y(\tau))$,

$$G(t_i, \ i \in P; \ y) = \begin{pmatrix} 2\,H(t_i)\,k_{i-1} & , & i \in P & \cdots & 0 \\ & \vdots & & \ddots & \vdots \\ \operatorname{diag}\left(y'\,H - c'\right)\mu(t_i)\,, & i \in P & \cdots & -H'(t_i)\,, & i \in P \end{pmatrix},$$

$\mu(t) = F(t^* - t)\,A\,b$, of the system (3.4) is non-singular at every part of the constant structure. Therefore there exists the unique continuous solution (3.3). The simplest computational method of constructing (3.3) consists in the following.

Let the solution $t_i(\tau_*), \ i \in P; \ y(\tau_*)$ be known at the initial point of a part, the solution (3.3) be constructed at the nodes $\tau = \tau_* + s\,h, \ s = \overline{0, k-1}$, and at the same time there occur the inequalities

$$(4.1) \qquad t_1(\tau) > \tau, \ t_i(\tau) < t_{i+1}(\tau), \ i = \overline{1, p-1}; \ t_p(\tau) < t^*$$

$$\tau = \tau_* + s\,h, \ s = \overline{0, k-1},$$

here $h > 0$ is a parameter of the method.

To calculate the elements (3.3) at $\tilde\tau = \tau_* + k\,h$ we construct the vectors $z^l = (t_i^l, \ i \in P: y^l), \ l = \overline{1, l_0}$:

$$(4.2) \qquad z^1 = (t_i^1 = t_i(\tilde\tau - h), \ i \in P; \ y^1 = y(\tilde\tau - h)),$$

$$z^l = z^{l-1} - G^{-1}(z^{l-1}) \begin{pmatrix} f(\tilde\tau; \ t_i^{l-1}, \ i \in P; \ x^*(\tilde\tau)) \\ \\ q_j(t_i^{l-1}, \ i \in P; \ y^{l-1}), \ j \in P \end{pmatrix}, \ l = \overline{2, l_0}.$$

Set $t_i(\tau_* + k\,h) = t_i^{l_0}, \ i \in P; \ y(\tau_* + k\,h) = y^{l_0}$. The number l_0 is selected from the condition

$$\| f(\tilde\tau; \ t_i^{l_0}(\tilde\tau, \ i \in P; \ x^*(\tilde\tau)) \| \leq \varepsilon,$$

$$\left\| \left(q_j(t_i^{l_0}(\tilde\tau), \ i \in P; y^{l_0}(\tilde\tau)), \ j \in P \right) \right\| \leq \varepsilon$$

where ε is required accuracy of calculating for the terminal constraints and the optimality conditions.

Suppose that at the moment $\bar{\tau}$ one of the relations (4.1) is violated (to within h). This means a possibility of change of the structure of the defining equations and the necessity of modifications of the rules of constructing a solution. We describe only those situations where modifications are required in (4.2).

Let $\bar{\tau} = t_1(\bar{\tau})$ and for $\tau \in T^+(\bar{\tau})$ the structure does not change (or $\bar{p} = p + 1$ i.e. $t_1(\bar{\tau})$ is a new zero of the co-control). In this case for constructing the totality (3.3) at $\tau = \bar{\tau} + h$ we modify only the rules of calculating the vector z^1:

$$z^1 = (t_1^1 = \bar{\tau} + h, \ t_i^1 = t_i(\bar{\tau}), \ i = \overline{2,p}, \ y^1 = y(\bar{\tau})).$$

In the case when $t_s(\bar{\tau}) = t_{s+1}(\bar{\tau})$ and for $\tau \in T^+(\bar{\tau})$ the structure of the equations (3.4) remains (or $\bar{p} = p + 2$ i.e. $t_s(\bar{\tau})$, $t_{s+1}(\bar{\tau})$ are new zeroes of the co-control) at the point $\tau = \bar{\tau} + h$ the vectors z^l, $l = \overline{1,l_0}$; $l \neq 2$, are constructed according to (3.5). The vector z^2 is constructed as $z^2 = z^1 + \Delta z^1$, where

$$\Delta z^1 = (\Delta t_i^1, \ i \in P; \ \Delta y^-), \ \Delta t_i^1 = \mu_i, \ i \in P\backslash\{s, s+1\},$$

$$\Delta t_s^1 = -\Delta t_{s+1}^1 = \mu_s/2,$$

$$(\mu_i, \ i \in K_s, \ \Delta y^1) = -G^{-1}(t_i^1, \ i \in K_s, \ y^1) \left(\begin{array}{c} f(\bar{\tau} + h; \ t_i^1, \ i \in P; \ x^*(\bar{\tau} + h)) \\ q_j(t_i^1, \ i \in P; \ y^1), \ j \in K_s \end{array} \right),$$

$$K_s = P\backslash\{s+1\}.$$

It can be shown that $\det G(t_i^1, \ i \in K_s; y^1) \neq 0$ for $H(t_s(\bar{\tau})) \neq 0$, $\partial \dot{\Delta}_{\bar{\tau}}^0(t)/\partial t|_{t=t_i^1} \neq 0$, $i \in K_s\backslash\{s\}$.

5   The case of discrete measurements

To simplify notation we suppose that measurements are conducted at the moments of correction of the control ϑ_k, $k = \overline{0, N-1}$. In this case in the process of control only the states $x^*(\vartheta_k)$, $k = \overline{0, N-1}$, are known, and the defining equations have to be solved only at the moments ϑ_k, $k = \overline{0, N-1}$. As compared with i.4 the problem gets comlicated because the adjacent values $x^*(\vartheta_k)$ can differ significantly now.

In this connection difficulties are possible with respect to (3.4) of the method used for solving the defining equations. The following way of overcoming these difficulties at the expense of increasing the volume of work at the moments ϑ_k, $k = \overline{1, N-1}$, is suggested. For solving the defining equations at the moments $\vartheta_k (k \geq 1)$ the vector $x^*(\vartheta_k) - x^*(\vartheta_{k-1})$ is devided into M parts and the defining equations are solved M times for the vectors

$$x^*(\vartheta_k) = x^*(\vartheta_{k-1}) + q(x^*(\vartheta_k) - x^*(\vartheta_{k-1}))/M, \; q = 1, 2, \ldots, M.$$

For M large enough the vectors $x_q^*(\vartheta_k)$, $x_{q+1}^*(\vartheta_{k-1})$ are close to each other so for every q the defining equations will be solved by the Newton method in a few number of steps. Having reached $q = M$ we obtain the solution to the defining equations needed to the optimal controller at the position $(\vartheta_k, \; x^*(\vartheta_k))$.

6 An algorithm for operating the optimal controller

The controller begins its work at the moment $t = 0$ from the information corresponding to the optimal program control $u^0(t)$, $t \in T$, of the problem (2.1) - (2.3). This information can be prepared beforehand.

At an arbitrary moment $\vartheta_k (k = \overline{0, N-1})$ the controller using the results of solving the definig equations constructs the control $u^0(t|\vartheta_k, \; x^*(\vartheta_k))$, $\vartheta_k \leq t < \vartheta_{k+1}$, where $x^*(\vartheta_k)$ is the state of the system at the moment ϑ_k under the influence of the control $u^*(t)$, $0 \leq t \leq \vartheta_k$, produced before and the realized perturbation $w^*(t)$, $0 \leq t < \vartheta_k$. The control entered the dynamic system in the next interval gets the form

$$u^*(t) = u^0(t|\vartheta_k, \; x^*(\vartheta_k)), \; \vartheta_k \leq t \leq \vartheta_{k+1} \; .$$

The optimal controller acting according to this rule produces piecewise continuous controls circulating in the system closed by the optimal program-positional feedback.

7 Example

As the illustration we consider the operating of the optimal controller for calming the oscillator with the minimal fuel consumption

$$\int_0^{4\pi} u(t)\, dt \quad \rightarrow \quad \min, \; \ddot{x} + x = u, \; x(0) = 1.0, \; \dot{x} = 0.4$$

$$x(4\pi) = \dot{x}(4\pi) = 0, \; 0 \leq u(t) \leq 1, \; t \in T = [0, 4\pi].$$

Let us regard that the perturbation $w(t) = \sin 3t$, $t \in T^0 = [0, 7.0]$, $w(t) \equiv 0$, $7.0 < t \leq 4\tau$, unknown to the controller is acting to the system.

The result of the program solving the problem are given in the first row of the Table 2. In Fig. 1 the corresponding phase trajectory is given.

	$t_i,\ i = \overline{1, p}$	$u(+0)$	I^0
	1.67868, 2.223925, 7.961865, 8.50711	0	1.090491
C1	1.86, 2.56, 8.686176, 9.201702	0	1.215527
C2	2.10, 2.72, 8.393965, 8.957373	0	1.183408

Table 2: Numerical results

Under conditions of perturbations the system was controlled by the optimal positional controller (C1) [3] and by the program-positional controller (C2) with correction of the control rule at the points $\vartheta_k = 1.4\,k$, $k = \overline{0, 5}$, under continuous measurement of the system state.

The results of acting of the controllers are presented in Table 2, where the following notation is used:
$t_i,\ i = \overline{1, p}$, are switch points of the control, $u(+0)$ is the value of the control in the first interval of constancy, I^0 is the value of the control criterion. Table 2 governed by the controller C1 is represented. The phase states corresponding to the moments ϑ_k, $k = \overline{0, 5}$, are marked by the points $0, \ldots, 5$. The corresponding data for the controller C2 are represented in Fig. 3.

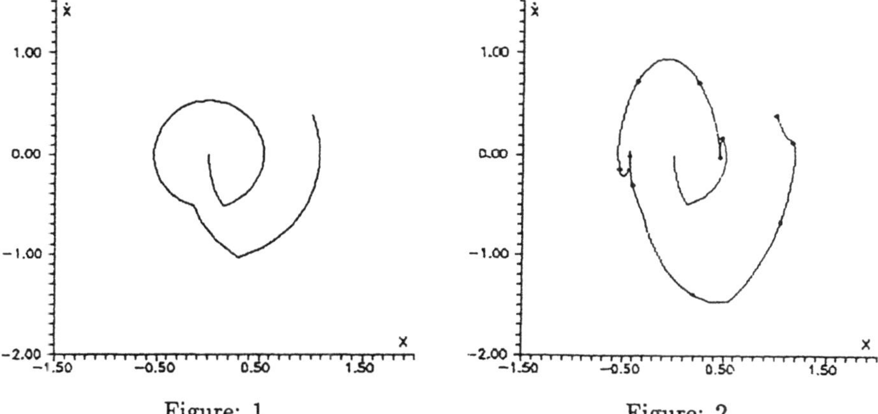

Figure: 1

Figure: 2

The defining equations were solved by the PC AT - 80286/287 with the tact frequency 16 MHz for the values $h = 0.02$, $\varepsilon = 10^{-6}$. In this case for solving the defining equation at 10 points $\tau + s\,h$, $s = \overline{0,9}$, it was required from 0.329670 sec to 0.439560 sec. The number of iterations of the Newton method needed for solving the defining equations to within ε depending on real time is given in Fig. 4.

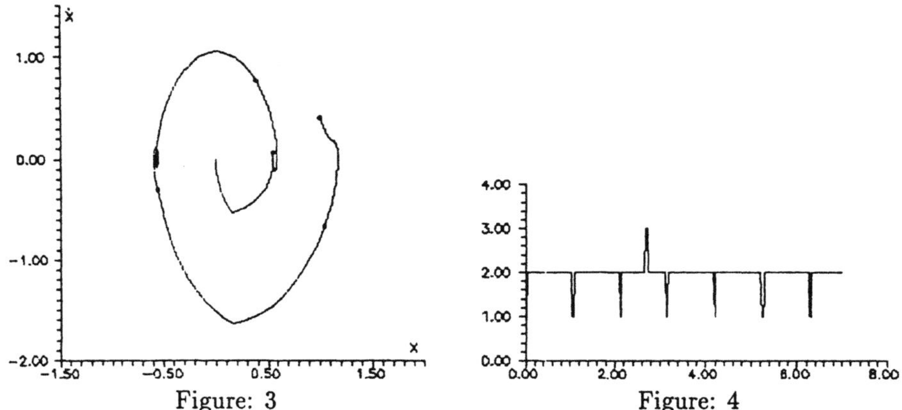

Figure: 3   Figure: 4

In Fig. 5 the behaviour of the value $\lg N_1(t)$ in the interval T^0 is presented, $N_1(t)$ is the norm of the discrepancy of the terminal constraints before solving the defining equations at the moment $t \in T^0$. The values $\lg N_2(t)$, $t \in T^0$, are given in Fig. 6, $N_2(t)$ is the norm of the discrepancy of the terminal constraints after solving the defining equations at the moment $t \in T^0$.

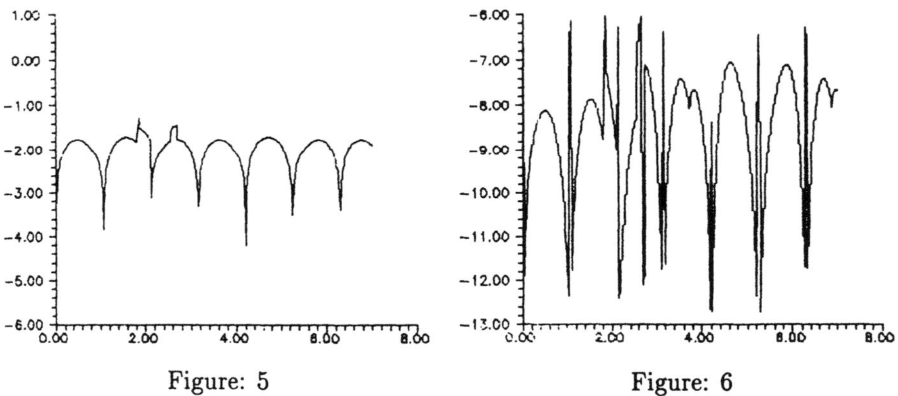

Figure: 5   Figure: 6

Using discrete measurements of the system states at the points $\vartheta_k = 1.4\,k$, $k = \overline{0,5}$, the points $x_q^*(\vartheta_k)$, $q = 1, 2, \ldots$, were selected in order to ensure convergence of the Newton method in 3 iterations for solving the defining equations.

In Table 3 the values of the norm of the discrepancy of the terminal constraints $N_1(\vartheta_2)$ accumulated to the moments ϑ_k, $k = \overline{0,5}$, under the influence of the perturbation and time required for calculating the control $u^0(t|\vartheta_k, x^*(\vartheta_k))$, $\vartheta_k \le t < \vartheta_{k+1}$, are given.

ϑ_k	$N_1(\vartheta)$	time
1.4	2.747253	1.069877
2.8	3.461538	1.504569
4.2	2.307692	1.265989
5.6	2.307692	1.205781
7.0	2.032967	0.675608

Table 3:

References

[1] A.A. Feldbaum, *Fundamentals of Theory of Optimal Automatic Systems.*
Fizmatgiz, Moscow, 1963.

[2] R. Gabasov, F.M. Kirillova, *Constructive Methods of Optimization.*
Part 2, University Press, Minsk, 1984.

[3] R. Gabasov, F.M. Kirillova and O.I. Kostyukova, *Constructing optimal feedback controls in linear problem.*
Doklady AN SSSR 320, No. 6 (1991), 1294-1299

Authors' addresses

R. Gabasov, N.V. Balashevich, Faculty of Applied Mathematics, Byelorussian State University, 220080 Minsk, Belarus.

F.M. Kirillova, Labority of Theory of Control Processes, Institute of Mathematics, 220604 Minsk, Belarus.

Synthesis of Bilinear Controlled Systems with Delay

Vladimir Kolmanovskii, Natalia Koroleva

Abstract

The optimal control problem of bilinear systems with delays in control and coordinates is considered. The construction of optimal control is reduced to the solution of initial boundary value problem for a system of ordinary and partial linear differential equations. Bilinear systems with delay only in coordinates were studied in [1].

1 Statement of the problem

Initially for the simplicity we shall consider the bilinear system

$$(1.1) \quad \dot{X}(t) = A_1(t)\, X(t-h) + A_2(t)\, u(t-h_1) + [A(t)\, X(t) + B(t)]\, u(t), \ 0 \le t \le T$$

where $X(t) \in R^n$, $u \in R^m$, matrices A, A_i, B with piecewise continuous bounded elements, delays $h \ge 0$, $h_1 \ge 0$ and instant $T > 0$ are prescribed. Initial conditions take the form

$$(1.2) \quad X_0 = \varphi \in C[-h, 0]$$

$$(1.3) \quad u_0 = \psi \in D[-h_1, 0]$$

where $C[-h, 0]$ is a space of continuous functions on $[-h, 0]$; $D[-h_1, 0]$ is a space of piecewise continuous bounded functions on $[-h_1, 0]$ with uniform metrics; functions φ, ψ are given

$$X(t) = X(t + \theta), \ -h \le \theta \le 0; \ u(t) = u(t + \zeta), \ -h \le \zeta \le 0.$$

The cost functional is

$$(1.4) \qquad J = X'(T)\, N_1\, X(t) + \int\limits_{t_0}^{T} [X'(T)\, N_2(t)\, X(t) + u'(T)\, N_0(t)\, u(t) + F(t, X_t, u_t)]\, dt$$

where prime is a transposition sign, matrices $N_i \geq 0$, $N_0 > 0$, matrices N_1, N_2 are picewise continuous and bounded. The continuous non negative functional F

$$F \,:\, R \times C[-h, 0] \times D[-h_1, 0] \,\to\, R$$

being derived by using of the generalized work principle for delay systems, see [2] , will be given below.

The choice of the concrete form of F should be fulfilled in such a way that the corresponding Bellman equation would become a linear one.

Admissible control is called any piecewise continuous function $u \,:\, [t_0, T] \,\to\, R$ such that for any $(t, \varphi, \psi) \in [t_0, T] \times C[-h, 0] \times D[-h1, 0]$ there exists the solution of Eq. (1.1) on $[t, T]$ with initial conditions $X_t = \varphi$, $u_t = \psi$ under control u. The set of all admissible control is denoted by W.

The problem is to find such admissible control that minimizes cost functional (1.4).

2   Optimality Conditions

Define the Bellman functional

$$V \,:\, [t_0, T] \times C[-h, 0] \times D[-h_1, 0] \,\to\, R$$

as follows. Let

$$X(\cdot; t, \varphi, \psi; u) \,:\, [t, T] \,\to\, R^n \quad (t \in [t_0, T],\ \varphi \in C[-h, 0],\ \psi \in D[-h_1, 0],\ u \in W)$$

be a solution of Eq. (1.1) under control u with initial conditions

$$X(t + \theta; t, \varphi, \psi; u) = \varphi(\theta),\ -h \leq \theta \leq 0,\ u(t + \zeta) = \psi(\zeta),\ -h_1 \leq \zeta \leq 0.$$

Designate

$$X_\tau(t, \varphi, \psi; u) = X(\tau + \cdot; t, \varphi, \psi; u) \,:\, [-h, 0] \,\to\, R^n.$$

Then

$$
\begin{aligned}
V(t,\varphi,\psi) \;=\; &\inf_{u\in W}\{X'(T;t,\varphi,\psi;u)\, N_1\, X(T;t,\varphi,\psi;u) \\
&+ \int_t^T [X'(s;t,\varphi,\psi;u)\, N_2(s)\, X(s;t,\varphi,\psi;u) + u'(s)\, N_0(s)\, u(s) \\
&+ F(s,X_s(t,\varphi,\psi;u),u_s)]\, ds\}.
\end{aligned}
$$

Introduce the operator

$$
L_u V(t,\varphi,\psi) = \overline{\lim_{\Delta\to 0+}}\,\frac{1}{\Delta}\,[V(t+\Delta, X_{t+\Delta}(t,\varphi,\psi;u), v_{t+\Delta}) - V(t,\varphi,\psi)].
$$

Here $v_{t+\Delta}$ is a control equal u on $[t, t+\Delta]$ and equal φ on $[t+\Delta - h, t]$. Remark that operator $L_u V$ is a full derivative of the functional $V(t,\varphi,\psi)$ along the trajectories of Eq. (1.1) under control u .

Application of dynamic programming method leads us to the following optimality conditions.

Theorem 2.1. Let there exist the functional

$$
V_0 \;:\; [t_0,T] \times C[-h,0] \times D[-h_1,0] \;\to\; R,
$$

satisfying the local Lipschitz condition and the functional

$$
u_0 \;:\; [t_0,T] \times C[-h,0] \times D[-h_1,0] \;\to\; R,
$$

$$
\begin{aligned}
(2.1)\quad \inf_{u\in R}&[L_u\, V_0(t,\varphi,\psi) + \varphi'(0)\, N_2(t)\, \varphi(0) + u'\, N_0(t)\, u + F(t,\varphi,\psi)] \\
&= L_{u_0}\, V_0(t,\varphi,\psi) + \varphi'(0)\, N_2(t)\, \varphi(0) + u_0'(t,\varphi,\psi)\, N_0(t)\, u_0(t,\varphi,\psi) \\
&\quad + F(t,\varphi,\psi) \\
&= 0
\end{aligned}
$$

$$
(2.2)\quad V_0(T,\varphi,\psi) = \varphi'(0)\, N_1\, \varphi(0).
$$

Then $u_0(t,\varphi,\psi)$ is an optimal control and $V_0(t,\varphi,\psi)$ a Bellman functional for the problem (1.1) - (1.4) .

Remark. Optimalty conditions formulated above will be valid for the case $u \in R^m$ as well. In this connection A is a tensor with components a_{jk}^i ; the product $A\,X\,u$ is a vector with components $\sum_{j,k} a_{jk}^i\, X_j\, u_k$, $i = 1,\dots,n$ and infimum in (2.1) is calculated with respect to the vector $u \in R^m$.

3   Construction of the solution

Let us try to seek the solution $V(t, \varphi, \psi)$ of the problem (2.1)- (2.2) in the form

$$
(3.1) \quad V_0(t, \varphi, \psi) = \varphi'(0)\, P(t)\, \varphi(0) + \varphi'(0) \int_{-h}^{0} Q(t, \tau)\, \varphi(\tau)\, d\tau
$$

$$
+ \varphi(0) \int_{-h}^{0} Q'(t, \tau)\, \varphi'(\tau)\, d\tau + \int_{-h}^{0} \int_{-h}^{0} \varphi'(\tau)\, R(t, \tau, \tau_1)\, \varphi(\tau_1)\, d\tau\, d\tau_1
$$

$$
+ \varphi'(0) \int_{-h_1}^{0} L(t, \rho)\, \psi(\rho)\, d\rho + \varphi(0) \int_{-h_1}^{0} L'(t, \rho)\, \psi'(\rho)\, d\rho
$$

$$
+ \int_{-h}^{0} \int_{-h_1}^{0} \varphi'(\tau)\, K(t, \tau, \rho)\, \psi(\rho)\, d\tau\, d\rho
$$

$$
+ \int_{-h}^{0} \int_{-h_1}^{0} \varphi(\tau)\, K'(t, \tau, \rho)\, \psi'(\rho)\, d\tau\, d\rho
$$

$$
+ \int_{-h_1}^{0} \int_{-h_1}^{0} \psi'(\rho)\, M(t, \rho, \rho_1)\, \psi(\rho_1)\, d\rho\, d\rho_1.
$$

All matrices in (3.1) are assumed to be piecewise continuously differentiable bounded functions and $P(t) \geq 0$. Substitute (3.1) into (2.1) and find the control u_0 minimizing (2.1). Then we get

$$
(3.2) \quad u_0(t, \varphi, \psi) = -N_0^{-1}(t)\{[(A(t)\, X + B(t))'\, P(t) + L'(t, 0)]\, X
$$

$$
+ \int_{-h}^{0} [K'(t, \tau, 0) + (A(t)\, X + B(t))'\, Q(t, \tau)]\, \varphi(\tau)\, d\tau
$$

$$
+ \int_{-h_1}^{0} [M'(t, \tau, 0) + (A(t)\, X + B(t))'\, L(t, \tau)]\, \psi(\rho)\, d\rho\}
$$

where $X = \varphi(0)$.

Take now F in (1.4) to be equal to

$$
(3.3) \quad f(t, X_t, u_t) = N_0^{-1}(t)\{[(A(t)\, X(t) + B(t))'\, P(t) + L'(t, 0)]\, X(t)
$$

$$
+ \int_{-h}^{0} [K'(t, \tau, 0) + (A(t)\, X(t) + B(t))'\, Q(t, \tau)]\, X(t + \tau)\, d\tau
$$

$$
+ \int_{-h_1}^{0} [M'(t, \tau, 0) + (A(t)\, X(t) + B(t))'\, L(t, \tau)]\, u(t + \rho)\, d\rho\}^2.
$$

To define the coefficients of the functional V_0 substitute (3.1) - (3.3) into (2.1). (2.2) and equate to zero the coefficients of corresponding quadratic forms of variables φ and ψ. As a result we obtain a system of equations.

$$(3.4) \qquad \dot{P}(t) + Q(t,0) + Q'(t,0) + N_2(t) = 0$$

$$\left(\frac{\partial}{\partial t} - \frac{\partial}{\partial \tau}\right) Q(t,\tau) + R(t,0,\tau) = 0$$

$$\left(\frac{\partial}{\partial t} - \frac{\partial}{\partial \tau} - \frac{\partial}{\partial \tau_1}\right) R(t,\tau,\tau_1) = 0$$

$$\left(\frac{\partial}{\partial t} - \frac{\partial}{\partial \tau} - \frac{\partial}{\partial \rho}\right) K(t,\tau,\rho) = 0$$

$$\left(\frac{\partial}{\partial t} - \frac{\partial}{\partial \rho} - \frac{\partial}{\partial \rho_1}\right) M(t,\rho,\rho_1) = 0$$

$$\left(\frac{\partial}{\partial t} - \frac{\partial}{\partial \rho}\right) L(t,\rho) + K(t,0,\rho) = 0$$

$$0 \leq t \leq T, \; -h_1 \leq \rho, \; \rho_1 \leq 0, \; -h \leq \tau, \; \tau_1 \leq 0.$$

Analogously equating to zero quadratic forms of variables $X(t-h)$ and $u(t-h_1)$ we get the boundary conditions

$$(3.5) \qquad P(T) = N_1; \; Q(T,\tau) = R(T,\tau,\tau_1) = L(T,\rho) = K(T,\tau,\rho) = M(T,\rho,\rho_1) = 0$$

$$-h < \tau, \; \tau_1 \leq 0; \; -h_1 < \rho, \; \rho_1 \leq 0$$

$$-Q(t,-h) + P(t)\, A_1(t) = 0, \; R(t,\tau,\rho) = R'(t,\rho,\tau)$$
$$-R(t,-h,\tau) - R'(t,\tau,-h) + 2\, A_1'(t)\, Q(t,\tau) = 0$$
$$-L(t,-h_1) + P(t)\, A_2(t) = 0$$
$$-K(t,-h,\rho_1) + A_1'(t)\, L(t,\rho_1) = 0$$
$$-K(t,\tau,-h_1) + Q'(t,\tau)\, A_2(t) = 0$$
$$-M(t,-h_1,\rho) - M'(t,\rho,-h_1) + 2\, A_2'(t)\, L(t,\rho) = 0$$
$$M(t,\rho,\rho_1) = M'(t,\rho,\rho_1)$$
$$0 \leq t \leq T; \; -h \leq \tau; \; \tau_1 \leq 0; \; -h_1 \leq \rho; \; \rho_1 \leq 0.$$

Under our assumptions by virtue of [3] there exists the unique solution of the problem
(3.4), (3.5) at the class of piecewise continuously differentiable bounded functions with
$P(t) \geq 0$. From here and [4] it follows the existence of a solution of the problem (1.1) -
(1.3) under control u in some neighborhood of the initial point $t = 0$. It can be proven
that this solution can be continued to the closed interval $[0, T]$. This means that the
control u_0 is admissible and optimal. The derivative of the functional V_0 along (1.1)
with u_0 is nonpositive. Hence

$$V_0(t, X_t, u_{0t}) \leq V_0(0, \varphi, \psi).$$

From here and (2.1) it follows that

$$\int\limits_0^t u_0'(s)\, N_0(s)\, u_0(s)\, ds \leq 2\, V_0(0, \varphi, \psi).$$

Here $u_0(s) = u_{(}s, X_s, u_{0s})$. But the matrix $N_0(s)$ is uniformly positive definite. Hence
for some positive C

$$\int\limits_0^t u_0^2(s)\, ds \leq C\, V_0(0, \varphi, \psi).$$

Thus at any interval $[0, t]$ where the solution of the problem (1.1) - (1.3) exists under
control u_0 this control along the solution is uniformly in t square integrable. Therefore the trajectory X corresponding u_0 can be interpreted as a solution of the linear
equation (1.1) (where u is changed into u_0) with coefficients square integrable on $[0, T]$.
Consequently the solution $X(t)$ of the problem (1.1) - (1.3) under control u_0 can be
continued at the closed interval $[0, T]$.

Thus, the solution of the initial optimal control problem being reduced to the boundary
problem (3.4), (3.5) is given by the formulae (3.1), (3.2).

Consider first of all the particular case for which a solution of the problem (3.4), (3.5)
can be represented analytically. Assume that (1.1) has delay h_1 only in control

$$\dot{X}(t) = [A(t)\, X(t) + B(t)]\, u(t) + A_2(t)\, u(t - h_1).$$

Then the solution of the corresponding Bellman equation (2.1), (2.2) is

$$\begin{aligned}
V_0(t, \varphi, \psi) \;=\; & \varphi'(0)\, P(t)\, \varphi(0) \\
& + \varphi'(0) \int\limits_{-h_1}^0 L(t, \rho)\, \psi(\rho)\, d\rho + \varphi(0) \int\limits_{-h_1}^0 L'(t, \rho)\, \psi'(\rho)\, d\rho \\
& + \int\limits_{-h_1}^0 \int\limits_{-h_1}^0 \psi'(\rho)\, M(t, \rho, \rho_1)\, \psi(\rho_1)\, d\rho\, d\rho_1 .
\end{aligned}$$

In view of (3.2) the optimal control $u_0(t, X_t, U_{0t})$ takes the form

$$U_0(t, X_t, U_{0t}) \;=\; -N_0^{-1}(t)\,[(A(t)\,X + B(t))'\,(P(t)\,X(t) + \int_{-h_1}^{0} L(t,\rho)\,U_0(t+\rho)\,d\rho)$$

$$+ \int_{-h_1}^{0} M(t,0,\rho)\,U_0(t+\rho)\,d\rho + L'(t,0)\,X(t)].$$

Matrices P, L, M satisfy the equation

$$(3.6) \quad \dot{P}(t) + N(t) = 0,\; 0 \le t \le T$$

$$\left(\frac{\partial}{\partial t} - \frac{\partial}{\partial \rho}\right) L(t,\rho) = 0 \qquad -h_1 \le \rho,\; \rho_1 \le 0$$

$$\left(\frac{\partial}{\partial t} - \frac{\partial}{\partial \rho} - \frac{\partial}{\partial \rho_1}\right) M(t,\rho,\rho_1) = 0$$

with boundary conditions

$$(3.7) \quad P(T) = N_1;\; L(T,\rho) = 0,\; M(T,\rho,\rho_1) = 0$$
$$0 < t \le T;\; -h_1 < \rho,\; \rho_1 \le 0$$

$$P(t)\,A_2(t) - L(t,-h_1) = 0$$

$$2\,A_2'(t)\,L(t,\rho) - M(t,-h_1,\rho) - M'(t,-h_1,\rho) = 0.$$

The solution of the problem (3.6), (3.7) is

$$P(t) = N_1 + \int_{t}^{T} N_2(s)\,ds,\; 0 \le t \le T$$

$$L(t,\rho) = P(t + \rho + h_1)\,A_2(t + \rho + h_1) \quad \text{for } t + r + h < T,$$

$$L(t,\rho) = 0 \quad \text{for } t + \rho + h_1 \ge T$$

$$M(t,\rho,\rho_1) = A_2'(t + \rho + h_1)\,P(t + \beta + h)\,A_2(t + \rho + h_1),\; \text{if } w < 0$$

and $M(t,\rho,\rho_1) = 0$ if $w \ge 0$

where $w = \max(t + \rho + h_1 - T, t + \rho_1 + h_1 - T),\; b = \max(\rho,\rho_1).$

Now let us turn to the general case (3.4), (3.5). Represent the solution of the problem (3.4), (3.5) as a sum of two solutions. First for $N_2 = 0$ and arbitrary matrix $N_1 \geq 0$ and second for $N_1 = 0$ and arbitrary matrix $N_2 \geq 0$. First the solution (for $N_2 = 0$) has a form

$$(3.8) \quad P(t) = b'(t)\, N_1\, b(t), \quad Q(t,\tau) = -b'(t)\, N_1\, \dot{b}(t+\tau),$$

$$R(t,\tau,\tau_1) = \dot{b}'(t+\tau)\, N_1\, \dot{b}(t+\tau_1), \ 0 \leq t \leq T, \ -h \leq \tau,\ \tau_1 \leq 0$$

$$L(t,\rho) = b'(t)\, N_1\, b_1(t+\rho+h_1),$$

$$K(t,\tau,\rho) = -\dot{b}'(t,\tau)\, N_1\, b_1(t+\rho+h_1)$$

$$M(t,\rho,\rho_1) = b_1'(t+\rho+h_1)\, N_1\, b_1(t+\rho_1+h_1)$$

where

$$b_1(t+\rho+h_1) = \begin{cases} b(t+\rho+h_1)\, A_2(t+\rho+h_1)\,, & t+\rho+h_1 \leq \min(T,\rho+h_1) \\ 0 & ,\quad t+\rho+h_1 > \min(T,\rho+h_1). \end{cases}$$

The matrix $b(t)$ is a solution of the Cauchy problem

$$\dot{b}(t) = -b(t+h)\, A_1(t+h), \ b(T) = I, \ b(s) \equiv 0, \ s > T.$$

To construct the second solution (for $N_1 = 0$) denote $t_i = T - i\,h, \ i = 0,1,\ldots$; and introduce matrices L_1, M_1, A_3, A_4. For $t_{i+1} \leq t \leq t_i$ the function

$$\begin{aligned} A_3(t+\tau+h) &= 0 & &\text{if } -t+t_{i+1} < \tau \leq 0 \quad \text{and} \\ A_3(t+\tau+h) &= A_1(t+\tau+h) & &\text{if } -h \leq \tau \leq -t+t_{i+1}. \end{aligned}$$

Analogously

$$\begin{aligned} A_4(t+\rho+h_1) &= A_2(t+\rho+h_1) & &-h_1 \leq \rho \leq -t+t_{i+1}, \\ A_4(t+\rho+h_1) &= 0 & &-t+t_{i+1} < \rho < 0,\ t_{i+1} \leq t \leq t_i. \end{aligned}$$

As last

$$\begin{aligned} M_1(t_i, t+\rho-t_i, \rho) &= 0 \\ L_1(t_i, t+\rho-t_i) &= 0 \\ K_1(t_i, \tau, t+\rho-t_i) &= 0 & &\text{if } -h_1 \leq \rho \leq -t+t_{i+1} \end{aligned}$$

and

$$\begin{aligned} K_1(t_i, \tau, t+\rho-t_i) &= K(t_i, \tau, t+\rho-t_i) \\ M_1(t_i, t+\rho-t_i, \rho) &= M(t_i, t+\rho-t_i, \rho) \\ L_1(t_i, t+\rho-t_i) &= L(t_i, t+\rho-t_i) & &\text{if } -t+t_{i+1} \leq \rho \leq 0,\ t_{i+1} \leq t \leq \end{aligned}$$

Then for $t_{i+1} \leq t \leq t_i$ we get

$$(3.9) \quad P(t) = P(t_i) + \int_t^{t_i} N_2(s)\, ds + \int_{t-t_i}^0 [Q(t_i, s) + Q'(t_i, s)]\, ds$$

$$+ \int_{t-t_i}^0 \int_{t-t_i}^0 R(t_i, s, \alpha)\, ds\, d\alpha$$

$$Q(t, \tau) = [P(t_i) + \int_{t+\tau+h}^{t_i} N_2(s)\, ds + \int_{t-t_i}^0 Q(t_i, s)\, ds + \int_{t-t_{i+1}+\tau}^0 Q'(t_i, s)\, ds$$

$$+ \int_{t-t_i}^0 d\alpha \int_{t-t_{i+1}+\tau}^0 R(t_i, s, \alpha)\, ds\,] A_3(t + \tau + h)$$

$$+ Q_1(t_i, t + \tau - t_i) + \int_{t-t_i}^0 R_1(t_i, s, t + \tau - t_i)\, ds.$$

Here Q_1, R_1 are defined by the relations

$$Q_1(t_i, \tau + t - t_i) = Q(t_i, \tau + t - t_i)$$
$$R_1(t_i, \tau + t - t_i, \tau_1) = R(t_i, \tau + t - t_i, \tau_1), \quad -t + t_{i+1} \leq \tau \leq 0$$

and

$$Q_1(t_i, \tau + t - t_i) = 0$$
$$R_1(t_i, \tau + t - t_i, \tau_1) = 0, \quad -h \leq \tau \leq -t + t_{i+1}; \quad -h \leq \tau_1 \leq 0, \quad t_{i+1} \leq t \leq t_i.$$

Further more for $t_{i+1} \leq t \leq t_i$ define

$$R_2(t_i, \tau + t - t_i, \tau_1 + t - t_i) = R(t_i, \tau + t - t_i, \rho + t - t_i)$$
$$\text{if } t_{i+1} \leq t \leq t_i, \quad -t + t_{i+1} \leq \tau, \quad \tau_1 \leq 0$$
$$\text{and } R_2(t_i, \tau + t - t_i, \tau_1 + t - t_i) = 0$$

even one of the arguments τ, τ_1 does not belong to the interval $[-t + t_{i+1}, 0]$. Then $R(t, \tau, \tau_1)$ has the form

$$(3.10) \quad R(t, \tau, \tau_1) = A_3'(t + \tau + h) \cdot [\ \int_{t+h+\max(\tau, \tau_1)}^{t_i} N_2(s)\, ds + P(t_i)$$

$$+ \int_{t-t_{i+1}+\tau_1}^0 Q'(t_i, s)\, ds + \int_{t-t_{i+1}+\tau}^0 Q(t_i, s)\, ds$$

$$+ \int_{t-t_{i+1}+\tau}^0 d\alpha \int_{t-t_{i+1}+\tau_1}^0 R(t_i, s, \alpha)\, ds] \cdot A_3(t + \tau_1 + h)$$

$$+A'_3(t + \tau + h) \cdot [Q'_1(t_i, t + \tau_1 - t_i)$$
$$+ \int\limits_{t-t_{i+1}+\tau}^{0} R_1(t_i, s, t + \tau_1 - t_i)\, ds]$$
$$+A_3(t + \tau_1 + h) \cdot [Q_1(t_i, t + \tau - t_i)$$
$$+ \int\limits_{t-t_{i+1}+\tau}^{0} R_1(t_i, s, t + \tau - t_i)\, ds]$$
$$+R_2(t_i, t + \tau - t_i, t + \tau_1 - t_i).$$

For $L(t, \rho)$ and $K(t, \tau, \rho)$ we get

$$(3.11) \quad L(t, \rho) = [P(t_i) + \int\limits_{t+\rho+h_1}^{t_i} N_2(s)\, ds + \int\limits_{t-t_i}^{0} Q'(t_i, s)\, ds + \int\limits_{t-t_{i+1}+\rho}^{0} Q(t_i, s)\, ds$$
$$+ \int\limits_{t-t_i}^{0} d\alpha \int\limits_{t-t_{i+1}+\rho}^{0} R(t_i, s, \alpha)\, ds \,] \, A_4(t + \rho + h)$$
$$+L_1(t_i, t + \rho - t_i) + \int\limits_{t-t_i}^{0} K_1(t_i, s, t + \rho - t_i)\, ds$$

$$(3.12) \quad K(t, \tau, \rho) = A'_3(t + \tau + h) \cdot [\int\limits_{t+h+\max(\tau+h,\rho+h_1)}^{t_i} N_2(s)\, ds + P(t_i)$$
$$+ \int\limits_{t-t_{i+1}+\tau}^{0} Q'(t_i, s)\, ds + \int\limits_{t-t_{i+1}+\rho}^{0} Q(t_i, s)\, ds$$
$$+ \int\limits_{t-t_{i+1}+\tau}^{0} d\alpha \int\limits_{t-t_{i+1}+\rho}^{0} R(t_i, s, \alpha)\, ds] \cdot A_4(t + \rho + h_1)$$
$$+A'_3(t + \tau + h) \cdot [L_1(t_i, t + \rho - t_i)$$
$$+ \int\limits_{t-t_{i+1}+\rho}^{0} K_1(t_i, s, t + \tau - t_i)\, ds]$$
$$+A_4(t + \rho + h_1) \cdot [Q'_1(t_i, t + \tau - t_i)$$
$$+ \int\limits_{t-t_{i+1}+\rho}^{0} R_1(t_i, s, t + \tau - t_i)\, ds]$$
$$+K_2(t_i, t + \tau - t_i, t + \rho - t_i).$$

At last the matrix M is

$$(3.13) \quad M(t, \rho, \rho_1) = A'_4(t + \rho + h) \cdot [\int\limits_{t+h_1+\max(\rho,\rho_1)}^{t_i} N_2(s)\, ds + P(t_i)$$

$$+ \int_{t-t_{i+1}+\rho}^{0} Q'(t_i, s)\, ds + \int_{t-t_{i+1}+\rho}^{0} Q(t_i, s)\, ds$$

$$+ \int_{t-t_{i+1}+\rho}^{0} d\alpha \int_{t-t_{i+1}+\rho_1}^{0} R(t_i, s, \alpha)\, ds] \cdot A_4(t + \rho_1 + h_1)$$

$$+ A_4'(t + \rho + h_1) \cdot [L_1(t_i, t + \rho_1 - t_i)$$

$$+ \int_{t-t_{i+1}-\rho}^{0} K_1(t_i, s, t + \rho_1 - t_i)\, ds]$$

$$+ A_4(t + \rho_1 + h_1) \cdot [L_1'(t_i, t + \rho - t_i)$$

$$+ \int_{t-t_{i+1}+\rho_1}^{0} K_1'(t_i, s, t + \rho - t_i)\, ds]$$

$$+ M_2(t_i, t + \rho - t_i, t + \rho_1 - t_i).$$

Here $\quad M_2(t_i, t + \rho - t_i, t + \rho_1 - t_i) = M(t_i, t + \rho - t_i, t + \rho_1 - t_i)$

$\qquad$ for $t_{i+1} \le t \le t_i$, $-t + t_{i+1} \le \rho$, $\rho_1 \le 0$

and $\quad M_2(t_i, t + \rho - t_i, t + \rho_1 - t_i) = 0,$

$\qquad$ if even one of the ρ, ρ_1 does not belong to the interval $[-t + t_{i+1}, 0]$.

Recurrent formulae (3.9) - (3.13) allows us to obtain the second solution of the problem (3.4), (3.5) sucessively on the interval $[t_{i+1}, t_i]$, $i = 0, 1, \ldots$. The sum of both solutions is a general one. In particular for $T - h \le t \le T$ we get

$$P(t) \;=\; \int_t^T N_2(r)\, dr + N_1$$

$$Q(t, \tau) \;=\; [\int_{t+\tau+h}^T N_2(r) + N_1] \cdot A_3(t + \tau + h)$$

$$R(t, \tau, \tau_1) \;=\; A_3'(t + \tau + h) \cdot [\int_{t+h+\max(\tau, \tau_1)}^T N_2(r)\, dr + N_1] \cdot A_3(t + \tau_1 + h)$$

$$L(t, \rho) \;=\; [\int_{t+\rho+h_1}^T N_2(r)\, dr + N_1] \cdot A_4(t + \rho + h_1)$$

$$K(t, \tau, \rho) \;=\; A_3'(t + \tau + h) \cdot [\int_{t+\max(\tau+h, \rho+h_1)}^T N_2(r)\, dr + N_1] \cdot A_4(t + \rho + h_1)$$

$$M(t, \rho, \rho_1) \;=\; A_4'(t + \rho + h_1) \cdot [\int_{t+h_1+\max(\rho, \rho_1)}^T N_2(r)\, dr + N_1] \cdot A_4(t + \rho_1 + h_1).$$

4   Some generalizations

The obtained results can be generalized for another type of controlled systems. Consider for example the equation

$$(4.1) \quad \dot{X}(t) = A_1(t)\, X(t-h) + A_2(t)\, u(t-h_1) + (A(t, X_t) + B(t))\, u(t), \ 0 \le t \le T$$

where the functional $A : R \times C[-h_2, 0] \to R$ is measurable, piecewise continuous in t, satisfies the Lipschitz condition in the second argument and $X_t = X(t+\tau)$, $-h_2 \le \tau \le 0$.

The solution of Eq. (4.1) is defined by initial conditions (1.3) and $X_0 = \varphi \in C[-\max(h_1, h_2), 0]$.

The cost functional takes the form (1.4) with F given by (3.3) where $A(t)\, X(t) + B(t)$ is changed by $A(t, X_t)$. The optimal control is given by (3.2) where $A(t)\, X(t) + B(t)$ is changed $A(t, \varphi)$. The Bellman functional (3.1) is just the same as earlier. In particular it means that the functional (3.2) does not depend on h_2 though optimal control and trajectory do depend on h_2. Hence for $h = 0$ the optimal value of the cost functional does not depend generally on the initial function $\varphi(s)$, $s < 0$ though the optimal trajectory and control depend on it essentially.

The proposed algorithm was applied to a biological reactor described by bilinear equations with delay.

References

[1] Kolmanovskij V.B., Koroleva N.I. *Optimal control of bilinear hereditary systems of their applications.* Applied Mathematic Letters 4, No. 2 (1991), 1-3.

[2] Andreeva E.A., Kolmanovskij V.B., Shaikhet L.E. *Control of systems with delay.* Nauka. Moscow Main Editorial board (1992).

[3] Kolmanovskij V.B., Nosov V.R. *Stability of functional differential equations.* N-Y., London, Academic Press, 1986.

[4] Kolmanovskij V.B., Mainzenberg T.L. *Optimal control of stochastic systems with aftereffect.* Avtomat., Telemekh., No. 1 (1973).

Authors' addresses

Prof., Dr. Vladimir Kolmanovskij, Department of Cybernetics, Moscow Institute of Electronic Engineering, Bolshoy Vuzovskij, 3/12, Moscow, Russia, 109028

Dr. Natalia Koroleva, Department of Cybernetics, Moscow Institute of Electronic Engineering, Bolshoy Vuzovskij, 3/12, Moscow, Russia, 109028

Constructing Feedback Control in Differential Games by Use of "Central" Trajectories

G. Sonnevend

Abstract

A new approach is developed for the solution of a large class of feedback control problems arising in differential games of stabilization, with pointwise bounded controls and measurement errors based on the notion of analytic centers and central trajectories.

1   Introduction

Differential games are one of the most natural idealizations of various problems in the area of feedback control of uncertain dynamical systems, where we have two (or more) different inputs, which are chosen by independent — in fact "unpredictable" or conflicting actors — and the problem of a worst case "strategy" (feedback control) for each of them arises. A distinctive feature of this idealization — as compared to other, more traditional approaches to the problems of feedback control — is the assumption that the actuator (control) input, the uncertain "disturbance" input and measurement error acting on the system are *bounded* (must satisfy "saturation bounds") i.e. belong, at each moment of time, to specified known closed sets, as well as the state vector of the system is also known (or desired) to be confined in a specified closed set; no probabilistic assumptions are made on the inputs (note that incorporating "integral" constraints on the inputs is always possible by transforming them to state constraints).

We shall restrict our study to linear, time invariant control systems

$$(1.1) \quad \dot{x} \;=\; Ax + Bu + Dv, \quad u(t) \in P, \quad v(t) \in Q,$$

$$(1.2) \quad y \;=\; Cx + Ge, \quad e(t) \in S,$$

where $x \in R^n$ is the state vector, y is the vector of measured outputs, u is the control (actuator) input, v is the disturbance, i.e. the uncertain control of the opponent or "nature", and e is a "measurement error" vector. The values u, v, e should resp. are assumed, i.e. to belong to closed sets P, Q, S. The matrices A, B, C, D, G are assumed to be known exactly. In fact, the methods we introduce here are (for most parts) applicable to more general classes of nonlinear, time variant,... systems.

Here we shall present the basic ingredients of these methods, they have been applied with success to (i.e. have been tested numerically) on a wide variety of nontrivial low dimensional problems and on a class of high dimensional "linear programming games", but here — due to lack of space — we can give just a few applications and refer to a larger report (see below on the first page) on results of extensive computer (simulation) tests, which also includes some of the source files, written in Matlab allowing the reader to make his own experiments. More specifically, we study here mainly the following:

Stabilization problem

Find an algorithm (esp. a positional feedback control) to compute $u(t)$, based on present and past values of $y(s), s \leq t$, and $x(0)$ such that, for given convex, quadratic functions $K_1, \ldots, K_m$ and real numbers $d_1, \ldots, d_m$,

$$(1.3) \qquad x(t) \in K(d_1, \ldots, d_m) := \{z | K_i(z) \leq d_i, i = 1, \ldots, m, \}$$

holds for all $t \geq 0$, whenever $x(0) \in K(d_1, \ldots, d_m)$.

Below we shall also consider a more general situation, where the inclusion (1.3) is to be satisfied by a suitable feedback control $u(\cdot)$ under the condition, that the numbers $d_i, i = 1, \ldots, m$, are replaced by uncertain functions $d_i(t)$ for which only the following a priori information is avaible

$$(1.4) \qquad |d_i'(t)| \leq \delta_i, \quad i = 1, \ldots, m, \quad t \geq 0,$$

—where $\delta_i, \ i = 1, \ldots, m$ are fixed, known constants. Of course for fixed sets P, Q, S it is not always possible to satisfy these bounds, and a more realistic problem is to analyze the interplay ("payoff" relations) between the magnitude of the achievable $\delta^m = (\delta_1, \ldots, \delta_m)$ and P, Q, S. Our aim is thus to deal with the analogons of the problems solved in the wellknown H^∞ theory, where only integral, quadratic bounds on the inputs and states are imposed; due to lack of space we study here in detail the example of a "surveillance" problem for the well known game of the "boy and the crocodile" and the problem of robust stabilization of an unstable pendulum.

The main tool, whose use and effectivity we try to illustrate here, is the notion of an *analytic center* of a finite or semiinfinite system of convex, analytic (here quadratic) inequalities in R^n, and that of *central paths* corresponding to oneparametric families of the latter.

This tool has proved itself to be very effective for the solution of linear (and more general convex, analytic) programs as one of the basic tools of a new class of recently developed "interior point" methods for the latter, especially in the case of high dimensions. Below we shall demonstrate that another advantage of this tool consist of the relative ease by which nearly optimal feasible solutions can be updated when the constraints change with time. In fact this good updatability is one of the basic new features and that hints to the effectivity of the method proposed below for feedback control. We shall present an analysis of a linear programming game which models the problem of scheduling an inertial production in case of quickly changing and not exactly measurable demands.

An important element of our approach is to exploit that the sets $H = P$, Q or S are algebraically simple; more precisely, that they are just — like the sets in (1.3) — finite intersections of ellipsoids (e.g. half spaces).

$$H = \{h|H_i(h) \leq \gamma_i, i = 1, \ldots, r\}$$

where H_i, $i = 1, \ldots, r$, are convex quadratic in h. Thus polyhedral sets are special cases in which all H_i are linear. The values of the feedback controls we build will be interior points of the set P, but in fact they will approximate bang-bang controls surprisingly well.

We stress that our interior point methodology is quite different from previous approaches to deal numerically with the given class of problems, which emphasized (and tried to implement numerically) general notions of "nondifferentiable convex analysis". We exploit that each elementary constraint is (individually) algebraically simple and reject to use local, nondifferentiable objects (since they usually lead to combinatorial blow up when the dimension of the state vector becomes large, which happens for most known methods connected with the computation of the fine structure of the value function or stable bridges, alternating integrals, etc...). In contrast to the theoretical approaches like [1], we are interested to construct and use only notions and tools which are implementable (also in *higher* dimensional situations). Of course, local conditions of optimality and stability (via subdifferentials, contingent cones, e.t.c.) are interesting, we are however not aware of any feasible numerical method which implements them and is based solely on these notions and the more or less well known relations governing them.

In [15] we outlined the applicability of interior point methods using analytical centers to a larger class of feedback control problems. In fact we think that interior point methods will turn out to be more efficient also for the much simpler problems of optimal control under state constraints and bounded control.

In order to understand the difficulties connected with a *local* approach using nondifferentiable techniques, it is enough to think about the problem of state or input reconstruction under observation errors, which might be arbitrary measurable or piecewise continuous functions: using the maximum principle to compute a "corner-optimal" recovery would lead to enormous (noncountable) number of switching points for the adjoint variables. For phase constraints complicated singular arcs appear as a rule,

..., see e. g. [4], and it is clear, that often in such an approach the structure is too complicated and unnecessarily fine.

In this moment we are not yet able to present such a full, complete theory, as exists for the games (1.1)-(1.2) with linear quadratic integral costs via H^∞ see e.g. [10]. It is known, that bounded control games are much more complicated than those with no bounds on the controls and states. Attempts to build up an analogous H^1 theory are described in [6], the *linear* feedback controllers proposed there seem not to be suitable to produce good approximations to optimal strategies (which should be of the bang-bang type, as our examples below will demonstrate). Informations about the present state of art could be drawn e.g. from the two proceedings volumes [2], [3]. A general, in a sense "complete" theory, in which basis facts about the existence, the characterization and the properties of the optimal strategies are clarified (and a lot of other, interesting, but more partial approaches) have been developed in the Soviet Union, see[7], [5] and [21] for a survey of these results and methods.

In a theoretical sense, differential games is now a "grown up" discipline, which is ready for its application to traditional fields of feedback control. The word traditional is used to stress that the differential game approach can be applied not only to pursuit evasion games, but to a wide variety of problems in the field of process control engineering, where in a worst case design approach nature is regarded as the uncertain "opponent". There is a relevant, large body of literature on "Robust Control" methods, see e.g. the proceedings [15] on more direct, but partial methods dealing with bounded "uncertainties". A classical, engineering approach was to use "loop shaping" in a quadratic optimal, linear filter and controller, i.e. to use suitable weight therein, determined by sucessive trials, for a number of different linearization points — to achieve desired performance and satisfy given saturation bounds. Studying minimax optimal control problems and resolving the arising two - point boundary problems in real time by homotopy and Newton's method techniques, see [4], can yield in specific examples satisfactory results. Some, not too complicated problems could be solved approximately by a suitable parametrization of the controller and subsequent application of nonlinear programming techniques to optimize a finite number of parameters, see e.g. [23],[24]. Recent advances in the theory of Hamilton - Jacobi equations, see [21], seem to be interesting, even if the numerical solution of these equation via finite difference or finite element techniques seem to be viable only for low dimensions.

One of the main aspect of the differences of the proposed method with the "traditional" approaches is that we do not need to use any analogon of the "backward" procedures used therein to compute the value function for all, possible positions (t, x), in this sense the primary object for us is the feedback control itself and the implicitly arising invariant set (or level set of the value function) is secondary. It turns out that with the proposed schemes one can find *several* feedback controllers which assure within arbitrary accuracy the *same, optimal* bounds in (1.1)-(1.4) without the need to know explicitly a Ljapunov function (corresponding to any of the slightly different ϵ optimal invariant sets, which of course lie within the same bounds).

We demonstrate, that using our approach it is often possible to find rather explicit ways to construct and parametrize invariant sets without solving a free boundary problem for a corresponding Hamilton-Jacobi equation or computing alternating integrals. The idea of the latter was used in [12] to compute the largest invariant set (for the "first player" u) within a given set L (in the case of complete observations); the difficulty to perform the "elementary" set theoretic operations on the arising convex sets in higher dimensions make this method interesting only theoretically, see also [17].

Our approach — of using a modification of the system of "characteristics", i.e. trajectories of a Hamilton system arising from the Euler Lagrange problem defining the "central" trajectories for constructing a feedback control is closest to an earlier method proposed by Pontrjagin, even if his system of characteristics is nonsmooth and is built on a different idea. Also a more recent paper [22] should be mentioned here , in which for *convex* Hamiltonians the nonsmooth generalized solution of the Hamilton Jacobi equation say the value function of a convex, optimal control problem is obtained using characteristics. In our case of game problems however the Hamiltonians are usually not convex. There is some similarity with the methods which build feedback controllers based on realtime solution (updating - via homotopies) of the two-point boundary value problems arising from the application of the maximum principle for the corresponding open loop (minimax) controls, see [4]. In order to reduce the order of the feedback system and arrive at a more simple feedback controller we exploit, that the stabilization of the arising Hamiltonian system proposed here naturally leads to small (large) parameters, so that a relatively simple application of known methods for the analysis of the asymptotics of equations with small parameters allows us to eliminate the conjugate variables from the feedback system. The curves arising from our modification of the central trajectories behave (with respect to the family of solutions of the two-point boundary value problems) similarly as the "magistral" trajectories known e.g. from the asymptotic theory of optimal control of economic growth.

The scope of the present paper does not allow to give a more detailed comparison of the method proposed below with existing other methods, in fact the present author is unaware whether the higher dimensional examples given below have been or could be attacked more succesfully with other existing methods, where the saturation bounds are also taken explicitly and rigorously into account, see [3] for other numerical approaches to compute the value function and maximal bridges for related problems.

There are ample reasons to expect that generally the positional feedback control, which is constructed with the proposed method is, for a suitable choice of a small number of free parameters, in fact *optimal*, i.e. it produces minimal guaranted values $d_1, \ldots, d_m$ for the sets P, Q, S. This we can prove in simpler cases, see Theorem 2 below. This is of course reassuring, we stress however that the proposed method is designed for the "interactive" (and perhaps only suboptimal) solution of complicated highdimensional and nonlinear games, where the usually defined optimal solution is beyond reach (at least for the traditional methods, but maybe for whatever methods). We emphasize a particularly interesting feacture of the method proposed below: the feedback law we construct is an explicitly computed algebraic expression, i.e. nonlinear function of the (reconstructed) state containing a few parameters. These parameters are just the parameters defining the game, i.e. the matrices $(A, B, C, D, \dots)$ the numbers $d_1, \dots, d_m$, the parameters of $K_1, \dots, K_m$, and a few further ones, which are the weights used for the different logarithmic penalty terms in the Lagrangian, (which define — as extremals — the central paths used here.) This might greatly help us in understanding the mechanism of a good feedback law (by an interactive construction of the Euler-Lagrange problem from which it is derived), and the influence of the different parameters on the performance. This does not seem to be possible with other existing approaches, where the feedback value $u(t)$ is not given by explicit, i.e. symbolycally computable formulae, but only indirectly, through hardly computable quantities. In fact the tuning of the various free parameters, which is always necessary in the final analysis and implementation, is greatly aided by this availability of explicit feedback formulae and sometimes even of equations for the boundary of the invariant set.

2 Analytic centers, central paths and variational problems associated to linear programs

Definition. Let $Q_1, \dots, Q_m$ be concave, analytic functions such that the feasible set of the inequality system $I(Q_1 \dots, Q_m)$

$$(2.1) \qquad F(Q_1, \dots, Q_m) = \{x | Q_i(x) \geq 0, i = 1, \dots, m, \ x \in R^n\}$$

is bounded. The analytic center of $I(Q_1 \dots, Q_m)$ is defined as

$$(2.2) \qquad c(Q_1, \dots, Q_m) = \arg\max\{\Phi(x) := \sum_{i=1}^{m} \log Q_i(x) | x \in F(Q_1, \dots, Q_m)\}$$

whenever this solution is unique. In fact, in the case when $Q_i, i = 1, \dots, m$, are quadratic (e.g. linear) in x the boundedness of $F(Q_1, \dots, Q_m)$ implies the uniqueness of the above maximum (see the next theorem). Note that $c(Q_1, \dots, Q_m)$ depends not only on the set $F(Q_1, \dots, Q_m)$ but on its description $I(Q_1, \dots, Q_m)$.

Theorem 1. For arbitrary quadratic, concave functions $Q_1, \dots, Q_m$, $c(Q_1, \dots, Q_m)$ is affine invariantly connected to $I(Q_1, \dots, Q_m)$ and provides a two sided ellipsoidal approximation for $F = F(Q_1, \ldots, Q_m)$:

(2.3)
$$c(Q_1, \ldots, Q_m) + \frac{1}{2\sqrt{m}} E(Q_1, \ldots, Q_m) \subseteq F \subseteq c(Q_1, \ldots, Q_m) + \sqrt{2m} E(Q_1, \ldots, Q_m)$$

where

(2.4)
$$E(Q_1, \ldots, Q_m) := \{z| \ -\frac{1}{2} < D^2\Phi(c(Q_1, \ldots, Q_m))z, z >\leq 1\}$$

The proof is given in [19]. A generalization of this Theorem for approximating reachable sets (or the sets of localization for the unknown state in (1.1)-(1.2) is given in [16]. The global quadratic Taylor approximality of $\exp(\Phi(Q_1, \ldots, Q_m))$ around $c(F_1, \ldots, F_m)$ — a function which is positive inside $F(Q_1, \ldots, Q_m)$ an vanishes on its boundary — implead by this theorem is responsible for the effectivity of Newton's method, i.e. for the existence of a large domain for its quadratic convergence. Precisely this is needed for efficient updatability of $c(Q_1, \ldots, Q_m)$ when the parameters of $Q_1, \ldots, Q_m$ change: even large perturbations of $Q_1, \ldots, Q_m$ cause only a moderate — and algorithmically easy to realise — change of $c(Q_1, \ldots, Q_m)$. This observation is crucial to understand the motivation for using homotopies of "central" solutions. For a survey of its application to convex optimization problems and its generalizations to other systems of inequalities (where e.g. constraints like $\det x > 0$ are used as a barrier for the class of positive definite matrices x) we can here only refer to [14], [15], [19] and [20] for a more detailed analysis. We note, see [14], that central solutions can be regarded as generalization or "specializations" of the notion of a *maximum entropy* solution (of a moment problem), moreover provided a crucial element of the H^∞ theory by which linear games with quadratic, integral constraints on u, v and e are solved, see [10], [15].

It will be rather instructive to consider the application of centers to linear programming problems:

(2.5)
$$\lambda^* := \min\{c^T z | a_i^T z \leq b_i, \quad i = 1, \ldots, m\}$$

The central path $z(\lambda)$ is the homotopy path of centers corresponding to the one parameter family of inequalities $\{\lambda \geq c^T z, \quad b_i \geq a_i^T z, \quad i = 1, \ldots, m\}$, that is

(2.6)
$$z(\lambda) := \arg\max\{\log(\lambda - c^T z) + \sum_{i=1}^m \log(b_i - a_i^T z)\},$$

which is the solution of the one parameter family of equations

(2.7)
$$E(\lambda, z) := \frac{c}{\lambda - c^T z} + \sum \frac{a_i}{b_i - a_i z} = 0, \quad \lambda \geq \lambda^*.$$

Introducing the new parameter $t = (\lambda - c^T z(\lambda))^{-1}$ and $r = t^{-1}$ we see that the central path is identical with the "log-barrier" path

(2.8)
$$z(r) = \arg\min(c^T z - r \sum_{i=1}^m \log(b_i - a_i z)).$$

It turns out that along this path one can write down an Euler-Lagrange variational problem P, such that $z(t)$ becomes the derivative of an extremal of P: Indeed consider the variational problem

$$(2.9) \qquad \max \int_0^T (c^T x + \sum \log(b_i - a_i^T x'(t)))dt,$$

for which the Euler-Lagrange equations are

$$c = \frac{d}{dt}(\sum \frac{-a_i}{b_i - a_i^T x'})$$

from which the extremals are immediately obtained as solutions of

$$d + ct + \sum \frac{a_i}{b_i - a_i^T x'(t)} \equiv 0, \quad \text{for some } d \in R^n$$

and, of course, for arbitrary $x(0)$. Note that adding the term $d^T z$ to the function maximized in (2.6), we get all the extremals, $d = 0$ corresponds to the fullfilement of a transversality condition at $t = 0$, (the necessary condition of optimality for (2.9) with $x(0)$ being free, since

$$(2.10) \qquad \frac{\partial L}{\partial x'} = \sum_{i=1}^m \frac{a_i}{b_i - a_i^T x'} = 0$$

is the equation for the center $z(\infty)$ of the feasible set. The extremals exist for the whole interval $t = [0, \infty)$; in fact for $t \to \infty$ they all, i.e. $x'(t)$ converge to a point, which for $d = 0$ is the centre) of the optimal face (i.e. to the optimal point of (2.5) when it is unique. The above connection between central path and the variational problem (2.9) was noticed by J. Lagarias, who observed that the corresponding Hamiltonian system is completely integrable, i.e. has n independent first integrals in involution.

Let us consider the optimal control problems — for fixed $T, x(0)$ and $x(T)$ —

$$(2.11) \qquad \max \int_0^T c^T x(t)dt, \quad x'(t) = u(t), \quad a_i^T u(t) \le b_i, \quad i = 1, \dots, m$$

Notice that

$$\int_0^T c^T x\, dt = \int_0^T -tc^T x'(t)dt + \text{const},$$

taking here the maximum poinwise in t it becomes clear that the log-barrier path for the derivatives now with $r = t^{-1}$ is an is an extremal of (2.9), the latter thus arises simply by applying the log-barrier technique, which originated with the introduction of the subproblems (2.8)) to (2.11). In [20] it is shown, that the dynamics of this system

is equivalent (up to affine equivalence in R^n) to the following universal differential
equation

$$M' = [M, [M, D(Me)]] \quad e = (1, \dots, 1) \in R^m$$

where $D(Me)$ denotes the diagonal matrix with elements Me evolving in the Grassmann manifold of projection matrices $M : R^m \to R^m$ of rank n. Similar equations
(like that of the Toda-lattice) arise in the theory of inverse scattering (for the solution
of classes of nonlinear differential equations, see e.g.[14]) in fact there are further similarities of our constructions to those in this theory. We included this remark in order
to indicate that the use of the logarithmic barrier function is not just one technique
among many alternative ones but it has deep theoretical grounds like the invariance
under scaling and duality and ramifications, see [15],[20].

Consider now the system of canonical (Hamiltonian) equations, for (2.11), see [11]

$$(2.12) \qquad \Psi' = c, \quad x' = U(\Psi),$$

where $U(\Psi)$ is the solution of

$$(2.13) \qquad \min_u (\Psi^T u - \sum \log(b_i - a_i^T u))$$

$$(2.14) \qquad \Psi + \sum_i \frac{a_i}{b_i - a_i u} = 0.$$

For a free value of $x(0)$ we get the transversality condition $\Psi(0) = 0$. Explicit solution
of (2.12) is in general not possible, but in order to compute the extremals one can use
the following trick of differentiating

$$x'' = u(\Psi)' = \frac{\partial u}{\partial \Psi} \Psi = \left(\frac{\partial \Psi}{\partial u}\right)^{-1} = -\left(\sum \frac{a_i a_i^*}{(b_i - a_i^T u)^2}\right)^{-1} c.$$

Introducing the "slack" variables $d_i = b_i - a_i^T u$, $i = 1, \dots, m$ the above system becomes

$$(2.16) \qquad d' = A^*(AD(d)^{-2}A^*)^{-1}c,$$

where $D(d)$ is the diagonal matrix formed from the elements or d and A is the $n \times m$
matrix formed from the vectors $a_1, \dots, a_m$.

Now the as a Hamiltonian system, (2.12) is — in general — unstable, i.e. its transition
matrix has eigenvalues both with positive and negative real parts, therefore during
a numerical integration procedure the equation (2.14) will be violated. A way of
stabilization is the following: select Ψ' so that for some positive constant λ (oder
constant diagonal matrix λ)

$$(2.17) \qquad L(t)' = -\lambda L(t), \quad \text{where} \quad L(t) = \Psi(t) + \sum_i \frac{a_i}{d_i(t)},$$

this leads to a modification of the system equations by "recentering" i.e. by replacing c with $c - \lambda(ct + \sum a_i d_i^{-1})$ in (2.16).

Our numerical experiments showed, that this is a crucial improvement; for an analysis of the computational complexity of first order extrapolation path following methods methods using Newton-type corrections, see [20].

We found that in this "stabilised" form of (2.16) we can allow quickly varying Lipschitz functions of time $c = c(t), 0 < t < T$, and still maintain feasibility and get small average value for $c^T x'(t)$ over $[O, T]$

We should emphasize here that — for constant c (2.16), (2.17) are analytic differential equations (in fact ones arising from a oneparametric, analytic system of equations) and for their numerical solution (more precisely for following their solution paths) appropriately constructed high order methods (with recentering) are more effective.

We have tested the feedback control method — proposed below in full generality in Section 3 — on the following class of "linear" programming games. Suppose that the objective vector is an unknown Lipschitz function of time with known Lipschitz constant L_0

$$|c'(t)| \leq L_0, \quad \text{for all } t \geq 0$$

may be satisfying some other constraints like $c_i(t) - c_i^0 \leq K_i$, for $i = 1, ...m$

Suppose that the functions $b_i(t)$, $i = 1, \ldots, m$, defining the feasible polyheda $P(t)$ are changing with time according to the law

$$b_i(t) = a_i^T \nu(t) + d_i(t) + d_i^0, \quad \tilde{d}_i'(t) = v_i \tilde{d}_i(t), \quad |v_i(t)| \leq \sigma_i, \quad \|\nu(t)\| \leq L, \quad i = 1, \ldots, m$$

here ν is an unknown Lipschitz function, $\sigma_1, \ldots, \sigma_m$ are constants. From this definition follows automatically, that $d_i \geq 0$; i.e. the polyhedron $P(t)$ never becomes empty, in fact it contains the translate of a fixed polyhedron.

The "state" vector x, for which we would like to get values for which $c^T x_1(t) < 1$, for all t, is supposed to be controllable according to (1); for simplicity we just looked to the case

$$x = (x_1, x_2), \quad x_1' = x_2, \quad x_2' = u, \quad \|u\|^2 \leq q.$$

In order to construct the required feedback control $u(\cdot)$ we propose to use the Lagrangian

$$
\begin{aligned}
(2.18) \quad L_3 = {} & P_0 \sum_i \log(b_i - a_i^T x) + P_1 \sum_i \log(b_i' - a_i^T x_2 - \sigma_i d_i) \\
& + P_1 \sum_i \log(\sigma_i d_i - b_i' + a_i^T x_2) + P_2 \log(1 - c^T(t) x_1) \\
& + P_3 \log(L^2 - |x_2|^2) + P_4 \log(d_0^2(1 - c^T(t) x_1)^2 - (c^T(t) x_2 + c'^T(t) x_1)^2).
\end{aligned}
$$

Here we have assumed that the derivatives b_i', $i = 1, \ldots, m$ and c' are available. (Note that having only noisy measurements of $b_i(t)$ and $c(t)$ we can reconstruct nearly equivalent 'disturbance" functions using the same method as in the observation loop of Example 1.

Note that in the above Lagrangian most of the terms are used to express our a priori knowledge about the bounds for the magnitude of the several inputs, Example 1 below will show that it is a good idea (in fact necessary for optimality) to restrict the maximum velocity of the point x to be only slightly larger than that of the unknown vector ν. The last constraint in L is an example of a restriction which imposed "a posteriori", i.e. to get a better scheme, d_0 should be chosen to depend on $L_0, c_1^0, \ldots, c_n^0, K_1^0, \ldots K_r^0$. To incorporate the knowledge of $L_0 c_1^0, \ldots, c_n^0, K_1^0, \ldots K_n^0$ is a more difficult problem and we shall return to it later.

3 The method for constructing and analysing the feedback

In accordance with the "separation principle" we shall consider the problems of control and dynamic observation separately. For each of these problems, the Lagrange functions $L_i(y, x, u, v)$, $i = 1, 2$ we shall use for building a feedback control, will have the following form

$$(3.1) \qquad \max_u \int_0^T (\sum_{j=1}^q Q_j \log G_j(x) + g \log(\varrho - |u|^2) + \sum_{i=1}^m P_i \log(d_i - K_i(x)))dt$$

where in (1.1) the function $v(.)$ is fixed and

$$(3.2). \qquad \max_v \int_0^T (P_0 \log(\varepsilon^2 - \|y - Cx\|) + a \log(R\sigma^2 - \|v\|^2) + \sum_{j=1}^r R_j \log F_j(x, y)))dt$$

where in (1.1)-(1.2) $u(.)$ and $y(.)$ are fixed. Here the "artificial" bounds $F_j(x, y) > 0$, $j = 1, \ldots, r$ and $G_j(x) \geq 0$, $j = 1, \ldots, q$ are introduced for enchancig performance, see below, or just to the examples to see what is ment here; an example of choosing these is provided by the last two terms of the integrand in (2.18).

Note that, by the positivity of weights not only the Legendre condition (of local optimality) is satisfied in both problems (since the signs of g and a are positive), but also these Lagrangians are convex "globally".

The main idea is to consider in the common system (1.1)-(1.2) for *fixed* functions $y(\cdot)$ and $v(\cdot)$ the maximizing, control $u^*(\cdot)$, and on the other side to consider for each fixed $y(\cdot)$ and $u(\cdot)$ the maximizing, control $v^*(\cdot)$, and modify the well known canonical systems of ordinary differential equations describing these (via the maximum principle) so that they could be updated "on line" as T changes. The latter, maximizing controls and the corresponding trajectories will be sometimes called as the "central" ones.

Our aim is thus to calculate u^* and v^* on line as feedback functions of y and — in the observation problem, an auxiliary function which models —the state vector x (not anticipating ,i.e. depending uncausally, on the fixed, opponents control function $v(.)$):

$$u^*(t) = U(x,t), \quad v^*(t) = V(y,x,t),$$

then we can put $v^*(t) = V(y,x,t)$ back into (1.1) to get a closed loop, causal feedback working out $u(t)$. Of course the optimal functions $u^* = u^*(t,T)$ and $v^* = v^*(t,T)$, $0 \le t \le T$ depend ("noncausally") on the chosen, fixed interval $[0,T]$ and of the fixed "other" control (i.e. $v(\cdot)$ and $u(\cdot)$); they are obtained from extremals of Hamiltonian systems with Hamiltonians $H_1(x,\Psi_1,u,v)$ and $H_2(x,\Psi_2,u,v)$

$$(3.3) \qquad x_1' = \frac{\partial H_1}{\partial \Psi_1}(x_1,\Psi_1,U(x_1,\Psi_1),v); \quad x_2' = \frac{\partial H_2}{\partial \Psi_2}(x_2,\Psi_2,u,V(x_2,\Psi_2,y))$$

$$(3.4) \qquad \Psi_1' = -\frac{\partial H_1}{\partial x_1}(x_1,\Psi_1,U(x_1,\Psi_1),v); \quad \Psi_2' = -\frac{\partial H_2}{\partial x_2}(x_2,\Psi_2,u,V(x_2,\Psi_2,y)),$$

where — as we shall see below $U(\cdot)$ and $V(\cdot)$ depend only on the Ψ variables and —

$$H_i(x_i,\Psi_i,u,v,y) = \Psi^T(Ax + Bu + Dv) + L_i(y,x,u,v)$$

which should —for all fixed T — satisfy the transversality conditions

$$(3.5) \qquad \qquad \Psi(0,T) = 0, \quad \Psi_i(T,T) = 0, \quad i = 1,2.$$

Thus, $u^*(t) = u^*(t,T)$ is obviously not constructed causally. The functions $U(\cdot)$ and $V(\cdot)$ are obtained — using the maximum principle, see [11] — from the solution of the following simple problems:

$$(3.6) \qquad \max_u(\Psi^T Bu + g\log(\varrho - |u|^2)) \quad \text{resp.} \quad \max_v(\Psi^T v + a\log(R\sigma^2 - |v|^2))$$

and can be computed explicitly

$$u = qB^*\Psi(g + \sqrt{g^2 + q\|B * \Psi_2\|^2})^{-1}$$

Our main idea is to modify the system of equations for Ψ in order to be able to propagate them causally (and taking care of the transversality conditions in an approximative sense). Note that differentiating (3.5) we see that

$$\frac{d}{dT}\Psi(t,T) + \frac{d}{dt}\Psi(t,T) \equiv 0, \text{ for } T = t,$$

thus from

$$\frac{d}{dt}\Psi(t,T) = -\frac{\partial H}{\partial x} \text{ we get } \frac{d}{dT}\Psi(t,T) = \frac{\partial H}{\partial x}, \text{ at } t = T$$

We are naturally led to the following law for propagating the conjugate variables:

$$(3.7) \qquad \Psi_i' = \frac{\partial H_i}{\partial x_i} - \Lambda_i \Psi_i, \quad \Psi_i(0) = 0, i = 1, 2,$$

where Λ_i are diagonal matrices with large positive elements. Since it turns out, that —under rather general and easily verified conditions for the Lagrangians $L_i(\cdot)$ — from (3.3)-(3.4) and (3.6) we can "eliminate" Ψ_1 Ψ_2, the values $u(t) = U(x_1(t), \Psi_1(t), y(t)) \to U(x, y)$ and $v(t) = V(x_1(t), \Psi_2(t), y(t)) \to V(x, y)$ can be used to get the final, closed loop system. This elimination is obtained by replacing the adjoint equations with their asymptotic equivalents regarding Λ^{-1} as a small parameter. This amounts to determine Ψ as a function of x and y, by solving the equations

$$0 = \frac{\partial H_i}{\partial x_i}(x_i, \Psi_i, u(x, \Psi_i, y), v, y) - \Lambda Psi_i \quad i = 1, 2.$$

Notice, that (by the linearity of (1.1)) the above system is linear in Ψ, and since there are no cross terms between x and u in the Lagrangian L — thus $\frac{\partial L}{\partial x}$ does not depend on u — its solution has the form

$$\Psi(x, y) = -(A^* - \Lambda I)^{-1} \frac{\partial L}{\partial x} \quad i = 1, 2$$

In order to find a suitable starting value, $x(0)$ we recall the relation (2.10), i.e. that the transversality condition $\Psi(0) = 0$, will be satisfied if we start from the analytic centre of the inequality system imposed for the moment $t = 0$, which is just the origine (by the assumed centralsymmetry of the constraint sets). In the case of the more observation problem we may have to replace the function $y(t)$ by $g(t)y(t)$, where $g(t)$ is a smooth, positive function monotonically ("quickly") tending to 1, satisfying $g^{(k)}(0) = 0$ for $k = 0, \ldots, n$.

The closed loop system thus has the following form: (1.1)-(1.2) are accompanied (i.e. augmented) by a model system "driven" by $y(\cdot)$

$$(3.9) \qquad \dot{x} = Ax + Bu(x, \Psi_1(x, y)) + Dv(x, \Psi_2(x, y))$$

so that the complete system is $2n$ dimensional and is driven by the inputs $v(\cdot)$ and $e(\cdot)$; the only difference between these is that v appears linearly on the right side of (1.1), while now $e(\cdot)$ from (1.2) is appearing in (3.9) in a *nonlinear* way. Here we let — just for generality — $\Psi_1(x, z)$ also depend on z since in a more general situation a reference curve $z(\cdot)$ — to be tracked — could have been also given (instead of the origine or "zero" curve corresponding to the constraints (1.3)).

Returning to the Lagrange problems (3.1) and (3.2), consider now to the problem of selecting the free constants, the weight coefficients and auxiliary artificial bounds. First of all note that the constants R and E should be selected larger than 1 in order that dynamic tracking of the output y be possible, of course, a minimal requirement for them is that the system must be *stable* when restricted to the case of zero disturbances and measurements errors. From the closed form solution for u follows that choosing g and a to be positive and small (so that near bang-bang controls could be realized). Note that if g is small the small change of sign of Ψ (in the onedimensional case) is able to change $U(g, q, \Psi)$ from near $\sqrt{q}$ to near $-\sqrt{q}$.

When the constraint $\|u\| \leq \varrho$ is replaced by more complicated ones e.g. of the form

$$a_i^T u \leq b_i, \quad i = 1, \dots, m$$

we use exactly as in (2.8) the sum of the elementary logarithmic barrier functions. The equation (3.8) is then, in general no more solvable explicitly and depending on the situation we can use (second order) Newton corrections like in (2.15)-(2.16) to resolve these equations (always from a near solution point obtained by extrapolation see [20]), or we could use more direct (less expensive but more crude first order) path following methods using variants of the "extremal aiming" (see [7]) to solve — say, for the control problem (3.1) —

$$\frac{\partial H(x(t), \Psi, u)}{\partial u} = 0$$

in order to get $u = u(x, \Psi)$ and

$$0 = E(t, \Psi) = \frac{\partial H}{\partial x}(x(t), \Psi, u(x, \Psi)) - \Lambda \Psi$$

from which $\Psi = \Psi(x)$ should be determined, leading finally to $u = U(x)$. The "a priori imposed" constraints, which can be present in both problems (3.1) and (3.2) $K_j(x) < d_j$, $j = 1, \dots$, give a term $\partial_x K_j(x)(d_j - K_j(x))^{-1}$ for the value of Ψ'. Indeed, the choice of the functions K_j, $j = 1, \dots, m$ in (1.3) (which are given a priori, independently of the dynamics (1.1)) to be the only ones selected in (3.1) and (3.2) is — in general — *not* sufficient to obtain an optimal solution.

At this point we recommend the study of the Example 1 in section 4 before proceeding further. In Example 1 we had to introduce the constraint $\|x_2\| \leq \gamma_2$ in addition to the original single constraint $\|x_1\| \leq \gamma_1$ the reason was to exploit the a priori knowledge about the disturbance function, and it turned out that this is the right thing to do: it leads already to the best possible result. In the same Example 1 however if we impose just one constraint (instead of the two selected similarly as here)

$$\|k_1 x_1 + k_2 x_2\|^2 \leq \gamma_0$$

with suitable *positive* k_1, k_2, the resulting feedback system also yields an optimal (minimal) invariant set, see below. It was only relative small work to find out in all our examples of linear games the right new or alternative, quadratic constraints to be imposed (based on the identification of the stable and unstable subspaces of A, and the

(A, B) (A, D) and (C, A, D), invariant subspaces of (1.1) and its dynamics outside it, see [13] for their importance and analysis). In fact the setting up and solving of a naturally related quadratic game (H^∞ problem, (see [16] for the connections between central paths and classical linear quadratic problems, identified Taylor approximations around these paths !) may also yield preliminary information about the form of the "imposed" constraints. A final tuning of free (weight and stabilization) parameters should be done in all cases. Of course this is an important problem, we go in more details when describing concrete examples of the solution of our problem to find minimal invariant sets around zero.

A disadvantage of the scheme (3.1)-(3.2) is that, even if it is possible to achieve stabilization with them within the set K, outside K they do not work, while often; like in Example 1, one can steer, at least assymptotically all points of R^n to this set. One way to find a remedy is to introduce a homotopy in the prescribed and imposed bounds, i.e. replace the constants d_j by monoton (but not too fast) decreasing functions of time $d_j(t)$.

A second scheme, i.e. type of Lagrange functions suitable for this more general problem can be obtained as follows, say for control loops:

$$(3.10) \qquad L(x) = \sum_j (-\alpha_j^1 \frac{d}{dt} K_j(x) - \alpha_0 K_j(x)) - \sum \beta_j G_j(x) + g \log(\varrho - |u|^2)$$

Here α_j^0, α_j^1, β, g are all positive (since the integral is to be maximized), $G_j(x)$ stand for the eventual, "later imposed" (convex, quadratic) constraints. This scheme is thus also always convex. The idea is to add — in the functional to be maximized — the "final" values $-K_j(x(T))$, $j = 1, \ldots, I$, (to the logarithmic terms for the inputs) to the integral of these values, to achieve — for a suitable selection of the weight coefficients — desired uniform bounds for these values. By writing out the differential quotient we see that the Lagrange function explicitly depends on the control variables u and v. We can "stabilize" the Hamiltonian system in the same way (3.6) or (3.6') but before we could make a further modification in the system (3.3) (3.4) by simply deleting all terms of the right side, which contain the disturbance input. Note that on the one part we assumed the set Q to be symmetrical with respect to the origine, secondly the knowledge of $v(t)$ i.e. its invertibility (from the observed output) might be a problem, even if sometimes we could work with "equivalent" inputs as described earlier for the solution of the observation problem. Surprisingly enough this modification of the Lagrange function (3.10) seems to yield — even for some nonadditively occuring disturbances — often (but not always) just a little weaker results than in the case where the terms containing v are also kept, i.e. when v is assumed to be reconstructible.

In these systems the feedback (optimal) control values u^* depend not only on the adjoint variables, but also on the state vectors which in turn has the consequence that the adjoint equations — on the right side of which now u^* appears — are no more linear in Ψ, so that the asymptotical analysis of them becomes more complicated. Note that the inclusion of a nonadditively occuring, quickly varying, ... function $v(\cdot)$ would make this analysis even more difficult

Finally let us turn to the derived, closed loop system. It can be analyzed in itself, without any reference to its origine and if we can prove rigorously the existence of invariant sets i.e. ones from which no controls $v(\cdot)$, $e(\cdot)$ can lead out by other arguments, then well, we can "forget" about some of the "nonrigorous" or not fully understood elements of its derivation. Of course this is, in general not the way to be followed, in fact it should now be clear that there are ample reasons for these choices, one of the main reason seems to be the *stability* and *easy updatability* of analytic centers and central paths. The mechanism for the achieved appropriate behavior of the conjugate variables is that in many cases — e. g. when there are no "cross terms" between x and u in the Hamiltonian — only their sign , more generally the direction of a component of them, is important -to define the correct choice of an approximate) extreme point of P, now the direction of a small vector is easily changed; this change occurs when at least one of the constraints is going to be violated (i.e. is closely approached), and then the influence of the corresponding term, which governs the derivative of the conjugate variable and is inverse proportional to the distance to the constraint boundary) overrides the stabilising term and changes the direction correctly. Note also that in the first class of schemes the value of the feedback control depends only on the direction of a "component" of the ψ vector and not on its norm (because g is chosen to be rather small).

Another intutitive explanation of the mechanism of our asymptotic scheme, i.e. why it can achive causality without loosing performance of minimax, i.e. uncausal optimal controls is to invoke the analogy with the "magistral property" observed earlier for some classes of optimal control problems (e.g. in economic growth theory): the tendency of the optimal trajectories to follow a magistral path (before returning to their final position -where the transversality conditions become active, i.e. shortly before the final time; this is implemented in the choice (3.7), but its validity also depends on the choice of the weights and additional functionals. There is a large amount literature on asymptotic analysis of optimal control (two point boundary value) problems containing small parameters, which could be relevant here for a justification of the proposed modification of the equation for the conjugate variables.

Note that the feedback system itself has several free parameters some of them small some large, so that a further analysis using methods of the theory of differential equations with small parameters could be made, see the paper of E.F. Mischenko and L.S.Pontrjagin in [11] for a survey on some of these methods.

This analysis is most simple for the case when we have exact observations and when both controls are onedimensional i.e. $v \in [-1, 1]$ and $u \in [-q, q]$ or we have a small number of independent onedimensional controls u and v. The first thing is look at the closed loop system with $v \equiv 1$ and $v \equiv -1$, which correspond to two vector fields $F_1(x), F_2(x)$, and in order to analyse the boundary of the arising invariant set, more precizely: to express the loss of controllability at the points of the boundary in terms of Lie Algebraic conditions, we have to look at the Lie Algebra generated by them. The situation is especially easy, and this is a common case in our examples under suitable choice of the free parameters, if each of these vector fields has just one zero, which is asymptotically *stable*. If the dimension of the state is two, then there are exactly two "separatrices" S_1, S_2 connecting these 2 points. In the twodimensional case they enclose a compact region which is a natural candidate for an invariant set, see e.g. Example 1. If on these lines the actual v controls are selected differently from what "corresponds" to their places, the state can be expected to move "inside" that region. There are known methods for checking this, we do not study this here in more detail. We will see that this is the case in all twodimensional examples, where in addition this domain is convex (its convexity can be proved from the fact, that in games (1.1)-(1.2) the largest invariant set in a given set L is always convex, as follows from its construction, see [12], [17]). It is interesting to note that in the latter examples the two stable points are also connected by an "interior", singular, "switching" line along which a slow "chattering" motion is realized.

A more general situation is, see Example 2, when one or both of the two zero points are unstable, but there exists a stable (from outside) limit circle around them. Then we have to find two "separatrices" connecting these limit circles C_1 C_2 and enclosing again a closed, compact invariant set. More precisely, to obtain this set we have to find four points P_1, P_2, P_3, P_4 such that at these points the two kind of trajectories (tending asymptotically to, resp. coming from the other limit cycle) are tangential. The number of such smooth pieces on the boundary of the invariant set increases (remaining finite) when replace the segments P, Q by polyhedrons (i.e. polygons). Of course the form even the existence of such invariant sets depends on the selected weight constants and other parameters, the tuning of the latters is an important "implementation" problem.

In the case of dimension 3 the two separatrices S_1, S_2, connecting the the stable points P_1, P_2 are defined similarly. They are expected to be the intersection of two "parts" of a surface S (which then enclose the desired invariant set) consisting of the points x, where the following condition is satisfied

$$(3.11) \qquad\qquad < [F_1(x), F_2(x)], F_1(x) * F_2(x) >= 0$$

Here $[F_1, F_2]$ is the Lie bracket of the two vector fields, $< \cdot, \cdot >$ the scalar, $*$ the vector product. Do to the fact that $F_1(x)$ and $F_2(x)$ are explicitly computed by our feedback construction this equation can be analysed more easily (e.g. by simbolic computations). The reason for (3.11) is that at each point of the boundary of an invariant set, (which is now just twodimensional) the Lie bracket of the two fields F_1 and F_2 (each of which being in the tangent space) must lie in the tangent space, in order that no controls could steer out the point from this set. In higher dimensions we should add more equations corresponding to the higher order brackets to characterize the boundary of invariant sets formed from two vector fields. That the two separatrices lie on the surface follows from the fact that boths "stable" points obviously lie on this set (because the equation is satisfied when either F_1 or F_2 is zero) and (from invariance) these trajectories remain on the surface (their tangents being in its tangent space). The two surfaces seem to correspond two families of trajectories starting from each of the separatrices, along which $v(.)$ takes the opposite sign than on the separatrix, so that all points on these surfaces can be reached — from the corner points P_1 and P_2 — with at most two switches.

Concerning the applicability of the proposed method to more general game problems the following should be made clear. The game problems we studied here are specific in the sense that no finite duration interval $[0, T]$ or end point condition was specified. It is clear that different methods are needed to deal with "less stationary" problems and especially with problems of evasion, where the minimum of the distances is to be maximized. It is known that, roughly speaking, in the latter for the evader no continuous state feedback function can be better than the optimal open loop control (a theorem of Subbotin connected to an earlier, partial result of Zaremba), see [7]. Of course, this "restriction" concerns not the applicability of central solutions (logarithmic penalty functionals) to evasion problems, but simply points to the necessity of additional, different strategy concepts. A simple example for these kind of evasion problems is that of keeping the phase point of a controlled pendulum inside a ball but outside a smaller, concentric ball, see Example 5,In order to indicate the potential applicability of our methods to this kind of problems we note that the main necessary modification is just to introduce *periods* at the beginning of which we define some now starting values for the conjugate variables.

An important pecularity of the problems considered here is that, for the second player v, there exists an open loop, maximizing (worst case) strategy (i.e. control function), which is at the same time closed loop optimal. This open loop strategy is usually not unique (if the time interval considered is large enough). This nonuniqueness makes it difficult to compute these optimal controls for the second player or — what is rather the same — the maximin value function. In fact we could formulate the problems (1.1)-(1.3) in a more traditional fashion, say for just one constraint, $m = 1$

$$(3.12) \qquad d^* = \min_{U(\cdot)} \max_{v(\cdot)} \max_{0 \le t \le T} K(x(t)), \quad x(0) = 0,$$

for a fixed interval of time, if T is large enough we get the same value as for $T = \infty$. We deliberately avoided this formulation partly because the optimal solutions for a fixed terminal time have a complicated behavior (especially if the terminal conditions are not stable) which is not connected — at least for this class of problems — with the overall optimum, i.e. it is irrelevant.

Concerning the observation problem we note that in the case of centralsymmetric constraints optimal recovery of linear functions of the state vector can be obtained already by a linear observer, as this is (a priori assumed) in [6]. In a similar context an analogous theorem has been first proved by N. S. Bahvalow, for the "approximation theoretic" problem of recovery of functions based on values of linear functionals over a centralsymmetrical compact class . The connection to our control problems is that the classes of function on which such recovery is studied is described often in terms of restrictions on the higher derivatives, like in a controlled process, e.g. for the class

$$(3.13) \qquad x^k(t) = u(t), \ |u(t)| \le \varrho$$

one studied the problem of approximating $x^{(j)}$, for some $j < k - 1$ based on measuring error corrupted values of $x(t) = y(t) - \varepsilon(t)$ at a finite number of points $t_i, i = 1, ...N$. We refer to [18] for a detailed analysis of the questions concerning the conditions for attaining optimality by a linear and causal recovery in the class of nonlinear and noncausal recovery (observation) algorithms especially for the above class of functions (with r arbitrary). In the last reference examples are provided that in the general case of noncentralsymmetric constraints the class of linear observers may not be optimal. Sequential methods of selecting these points are analogous to closed loop strategies. The method proposed here is obviously nonlinear and applicable for nonlinear dynamics as well.

An other analogy — e.g. to indicate why the optimal open loop control of the second player is at the same time closed loop optimal — is with the problem of constructing optimal sequential (versus passive) algorithms for maximizing convex functions belonging to a suficiently rich centralsymmetrical set, noting that the distance function is convex in the controls. In fact we expect that a general argument exploiting just the last fact, i.e that the second player has to maximize (the maximum of) convex funtion suffices to check this equality of open and closed loop performance, the well known "conditions of regularity" studied, e.g. in ch.5 of [7], are less straightforward to check.

4 Description and analysis of examples

Example 1. Here $x \in R^n, n$ arbitrary (test results are given for $n = 1$ and $n = 2$.)

$$(4.1) \qquad x_1' = x_2 + v \qquad |x_1 + \tilde{x}_1| \le \varepsilon$$
$$(4.2) \qquad x_2' = u \qquad \|u\| \le \varrho, \|v\| < \sigma$$

The problem here is following ("tracking") a quickly varying object $z_1(z_1' = v)$ by an inertial one $z_2, (z_2'' = u), x_1 := (z_2 + z_1), x_2 := x_2'$, i.e. making $\qquad |x_1 - z_2|$ small for all time, under the condition, that the coordinate(s) of the second object can be measured only within a known error bound ($|z_1 - \tilde{z}_1| \le \epsilon$). The associated variational problems of control and observation are selected as follows

$$(4.3) \qquad \max \int_0^T (g \log(q - |u|^2) + P \log(D - |x_1|^2) + p \log(S - |x_2|^2)) dt.$$

$$(4.4) \qquad \max \int (P_0 \log(E\epsilon - |x_1 - \tilde{x}_1|^2) + a \log(R - |v|^2)) dt$$

Here D is a constant (desired upper bound for $\|x_1\|$) to be minimized while the constant S is introduced (giving an example of an "artifically imposed constraint in (3.1)) to incorporate the apriori bound for $\|v\|$ (as we shall see below), of course D and S as well as the weight constants P and p should be selected depending on both σ, q and ϵ, and eventual further constraints.

The positive constants $P_0, E > \epsilon^2, a$ and $R > \sigma^2$ are depending only on ϵ and σ. The resulting loops of observation and control with state (x_1, x_3) and (x_1, x_2) are separated, more precisely the "reconstructed " (equivalent) input disturbance (computed in the observation loop) is used as the (unknown) input in the control loop, where thus all coordinates are measured exactly. Note that here observation is in fact "smoothing", it is necessary if we would use the corrumpted values $\tilde{x}_1$ (instead of x_3) the system would be break down (i.e. "blow" up).

The system of modified canonical equation is now

$$(4.5) \qquad x_3' = x_2 + R\Psi_3(a + \sqrt{a^2 + R\|\Psi_3\|^2}), \quad x_1' = x_2 + v$$

$$\Psi_3' = \frac{P_0(\tilde{x}_1 - x_3)}{E - (\tilde{x}_1 - x_3)^2} - l_0\Psi_3, \quad \tilde{x}_1 = x_1 + d$$

$$(4.6) \qquad x_2' = q\Psi_2(g + \sqrt{g^2 + q\|\Psi_2\|^2})^{-1} \quad \Psi_2' = \Psi_1 - 2P\frac{x_2}{S - \|x_2\|^2} - l\Psi_2$$

$$(4.7) \qquad \Psi_1' = -2P\frac{x_3}{D - \|x_3\|^2} - k\Psi_1, \quad x = 0, \quad \Psi = 0$$

After excluding the fast variable Ψ_2 (and Ψ_1) we get the following explicit feedback system:

$$
\begin{aligned}
x_3' \ &= x_2 + R\Psi_3(\tilde{x}_1, x_3)(a + \sqrt{a^2 + R\Psi_3(\tilde{x}_1, x_3)}) \quad \Psi_3(\tilde{x}_1, x_3) = \frac{P_0(\tilde{x}_1 - x_3)}{l_0(E\epsilon^2 - |\tilde{x}_1 - x_3|^2)} \\
x_1' \ &= x_2 + v, \quad \Psi_2(x_1, x_2) = -(\tfrac{P}{k}\frac{y_1}{D - \|x_1\|^2} + p\frac{x_2}{S - \|x_1\|^2})l^{-1}, \quad \tilde{x}_1 = x_1 + d \\
x_2' \ &= q\Psi_2(x_3, x_2)(g + \sqrt{g^2 + q\|\Psi_2(x_3, x_2)\|^2})^{-1}.
\end{aligned}
$$

We see now, that the number of "free" parameters can be reduced, the feedback system is dependent only qP/gkl, qp/gl and RP_0/al_0. The analysis of this system (with two bounded disturbance "controls" v and d — to find invariant sets — is relatively simple in the "onedimensional" case.

A little thought on this game leads us to the observation, which was explained in a more general form above that for the case of exact measurement $\varepsilon = 0$, and $n = 1$ an inclusion minimal invariant set can be obtained from considering two special trajectories of the system. Indeed note, that for the constant control values $u = -\varrho, v = \sigma$ the trajectory of (1.2) in the time interval $\delta = \sigma/\varrho$ goes from the point $P_1 = (x_1 = -\sigma^2/\varrho, x_2 = \sigma)$ to the point $P_2 = (x_1 = \sigma^2/\varrho, x_2 = -\sigma)$ while the motion with opposite sign of the inputs move P_1 into P_2 during the same time interval. These two motions correspond to 2 curves, which enclose a convex smooth region $G(P_1, P_2)$, with just two corners (at P_1 and P_2). We prove now that an arbitrary near approximation G of this set can be constructed using the above scheme, i.e. when the parameters P_0, D, p, S, k, l are selected appropriately. To see this we have to select these parameters so that P_0 and P_1 become the zero points of the (x_3, x_2) system with σ replaced by $\sqrt{R}$. In the simulation we set

$$(4.8) \qquad D = 1.44\sigma^4/q, \quad S = 1.44\sigma^2$$

and then for fixed values of P/k, p is determined uniquely. Now it is easy to check that the obtained zero points are stable, however a further careful selection of the constants is necessary since the numerical integration — by which the feedback is implemented — has inherent errors. Without going into much details (concerning the requirements for a good ODE solver we just mention that higher order Runge Kutta methods often failed: there are examples when even the 2-3 order R-K-F ODE of Matlab (used with some modifications in many examples below) could not be used (because difficulties arise at the corners of the invariant set and on its boundary but also in its interior) we just used a simple 1-2 order (Euler+ "midpoint") method. On the other hand, in some problems it was better to use a R-K-F method of order 7-8 to obtain the needed local accuracy, mostly however we just used an Euler type method without adaptive stepsizes. One of the problems is that, because of the presence of arbitrary measureable, e. g. piecewise continuous inputs, the integration (esp. in the observation problem) could and should not be too accurate, but accurate enough to work out the rather sensitive nearly optimal feedback controls. The observation loop in this example is a rather trivial, ("negative positional") feeedback system, interesting is that, therefore the observer seems to "wish" to remain as far from the unprecisely observed point as the prescribed observation error bound allows. The latter is selected in our simulations to be 1.5 times the magnitude bound for the disturbance, of course 1.5 could be theoretically replaced by $1 + eps$, this is just to cope with errors of integration and the consequence of selecting a small, but different from zero. It is not difficult to see the minimality of this invariant set K with respect to inclusion and also in the sense that the functional

$$c(K) = \max\{\|z\| \mid z \in K\}$$

will be not smaller for any other invariant set. The second player v has always the possibility to force the first player to reach a velocity $|x_2(t_0)| \geq \sigma - \varepsilon_0$ for arbitrary ε_0 (otherwise the value of $\|x_1\|$ could be made arbitrary large. If at this moment $|x_1(t_0)| \leq \frac{\sigma^2}{\varrho}$, then using the constant control $v \equiv \sigma \operatorname{sign}(-x_1(t_0))$ at moment $t_0 + \sigma/\varrho$, the relation $|x_1(t_0)| \geq \sigma^2/\varrho - \operatorname{const} \varepsilon_0$ should hold. We have thus proved the following

Theorem 2. In Example 1 the proposed feedback scheme yields for each $\varepsilon_0 > 0$ and for suitably selected parameters, described above an invariant set which is ε_0-optimal with respect to the functional $c(K)$.

The worst case control of the second player is not unique at all, any control function which takes only the values σ and $-\sigma$ with sufficiently large intervals of constancy generates the boundary of the invariant set (this is often valid in the examples below, more precisely it holds when the ideal invariant set has just two corners, if its boundary contains parts of limit cycles, of course the moment (place) of swithing towards the other cycle has to be selected approprately).

In fact we expect that not only in the higher dimensional variant of this example (which in fact can be easily reduced — due to the sphericity of the constraints — to the onedimensional case) but in all of the 2, 3 and 4 dimensional examples studied here a similar statement as Theorem1 holds (we even see no reason why this might not be true in general, i.e. in higher dimensions and for the class of games studied here).

To take the above two terms in the control loop Lagrangian is not a necessary in the sense that we could replace the two terms by just one term of the type $P_0 \log(D - \|kx_1 + x_2\|^2)$

$$u = -\sqrt{q}\,\operatorname{sign}(kx_1 + x_2), \quad k = \sqrt{q}\sigma^{-1}$$

Note that the problem (2.18) above with $\sigma_1 = \ldots = \sigma_m = 0$, is a highdimensional generalization of Example 1 and shows a similar "switching from corner to corner" under "optimal", (worst case) choices of the function $v(.)$.

Let us now look at the realization of the scheme (3.10). The Lagrange function is here

$$L(x, u, v) = g\log(q - |u|^2) - P_1 x_1^2 - P_2 x_2^2 - 2r_1 x_1(x_2 + u) - 2r_2 x_2 v,$$

and the resulting system of (modified canonical) equations is — after deleting the terms containing the fonction $v(.)$ (assumed to be inaccesible) from the adjoint equations

$$\begin{aligned} x_1' &= x_2 + v; \\ x_2' &= u(x_1, x_4); \end{aligned}$$

$$\begin{aligned}
x_3' &= -2P_1 x_1 - k x_3 - 2r_1 x_2 - 2r_1 u(x_1, x_4); \\
x_4' &= -2P_2 x_2 - l x_4 + x_3 - 2r_1 x_1; \\
u(x_1, x_4) &= -q(x_4 - 2r_1 x_1)(g + \sqrt{g^2 + q(x_4 - 2r_1 x_1)^2})^{-1}.
\end{aligned}$$

We denoted here the conjugate variables Ψ_1, Ψ_2 just by x_3, x_4.

Of course, by introducing a smoothing procedure, like the one in (4.4) we could compute and use an "equivalent" $v(.)$. The system now asymptotically stabilizes itself to the same invariant set from *arbitrary* initial position $(x_1(0), x_2(0), 0, 0)$, the conjugate variables find their "right values" after a short time ! However for this to hold the free parameters should be chosen carefully, roughly speaking P_1 must be large, r_1 and r_2 small: in fact, under suitable choices of these, the laws

$$u(t) = \varrho sign(x_1 + k x_2),$$

–with different $k > 0$ — can be realized. Note that this scheme works already with $r_1 = 0$, $r_2 = 0$, (which choice is then nearly equivalent to select D and S to be large in (4.3)), in fact each of these schemes has its distinctive positive (or "negative") features. These schemes, even is significally different, can produce the same invariant set K for this they have to yield identical feedback values only on the boundary of K.

Example 2. The system is describing — after well known transformations — the "direct" control of an unstable pendulum (complete observations are assumed). Below in Example 4 we shall study a more difficult case of "indirect " control.

$$x_1' = w x_2 + u, \quad \|u\|^2 \le q, \quad w > 0$$

$$x_2' = w x_1 + v, \quad \|v\| \le \sigma, \quad x(0) = 0$$

The aim is to stabilize around $x(0) = 0$. Let us look first at the case of a Lagrange function with <u>two</u> imposed quadratic constraints on the state,

$$P_1 \log(D_1 - \|x_1\|^2) + P_2 \log(D_2 - \|x_2\|^2) + g \log(q - |u|^2).$$

The resulting system is

$$\begin{aligned}
x_1' &= w x_2 + u(x); \\
x_2' &= w x_1 + v; \\
p_1 &= -2P_1 x_1 (D_1 - x_1^2)^{-1}; \\
p_2 &= -2P_2 x_2 (D_2 - x_2^2)^{-1}; \\
A &= \begin{bmatrix} -k & w \\ w & -l \end{bmatrix}; \quad z = -A^{-1} \begin{bmatrix} p_1 \\ p_2 \end{bmatrix}; \\
u(x) &= q z_1 (g + \sqrt{g^2 + q z_1^2})^{-1};
\end{aligned}$$

The minimal difference to the derivation of the control loop in Example 1 is here that the solution of (3.7) really requires the inversion of the (2×2) matrix A. For given q and σ the right choice of the values for D_1, D_2, p_1, p_2 can be done as in Example 1 and in the following examples by analysing the fixed (stable) points (limit circles) and their stability, in the sources of the DFG report cited on p. 1 the rules of choice of some of the parameters are written out explicitly . A compact invariant set exists whenever $\sigma \leq \sqrt{q}$, e.g. here — for $\sqrt{q}$ near to σ and w large —we get with the selection rules similar to (4.8)

$$D_1 = 1.44\sigma^2/w^2 \text{ and } D_2 = 1.44q/w^2,$$

we get an invariant set bounded by two smooth curves (corresponding to special motions). For larger values of $\sqrt{q}/\sigma$ the invariant set is bounded by four pieces of special motions, suitable values for $D_i, i = 1,2$ can be taken from (4.10) below. The two terms here also could be as in Example 1, replaced by just one positive quadratic term, $P_0 \log(D - \|k_1 x_1 + k_2 x_2\|^2)$ with $k_1 = 1$, $k_2 = r$, which then ensures that

$$(4.10) \qquad \qquad \|x_2\| \leq \sigma/wr \text{ and } \|x_2\| \leq \sigma/w$$

for arbitrary parameters and $(2r + r^{-1})\sigma = \sqrt{q}$

Example 3. Consider now a more complicated observation problem (dynamic path following game for objects of the same type)

$$(4.11) \quad x_1' \;=\; x_2, \quad y = x_1 + e, \quad \|e\| \leq d,$$
$$(4.12) \quad x_2' \;=\; v, \quad \|v\|^2 \leq \sigma^2$$
$$(4.13)$$

At least at this moment it is clear that the problems (3.1) and (3.2) are almost equivalent which is of course to be expected by the known duality relations between observation and control, see [ ] for further details. A simple scheme is obtained from the following observation Lagrangian

$$(4.13) \quad g \log(R\sigma^2 - |v|^2) + P_0 \log(Ed^2 - |x - y|^2) + p \log(Dd_1^2 - \|x_2 - \frac{y(t) - y(t - h)}{h}\|^2),$$

where h is the optimal delay for approximating the derivative x_2 (in the uniform norm) of a function x_1 measured within accuracy d if it is known that $\sqrt{q}$ is a Lipschitz constant for x_1 (a problem solved in "approximation" theory, see [8] where a partially similar procedure is proposed)

$$h = 2\sqrt{d/\sqrt{q}},$$

the error being

$$d_1 = 2\sqrt{2d\sqrt{q}}.$$

Again E and D are selected to be larger than 1. The invocation of the Lagrange function of Example1 here has a pitfall; while the divided difference approximates x_2

"optimally", it is <u>not</u> a Lipschitz function with Lipschitz constant $\sqrt{q}$! The remedy is to introduce an extra loop for following the divided difference function with a Lipschitz function within the known error bound d_1 exactly as in the observation loop of Example 1. After this the remaining observation problem is identical with the control and observation problem of Example 1.

Example 4. This is the third order generalization of Example 1 under the simplification of assuming exact observations.

$$
\begin{aligned}
x_1 &= x_2 + v \quad \|v\| \le \sigma \\
x_2' &= x_3 \\
x_3 &= u \quad \|u\|^2 \le q
\end{aligned}
$$

We propose to use the following state dependent terms in the Lagrangian

$$
L_3 = P_1 \log(D_1^2 - |x_1|^2) + P_2 \log(D_2^2 - |x_2|^2) + P_3 \log(D_3^2 - |x_3|^2).
$$

The state feedback system is rather simple

$$
\begin{aligned}
x_1' &= x_2 + v, \quad x_2 = x_3 \\
x_3' &= q z_6 (g + \sqrt{g^2 + q z_6})^{-1}
\end{aligned}
$$

where

$$
\begin{aligned}
z_6 &= (z_5 - 2P_3 x_3 (D_3 - \|x_3\|^2))^{-1} \Lambda_3^{-1}, \\
z_5 &= (z_4 - 2P_2 x_2 (D_2 - \|x_2\|^2))^{-1} \Lambda_2^{-1}, \\
z_4 &= -2P_1 x_1 (D_1 - \|x_1\|^2)^{-1} \Lambda_1^{-1}
\end{aligned}
$$

The description of the structure of the invariant set as given at the end of section 3 is demonstrated by this example. This example and several others are analysed in more detail in the DFG Report cited on page 1.

Acknowledgement. The authors thanks the DFG for financial support and J. Stoer for manifold help which he gave during the long period of work on these and related topics.

References

[1] Aubin J.P., A survey of viability theory, SIAM J. on Control and Optimization, 1990, vol.28, pp.749-788, see also pp. 1294-1320.

[2] T. Basar, P. Bernhard, Differential games, Lecture Notes in Control and Information Sciences, vol. 135, Springer 1989.

[3] N.D. Botkin te al., Numerical solution of differential games, in Proc. IV. Intern. Symposium on Differ. Games and Applications, Helsinki, 1990.

[4] R. Bulirsch, F. Montrone, H.J. Pesch, About Landing in the Presence of Windshear as a Minimax Optimal Control Problem, Parts I-II JOTA, vol 70, No. 1 pp. 1-23, No. 2 pp. 223-254.

[5] Control in Dynamical Systems eds. A.I. Subbotin, V.N. Usakhov, Ekaterinenburg (Swerdlowsk), 1990, 126p. (in Russian)

[6] M. A. Dahleh, J.B. Pearson, Optimal Rejection of Bounded Disturbances, IEEE Trans. Aut. Contr., vol. 33, N.8 (August 1988),pp 722-731.

[7] Krasovskii, A.I. Subbotin, Game Theoretical Control Problems, Springer, New York, 1988.

[8] Krjazhimskii, A.V., Osipov Ju. S. On the best approximation of the differentiaton operator within the class of nonanticipatory operators, Matem. Zametki 1985, vol. 37 No. 2 pp. 192-199 in Russian.

[9] Ju.A. Mitropolski and others, Integrable dynamical systems; Spectral and Diff. geometrical Aspects, in Russian, Naukova Dumka, Kiev,1987.

[10] D. Mustafa, Minimum Entropy H_∞ Control, Dissertation, St. John's College, Cambridge, 1989.

[11] L.S. Pontrjagin, Collected Works, vol. II., Moskow, Nauka, 1988, (Gordon - Beach Publ. Comp., New York, 1990).

[12] G. Sonnevend, Structural problems in the theory of bounded control for linear systems, in "Survey of Mathematical Programming", ed. by A. Prekopa, Akademia Kiado, Budapest, 1976, pp. 217-232.

[13] G. Sonnevend, Output regulation in partially observed linear systems under disturbances, in Lecture Notes in Control and Information Sciences, vol. 6 (1978), pp 214-229.

[14] G. Sonnevend, Applications of Analytic Centers, NATO ASI ser., F vol. 70, "Numerical Linear Algebra and Digital Signal Processing" (P. van Dooren, and G.Golub eds.), Reidel 1988, 8p,

[15] G. Sonnevend, Application of analytic centers to feedback design for systems with uncertainties, in "Control on Uncertain Systems", eds. D. Hinrichsen, B. Martenson, Birkhäuser, 1989, 13p.,

[16] G. Sonnevend, Construction of ellipsoidal observers in linear games with bounded controls and measurement noise, in Lect. Notes in Contr. and Inf. Sci., vol 143 (1989), 413-423,

[17] G. Sonnevend, Existence and numerical computation of extremal invariant sets in linear differential games, Lect.Notes in Contr. and Inf. Sci.,vol.22, (1981) pp. 251-260.

[18] G. Sonnevend, J. Stoer, Global ellipsoidal approximations and homotopy methods for smooth, convex, analytic programs, in Applied Math. and Appl., 21(1980) pp. 139-165.

[19] G. Sonnevend, Sequential algorithms of optimal order global error for the uniform recovery of functions with monotone (r-1)-th derivatives, Analysis Mathematica, 10(1984) pp. 311-335.

[20] G. Sonnevend, J. Stoer, G. Zhao, On the complexity of following the central path by linear extrapolation in linear programs, Mathematical Programming, 1991,series B , 31p.

[21] A.I. Subbotin, Minimax Inequalities and the Hamilton Jacobi Equations, in Russian, Nauka, Moscow, 1992.

[22] N.N. Subbotina, Cauchy's Method of Characterics and Generalized Solutions of the Hamilton-Jacobi Equation, submitted to USSR Dokl. Akad. Nauk.

[23] Uncerainty and control, ed. by J. Ackermann, Lect. Notes in Control and Information Sciences, vol. 80, Springer V., Berlin, 1985.

[24] K.H. Well et al., Optimization of tactical aircraft maneuvers utilizing high angles of attack, Journal of Guidance, Control, Dynamics, Vol.5, No 2., pp.131-137.

Author's address

Prof. Dr. G. Sonnevend
Department of Numerical Mathematics
Eötvös University
Muzeum Körut 6 - 8
H-1088 Budapest
Hungary

Applications to Mechanical and Aerospace Systems

Singular Perturbation Time–Optimal Controller for Disk Drives

Mark D. Ardema, Evert Cooper

1. Introduction

One of the most important performance goals of a disk file actuator is rapid access time. The disk file actuator as incorporated within its magnetic head/disk assembly is a high-order, flexible dynamic system with unpredictable behavior due to manufacturing tolerances and temperature variations. The controller must operate within demanding limits of response time, power consumption, and storage capacity. Although the goal of rapid access time argues for a time optimal control law, the high system order necessitates approximation techniques. Further, the desire to maintain near time-optimality in the presence of temperature fluctuations and time-varying unmodeled dynamic effects motivates the need for an adaptive approach. The disk drive control problem is summarized in Reference 1.

Figure 1 shows the over-all architecture of a possible disk drive actuator controller. The time-optimal controller computes a control signal based on a reduced-order reference model of the true system. This control signal is augmented by a signal produced by an adaptive control loop. Only the optimal controller and its associated reference model are of interest in this paper.

It is highly desirable to construct a reference model closely representative of the true plant to minimize the efforts of the adaptive loop and to make the performance near-optimal. The main problem is that the system order must be severely limited. Although there are general algorithms for computing time-optimal controls for high-order systems (Ref. 2) and approximate methods for third-order systems (Ref. 3), these are impractical for our purpose, and analysis must be restricted to second order systems.

An attractive method for dealing with prohibitively high-order systems is to use singular perturbation theory. This method separates a dynamic system into reduced-order subsystems on different time scales, and provides a way to synthesize a near-optimal controller by combining the two reduced system controllers (see, for example, Ref. 4 and 5). This approach has been investigated for linear time-optimal systems in References 5 and 6. It is found that the optimal control of the full system is much like the optimal control of the reduced system (perturbation parameter set to zero) except

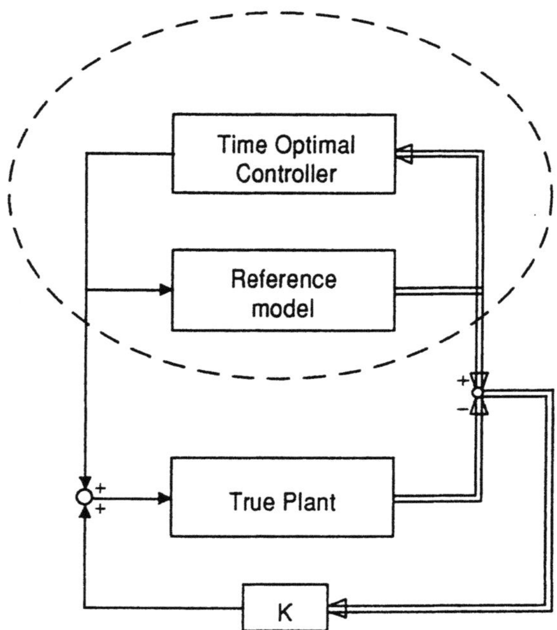

Figure 1. Schematic of disk drive actuator control system.

that there are rapid control switches near the end of the process to bring to rest the dynamics neglected by the reduced solution.

In this paper, we analyze a third-order disk drive actuator model as a singular perturbation problem. The perturbation parameter appears in such a way as to make the reduced problem a simple damped double integrator, for which the time optimal control is well-known and easily implementable. We then consider the boundary layer problem to obtain corrections to the reduced solution. Of particular interest is determination of the additional control switch time that appears in the third-order system but not in the reduced system. The analysis is restricted to zero order terms in the perturbation parameter.

Finally, we simulate the performance of the singular perturbation open-loop controller for typical values of the system parameters. As expected, the controller performs very well if the boundary layers are sufficiently small.

2. Problem Formulation and Reduced Problem

A mathematical model of a contemporary disk drive actuator control system is as follows:

$$
(2.1)\quad
\begin{bmatrix} \dot{\psi}_1 \\ \dot{\psi}_2 \\ \dot{I}_1 \\ \dot{I}_2 \\ \dot{x}_5 \\ \dot{x}_6 \end{bmatrix}
=
\begin{bmatrix}
-2\zeta\omega_n & -\omega_n^2 & 0 & 0 & 0 & 0 \\
1 & 0 & 0 & 0 & 0 & 0 \\
0 & \dfrac{L_2+L_3}{D} & -\dfrac{R_1(L_2+L_3)}{D} & -\dfrac{R_2 L_3}{D} & -\dfrac{K_f(L_2+L_3)}{D} & 0 \\
0 & \dfrac{L_3}{D} & -\dfrac{L_3 R_1}{D} & -\dfrac{R_2(L_2+L_3)}{D} & -\dfrac{K_f L_3}{D} & 0 \\
0 & 0 & \dfrac{K_f}{M} & 0 & 0 & 0 \\
0 & 0 & 0 & 0 & 1 & 0
\end{bmatrix}
\begin{bmatrix} \psi_1 \\ \psi_2 \\ I_1 \\ I_2 \\ x_5 \\ x_6 \end{bmatrix}
$$

$$
+
\begin{bmatrix} \omega_n^2 \\ 0 \\ 0 \\ 0 \\ 0 \\ 0 \end{bmatrix}
u
$$

The first two equations of this model represent the power amplifier, the second two the actuator currents, and the last two the dynamics of the actuator mass. In these equations, L_2, L_3, R_1, R_2 and D are electrical constants, K_f is the constant relating current to force, and M is the actuator mass. It is assumed that the actuator mass, as well as the disk drive housing, are rigid.

The model (2.1) is useful for design studies but is too complicated for use in a real-time controller. For this latter purpose, a third order model is adopted:

$$
(2.2)\quad
\begin{aligned}
\dot{x} &= y \\
\dot{y} &= z \\
\varepsilon \dot{z} &= -az - by + u
\end{aligned}
$$

The first two of these equations model the motion of the acuator mass and the third its electronic controller. The $-by$ term is the back *emf*; a third order system without this term was studied in Reference 7.

We wish to transfer the system from one rest position to another, the latter taken to be the state space origin,

$$
(2.3)\quad
\begin{aligned}
x(0) &= x_0, & x(t_f) &= 0 \\
y(0) &= 0, & y(t_f) &= 0 \\
z(0) &= 0, & z(t_f) &= 0
\end{aligned}
$$

while minimizing the transfer time:

$$
(2.4)\qquad J = \int_0^{t_f} dt
$$

The control u is bounded, $|u| \le 1$.

The necessary conditions for optimal control are easily stated. For this purpose, form the Hamiltonian function

$$(2.5) \qquad H = -1 + y\lambda_x + z\lambda_y + (-az - by + u)\lambda_z$$

and the adjoint equations

$$(2.6) \qquad \begin{aligned} \dot{\lambda}_x &= 0 \\ \dot{\lambda}_y &= -\lambda_x + b\lambda_z \\ \epsilon\dot{\lambda}_z &= -\lambda_y + a\lambda_z \end{aligned}$$

The Maximum Principle (Ref. 8) then gives the optimal control as

$$(2.7) \qquad u = sgn\lambda_z$$

Lemmas 14.1 and 14.2 of Ref. 8 establish that **extremal** control is bang-bang for this problem and that, for ϵ sufficiently small (specifically $\epsilon < a^2/4b$), there are at most two control switches. From (2.7) these switches occur at the zeros of λ_z. The adjoint equations (2.6) may be integrated to give

$$(2.8) \qquad \begin{aligned} \lambda_x &= C_1 \\ \lambda_y &= C_2 e^{s_1 t} + C_3 e^{s_2 t} + \frac{aC_1}{b} \\ \lambda_z &= \frac{C_2 s_1}{b} e^{s_1 t} + \frac{C_3 s_2}{b} e^{s_2 t} + \frac{C_1}{b} \end{aligned}$$

for ϵ sufficiently small, where

$$(2.9) \qquad s_{1,2} = \frac{a \pm \sqrt{a^2 - 4b\epsilon}}{2\epsilon}$$

Thus the control switch times depend on the constants C_1, C_2, and C_3, but determining these constants is nontrivial.

To obtain the reduced problem associated with system (2.2), set $\epsilon = 0$.

$$(2.10) \qquad \begin{aligned} \dot{x}_r &= y_r \\ \dot{y}_r &= -\frac{b}{a}y_r + \frac{1}{a}u_r \\ z_r &= -\frac{b}{a}y_r + \frac{1}{a}u_r \end{aligned}$$

with

$$(2.11) \qquad \begin{aligned} x_r(0) &= x_0, \quad x_r(t_{f_r}) = 0 \\ y_r(0) &= 0, \quad y_r(t_{f_r}) = 0 \end{aligned}$$

and

$$(2.12) \qquad J = \int_0^{t_{f_r}} dt$$

The solution to this problem is elementary. Assuming that $x_0 < 0$, there is one switch, at say $t = t_{s_r}$ from $u = +1$ to $u = -1$.

The solutions for t_{s_r} and t_{f_r} are

(2.13)
$$t_{s_r} = -\frac{a}{b}\ln(1 - K)$$
$$t_{f_r} = \frac{a}{b}\ln\left(\frac{1 + K}{1 - K}\right)$$

where $K = \sqrt{1 - e^{\frac{b^2 x_0}{a}}}$, and the values of the states at t_{s_r} are

(2.14)
$$x_{s_r} = x_0 - \frac{a}{b^2}[K + \ln(1 - K)]$$
$$y_{s_r} = \tfrac{1}{b}K$$

The adjoint variables, $0 \le t \le t_{f_r}$ are found to be

(2.15)
$$\lambda_{x_r} = \frac{b}{K}$$
$$\lambda_{y_r} = \frac{a}{K}(K - 1)e^{\frac{b}{a}t} + \frac{a}{K}$$

It is of interest to compare the switching function of the reduced problem (2.15b), with the switching function of the full problem, (2.8c). The latter, for ϵ sufficiently small, consists of two time-varying terms of approximately the forms $e^{at/\epsilon}$ and $e^{bt/a}$. Thus the two switching functions differ by a boundary layer type term, as expected.

3. Asymptotic Analysis and Control Law

The asymptotic analysis proceeds by dividing the motion into five segments as follows: (1) an initial boundary layer in which the z state variable rapidly and asymptotically approaches its equilibrium value; (2) an outer region ending at the first switch time; (3) an interior boundary layer beginning at the first switch time in which z approaches its new equilibrium value; (4) an outer region ending at the second switch time; and (5) a terminal boundary layer.

To zero order, the boundary layer motions, which are asymptotically stable, take place in zero time while the slow variables remain frozen at their boundary values. Consequently, the zero-order solution for the slow variables is exactly the same as for the reduced problem. The only effect of the boundary layer motions, to zero-order, is to bring the fast state z to zero after the slow states have already been brought to their final values. Thus there is a second switch at time t_{f_r} and the process ends a time increment of order ϵ later. There is no change in the first switch time due to the boundary layers.

To derive an expression for t_f, we consider the terminal boundary layer. The boundary layer is obtained by stretching time-to-go by introducing a new independent variable $\alpha = (t_f - t)/\epsilon$ in (2.2) and (2.6) and then setting $\epsilon = 0$ and $u = 1$:

$$(3.1) \qquad \begin{aligned} \frac{dx_b}{d\alpha} &= 0 & \frac{d\lambda_{x_b}}{d\alpha} &= 0 \\[2mm] \frac{dy_b}{d\alpha} &= 0 & \frac{d\lambda_{y_b}}{d\alpha} &= 0 \\[2mm] \frac{dx_b}{d\alpha} &= az_b + by_b - 1 & \frac{d\lambda_{z_b}}{d\alpha} &= \lambda_{y_b} - a\lambda_{z_b} \end{aligned}$$

$$(3.2) \qquad x_b(0) = 0, \qquad y_b(0) = 0, \qquad z_b(0) = 0$$

and the solution is

$$(3.3) \qquad \begin{aligned} x_b &= 0, & y_b &= 0, & z_b &= \tfrac{1}{a} - \tfrac{1}{a}e^{a\alpha} \\[1mm] \lambda_{x_b} &= C_4, & \lambda_{y_b} &= C_5, & \lambda_{z_b} &= C_6 e^{-a\alpha} + C_5 \end{aligned}$$

The constants of integration C_4, C_5, and C_6 are found by matching with the second outer solution. The result is, for λ_z,

$$(3.4) \qquad \lambda_{z_b} = 2e^{-a\alpha} - 1$$

Setting this to zero gives the second switch time, t_{s_2},

$$(3.5) \qquad \lambda_{z_b}(t_{s_2}) = 2e^{-a\alpha_{s_2}} - 1 = 0$$

Thus $\alpha_{s_2} = \tfrac{1}{a}\ln 2$ and, solving for t_f

$$(3.6) \qquad t_f = \frac{a}{b}\ln\left(\frac{1+K}{1-K}\right) + \frac{\epsilon}{a}\ln 2$$

The zero-order, open-loop control algorithm is now easily stated. First, precompute $t_{s_1}(= t_{s_r})$, $t_{s_2}(= t_{f_r})$, and t_f from (2.13) and (3.6). Then begin the process with control $u = 1$, and when $t = t_{s_1}$, switch the control to $u = -1$. When $t = t_{s_2}$, switch the control back to $u = +1$ and end the process at $t = t_f$.

An alternative to the open-loop controller just stated, would be to construct a zero-order composite representation of the switching function λ_z. The switching function in the second outer region is given by:

$$(3.7) \qquad \lambda_{z_0} = \frac{K-1}{K}e^{\frac{bt}{a}} + \frac{1}{K}$$

and that in the terminal boundary layer is given by (3.4). Taking limits

$$(3.8) \qquad \lim_{t \to t_f} \lambda_{z_0} = -1 = \lim_{\alpha \to \infty} \lambda_{z_b}$$

shows that these functions match and that the common part is -1. Consequently, the composite function is

$$\lambda_{z_c} = \lambda_{z_0} + \lambda_{z_b} - \text{common part}$$

$$(3.9) \qquad \lambda_{z_c} = \frac{K-1}{K} e^{\frac{bt}{a}} + \frac{1}{K} + 2 \left(\frac{1-K}{1+K} \right)^{\frac{a^2}{\epsilon b}} e^{\frac{at}{\epsilon}}$$

The two switch time t_{s_1} and t_{s_2} are then the roots of the equation:

$$(3.10) \qquad 0 = \frac{K-1}{K} e^{\frac{bt_s}{a}} + \frac{1}{K} + 2 \left(\frac{1-K}{1+K} \right)^{\frac{a^2}{\epsilon b}} e^{\frac{at_s}{\epsilon}}$$

The zero-order solution will be numerically accurate if the fast motion of the variable z is negligible just prior to the switch times. For the interval $t \in [0, \, t_{s_1}]$, this will be true if $e^{-at_{s_1}/\epsilon} \ll 1$, or

$$(3.11) \qquad (1-K)^{\frac{a^2}{\epsilon b}} \ll 1$$

This shows that the numerical accuracy of the zero order approximation depends on the initial condition x_0 as well as the system parameters a, b, and ϵ.

4. Numerical Examples

In this section, we apply the open-loop, zero-order control algorithm just derived to a simulation program that numerically integrates (2.2) subject to initial conditions (2.3). The nominal values of the parameters are $a = 0.00145$, $b = 0.332$, and $\epsilon_0 = 0.702 \times 10^{-7}$; these values are for a typical contemporary disk drive and meet the requirement $\epsilon < a^2/4b$.

We first consider a move of ten tracks on the disk (approximately 1.13×10^{-4} meters) with $\epsilon = \epsilon_0$. For this case $K = 0.0925$ and the inequality (3.11) is $0.00016 \ll 1$, indicating a very high degree of time-scale separation. This is shown on Figure 2 which plots the time histories of the state variables x, y, and z. It is seen that z has attained its steady state values to a high degree of accuracy before the control switches, and that all states are near zero at the final time. Figure 3 shows the case $\epsilon = .1\epsilon_0$ for a move of ten tracks; as expected this case shows an extremely high degree of accuracy.

Simulations for several values of ϵ were run for a move of ten tracks on the disk, and the resulting final values of the state variables are plotted on Figure 4. The error in the mechanical variables x and y at the final time approach zero as ϵ approaches zero, but there is a divergence in z; this is due to numerical error caused by the increasing stiffness of the system. All the errors shown on Figure 3 are very small and acceptable in actual systems.

Shorter moves on the disk are the most challenging for any control system, and we next consider the shortest move, a move of one track, or about 1.13×10^{-5} meters. In this case, $K = 0.0293$ and inequality (3.11) is $0.068 \ll 1$, indicating a moderate degree of time-scale separation. Consequently, the error in the states at the final time is greater than for moves of 10 tracks (Figure 5). For $\epsilon = .1\epsilon_0$, the accuracy is again quite good (Figure 6). The error in the final values of three states is shown on Figure 7.

In Figures 8 and 9 we investigate using the zero-order composite switching function (3.9) to determine switch times t_{s_1} and t_{s_2}. For a move of ten tracks (Figure 8), the first switch time is independent of ϵ for $0 < \epsilon \leq \epsilon_0$, and the second switch occurs at a time increment before the final time proportional to ϵ. For a move of one track (Figure 9), λ_z has no zeros for ϵ greater than about $.9\epsilon_0$, indicating an inadequate approximation. For $0 < \epsilon < .5\epsilon_0$, the switch times have the same characteristics as for the ten track move. Note that the final time predicted by this method is the same as the reduced problem final time.

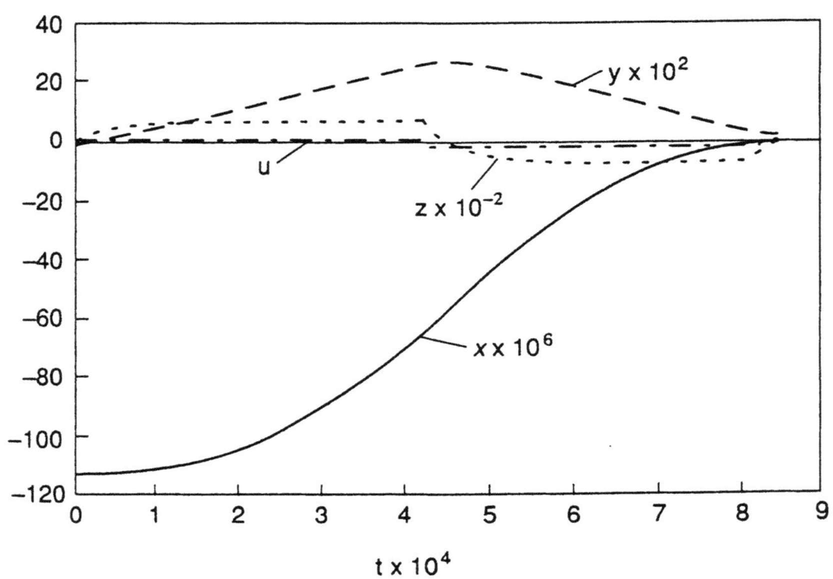

Figure 2. Time Histories of State Variables for x_0 = ten tracks and $\varepsilon = \varepsilon_0$.

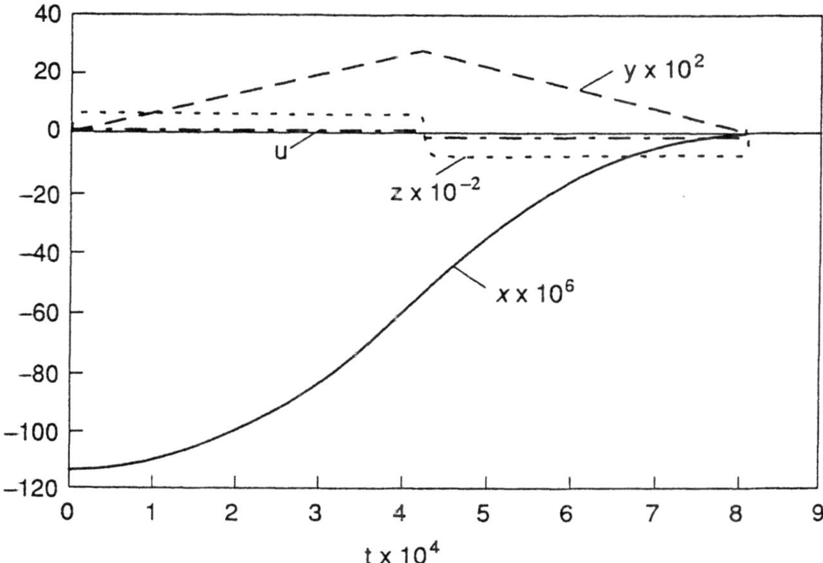

Figure 3. Time Histories of State Variables for $x_0 =$ ten tracks and $\varepsilon = 0.1\varepsilon_0$.

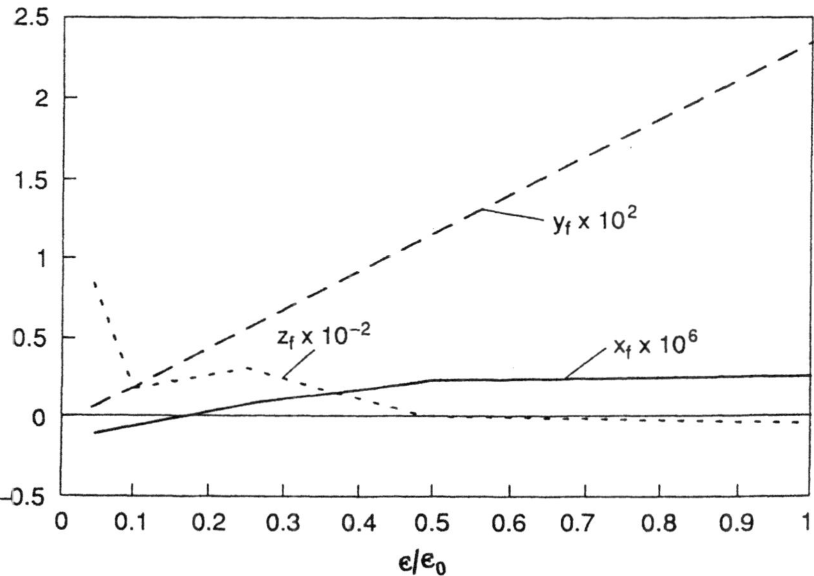

Figure 4. Effect of ε for $x_0 =$ ten tracks.

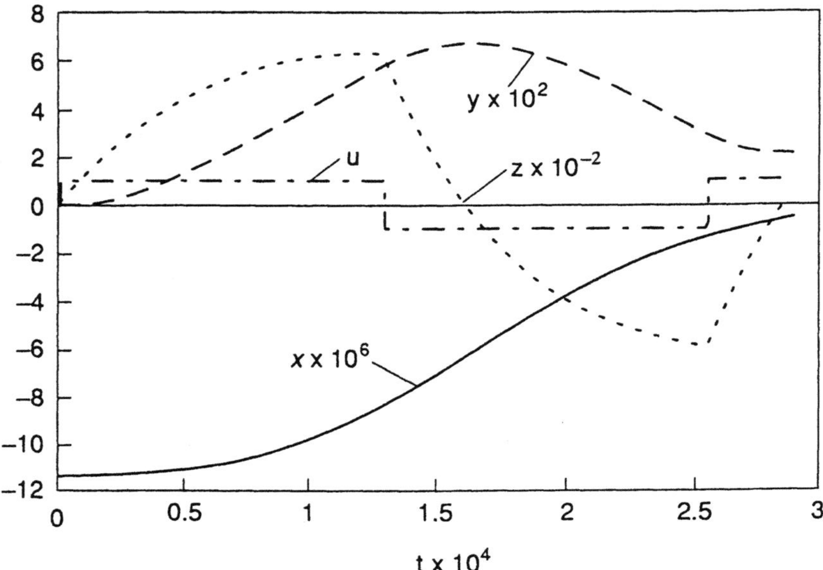

Figure 5. Time Histories of State Variables for x_0 = one track and $\varepsilon = \varepsilon_0$.

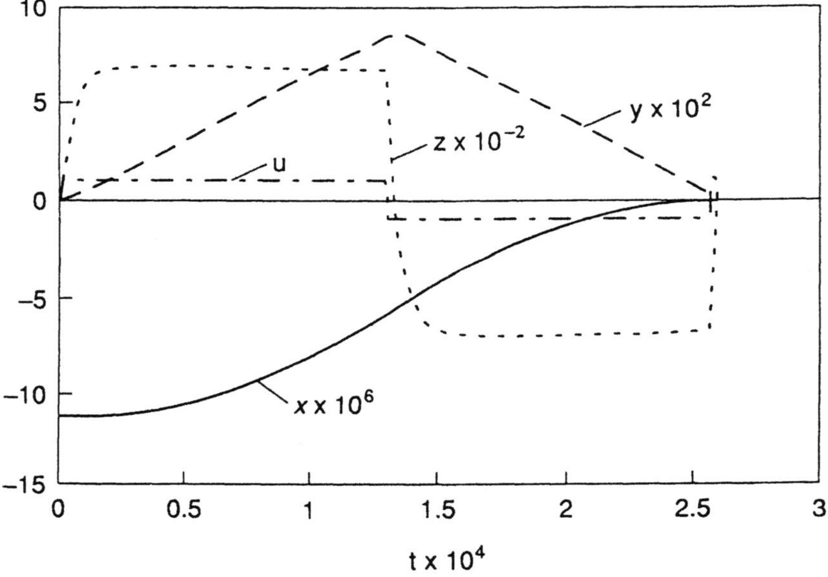

Figure 6. Time Histories of State Variables for x_0 = one track and $\varepsilon = 0.1\varepsilon_0$.

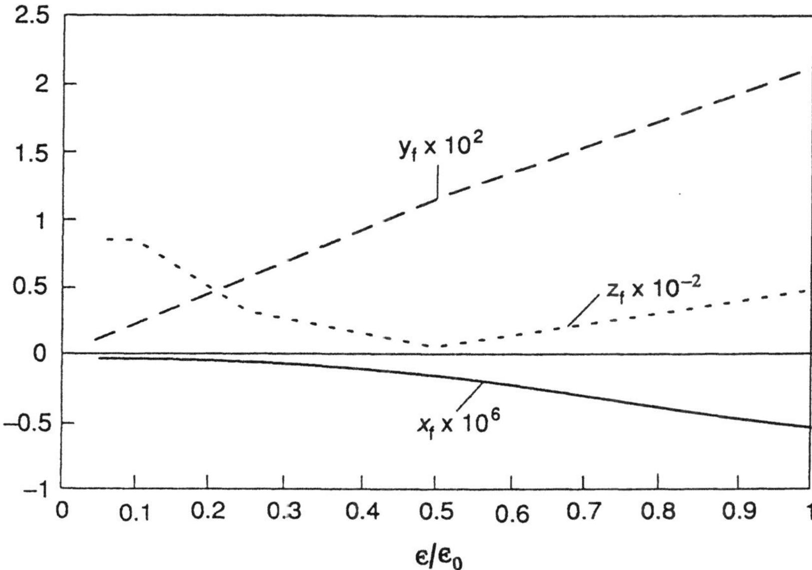

Figure 7. Effect of ϵ for $x_0 =$ one track.

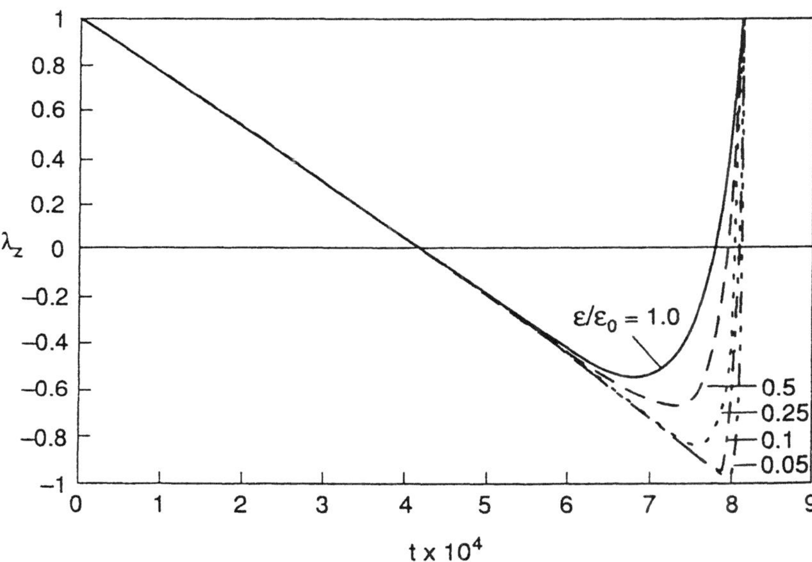

Figure 8. Composite Switching Function for $x_0 =$ ten tracks.

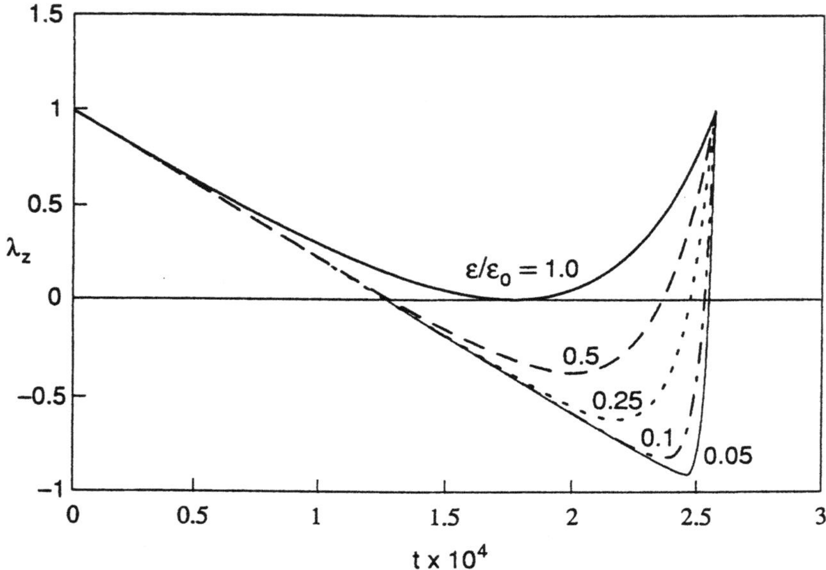

Figure 9. Composite Switching Function for $x_0 =$ one track.

5. Concluding Remarks

The use of singular perturbation methods for deriving near-time-optimal control laws for disk file actuator systems has been investigated. A typical disk drive actuator was modeled as a third order system with a small parameter. A control law giving the two control switch times was constructed from analysis of a sequence of reduced order problems. The control law is easy to implement and numerical simulations have shown that it gives good accuracy, even for short moves of the actuator.

References

1. Cooper, E., *Minimizing Power Dissipation in a Disk File Actuator*, IEEE Trans. on Magnetics, Vol. 24, No. 3, May 1988.

2. Yastreboff, M., *Synthesis of Time-Optimal Control by Time Interval Adjustment*, IEEE Trans. Auto. Control, Dec. 1969, pp. 707.

3. Kassam, S.A., Thomas, J.B., and McCrumm, J.D., *Implementation of Sub-Optimal Control for a Third-Order System*, Comput. Elect. Engng., Vol. 2, 1975, pp. 307.

4. Ardema, M.D., *An Introduction to Singular Perturbations in Nonlinear Optimal Control*, Singular Perturbations in Systems and Control, M.D. Ardema, ed., International Centre for Mechanical Sciences, Courses and Lectures No. 280, 1983.

5. Kokotovic, P.V., Khalil, H.K., and O'Reilly, J., *Singular Perturbation Methods in Control: Analysis and Design*, Academic Press, 1986.

6. Kokotovic, P.V. and Haddad, A.H., *Controllability and Time-Optimal Control of Systems with Slow and Fast Modes*, IEEE Trans, Auto Control, Feb. 1975, pp. 111.

7. Ardema, M.D. and Cooper, E., *Perturbation Method for Improved Time-Optimal Control of Disk Drives*, Lecture Notes in Control and Information Sciences, Vol. 151, J.M. Skowronski et. al. (eds), Springer-Verlag, 1991, pp. 37.

8. Leitmann, G., *The Calculus of Variations and Optimal Control*, Plenum, 1981.

Authors' addresses

Mark D. Ardema, Department of Mechanical Engineering,
Santa Clara University, Santa Clara, CA 95053, U.S.A.

Evert Cooper, Department of Electrical Engineering,
Santa Clara University, Santa Clara, CA 95053, U.S.A.

Optimal Design of Elastic Bars

Leszek Mikulski

Abstract

The paper concerns a strength optimization of elastic bars with variable cross-section. Bars are subjected to a dead weight and useful external load. The volume of the element or the deflection at the chosen point have been assumed as the cost function. Width of the cross-section is the control variable. The added constraints concern strength and geometrical restrictions. The multiple-shooting method is then used for the numerical solution of the problem.

1   Introduction

There are many reasons for the great importance of optimization in mechanics. The first reason is of economic nature. Beams with minimal weight and bars with minimal material expenditure are examples of this type.

The second reason has a purely mechanical qualitative character. Solutions, which from the mechanical point of view guarantee the best features of a system, are worth the trouble. Investigations of optimal design are undertaken in the hope that the obtained optimal elements can be applied in practice. However, even when this is impossible, they can be a bench mark of efficiency for practical design. Today, the number of papers in the fields of optimal design of construction exceeds 5,000. Many of them are discussed in detail in various monographs and handbooks.

In the present paper optimal control theory is used, by which the optimization problem can be reduced to a multipoint boundary value problem for ordinary differential equations. The multipoint shooting method [1] can then be applied for the numerical solution of these problems. The size of the multipoint boundary value problem (MP-BVP) is usually an essential numerical limitation.

In some papers, such a limitation of the size is due to sophisticated substitutions. There are problems of optimal design with respect to stability [2].

However, it turns out that only a few papers among such a great number deal with optimal design under constraints of the state variables [5].

The usefulness of developments of optimal design for bars with state variable constraints is manifest because constraints are frequently imposed upon deflections or stresses. Therefore, the knowlege of optimal design when constraints are active is very useful.

The present paper contains a series of concrete solutions of optimization problems for various bar elements. These solutions have been obtained by means of the BNDSCO [3] software package. The constraints of the state variables play a significant role in the problems considered. The constraints describe some side conditions which must be fulfilled, for example: conditions of constant bar volume, geometric constraints of control and constraint of stresses and deflections. Either the width of the bar is chosen as the control (when the height is fixed) or the height (when the width is fixed). The weight of the element or the deflection at defined point are the cost functions.

2   Formulation of the Task

In this section the particular solutions of optimal design problems for bars subjected to bending and tension will be shown. The bars with various cross-sections are designed for a minimum of weight or of deflection at a defined point. The height or the width of the bar with rectangular cross-section is taken as the control [6].

The side conditions restrict the control, normal stresses and deflections. The optimal design task for bars can be comprised in the following formula:

- State Equations

$$
\begin{aligned}
y_1' &= y_2 \\
y_2' &= \frac{(y_3 - ny_1)(1 + y_2{}^2)}{S_1 S_2{}^3} \\
y_3' &= y_4 \\
y_4' &= -A_1 - A_2 S_1 S_2
\end{aligned}
\tag{2.1}
$$

$$
A_1 = \frac{q_0 l^3}{EI_0} \quad , \qquad A_2 = \frac{p b_0 h_0 l^3}{EI_0}
$$

$$
S_1 = \frac{b}{b_0} \quad , \qquad S_2 = \frac{h}{h_0} \quad , \qquad \xi = \frac{x}{l} \quad , \qquad n = \frac{N l^2}{EI_0} \quad ,
$$

$$y_1 = \frac{\overline{y_1}}{l} \quad , \qquad y_3 = \frac{Ml}{EI_c} \quad , \qquad y_4 = \frac{Ql}{EI_0}$$

$\overline{y_1}$ – deflection, y_2 – angle of rotation, M – bendig moment, Q – shearing force, N – longitudinal force, I_0 – moment of inertia.
Boundary conditions for a beam with free ends

$$(2.2) \qquad \begin{aligned} y_1(0) &= 0 \quad , \qquad y_3(0) = 0 \quad , \\ y_1(1) &= 0 \quad , \qquad y_3(1) = 0 \quad . \end{aligned}$$

for fixed beam

$$(2.3) \qquad \begin{aligned} y_1(0) &= 0 \quad , \qquad y_2(0) = 0 \quad , \\ y_1(1) &= 0 \quad , \qquad y_2(1) = 0 \quad . \end{aligned}$$

- Cost Function

  - deflection at a fixed point of the bar

    $$(2.4) \qquad I(S) = y_1(\xi_k)$$

  - weight of the element

    $$(2.5) \qquad I(S) = \int_0^1 G_1 S_1 S_2 d\xi \quad , \qquad G_1 = \gamma b_0 h_0$$

- Constraints

  - geometrical

    $$(2.6) \qquad U_1 \leq S_i \leq U_2 \qquad i = 1,2 \qquad q_1 = S_1 - U_1 \geq 0$$

  - of strength

    $$(2.7) \qquad q_2 = \frac{n A_3{}^2 A_4}{12 S_1 S_2} + \frac{y_3 A_3 A_4}{2 S_1 S_2{}^2} - 1 \leq 0$$

    $$A_3 = \frac{h_0}{l} \quad , \qquad A_4 = \frac{E}{\sigma_0}$$

  - of displacements

    $$(2.8) \qquad y_1 \leq F(x)$$

  - constant volume

    $$(2.9) \qquad \int_0^1 S_1 S_2 d\xi = 1 \quad .$$

3   Necessary Conditions

Necessary conditions for the optimal solution are given in [4].
The Hamilton function has the form:

(3.1)
$$H = \lambda_0 S_1 S_2 + \lambda_1 y_2 + \lambda_2 \frac{(y_3 - n y_1)(1 + y_2{}^2)^{\frac{3}{2}}}{S_1 S_2{}^3}$$
$$+ \lambda_3 y_4 + \lambda_4(-A_1 - A_2 S_1 S_2) \quad .$$

The control results from the condition

(3.2)
$$\frac{\partial H}{\partial S_1} = -\frac{\lambda_2(1 + y_2{}^2)^{\frac{3}{2}}(y_3 - n y_1)}{S_2{}^3 S_1^2} + \lambda_0 S_2 - \lambda_4 A_2 S_2 = 0$$

(3.3)
$$\frac{\partial H}{\partial S_2} = -\frac{3\lambda_2(1 + y_2{}^2)^{\frac{3}{2}}(y_3 - n y_1)}{S_1 S_2{}^4} + \lambda_0 S_1 - \lambda_4 A_2 S_1 = 0$$

(3.4)
$$S_1{}^2 = \frac{\lambda_2(1 + y_2{}^2)^{\frac{3}{2}}(y_3 - n y_1)}{\lambda_0 - \lambda_4 A_2} \quad , \qquad S_2 = 1.0$$

(3.5)
$$S_2{}^4 = \frac{3\lambda_2(1 + y_2{}^2)^{\frac{3}{2}}(y_3 - n y_1)}{\lambda_0 - \lambda_4 A_2} \quad , \qquad S_1 = 1.0$$

The Legendre-Clebsch condition is fulfilled if

(3.6)
$$\lambda_2(y_3 - n y_1) \geq 0 \quad .$$

The corresponding system of the adjoint equations has the form:

(3.7)
$$\lambda_1' = \frac{\lambda_2 n (1 + y_2{}^2)^{\frac{2}{3}}}{S_1 S_2{}^3} \quad ,$$
$$\lambda_2' = -\lambda_1 - \frac{3\lambda_2 y_2 (y_3 - n y_1)(1 + y_2{}^2)^{\frac{1}{2}}}{S_1 S_2{}^3} \quad ,$$
$$\lambda_3' = -\frac{\lambda_2(1 + y_2{}^2)^{\frac{3}{2}}}{S_1 S_2{}^3} \quad ,$$
$$\lambda_4' = -\lambda_3 \quad .$$

Using the transversality conditions, one obtains

$$(3.8) \qquad \begin{array}{ll} \lambda_2(0) = 0 \ , & \lambda_4(0) = 0 \ , \\ \lambda_1(\tfrac{1}{2}) = -1 \ , & \lambda_3(\tfrac{1}{2}) = 0 \ . \end{array}$$

The constant volume is taken into account by an additional equation

$$(3.9) \qquad \begin{array}{c} y_0' = S_1 S_2 \ , \\ y_0(0) = 0 \ , \qquad y_0(\tfrac{1}{2}) = 0.5 \ . \end{array}$$

The state equations (1), (18) the adjoint equations (16) with the boundary condi-
tions (2), (3), (17) form a so-called two-point boundary-value problem for the variables
$(y_0, y_1, y_2, y_3, y_4, \lambda_0, \lambda_1, \lambda_2, \lambda_3, \lambda_4)$.
Expressions for the optimal control are given by (13), (14).
• Geometrical and Strength Constraints
The Hamilton function contains additional terms

$$(3.10) \qquad H_1 = H + \nu_1(S_1 - U_1) + \nu_2\Big(\frac{nA_3^2 A_4}{12 S_1 S_2} + \frac{y_3 A_3 A_4}{2 S_1 S_2^2} - 1\Big) \ .$$

If the constraint U_1 is active then $\nu_1 \geq 0$. In this case, the adjoint equations need not
be modified and the constraint is active in certain intervals.
If the constraint of the strength is active then $\nu_2 \geq 0$ and the control

$$(3.11) \qquad S_1 = \frac{nA_3^2 A_4}{12 S_2} + \frac{y_3 A_3 A_4}{2 S_2^2} \ ,$$

is active in certain intervals. In this case, however, the adjoint equation undergoes a
modification,

$$(3.12) \qquad \lambda_3' = -\frac{\lambda_2(1 + y_2^2)^{\frac{3}{2}}}{S_1 S_2^3} - \nu_2 \frac{A_4 A_3}{2 S_1 S_2^2} \ .$$

In the case of geometrical and strength constraints the optimal design has the form:

$$(3.13) \qquad S_1 = \begin{cases} \sqrt{\dfrac{\lambda_2(1 + y_2^2)^{\frac{3}{2}}(y_3 - n y_1)}{\lambda_0 - \lambda_4 A_2}} \ , & q_i < 0, \ i = 1, 2 \\[2ex] U_1 \ , & q_1 = 0 \\[2ex] \dfrac{nA_3^2}{12 S_2} + \dfrac{y_3 A_3 A_4}{2 S_2^2} \ , & q_2 = 0 \ . \end{cases}$$

Hence, we have obtained a multipoint boundary-value problem given in $(1, 16)$ with piecewise-defined right-hard sides (22) with switching function and expressions for the optimal design. The number of the switching points ξ_1 must be known ahead (a priori). The multiple shooting method can now be used to solve this MPBVP.

The results obtained by means of the program BNDSCO [3] is shown in graphs (1).

4 Conclusion

Minimum cost solutions for bars with variable cross-section are presented. These solution show that the inequality constraints associated with the geometry and the strength of the bars play an important role.

The obtained multipoint boundary-value problem has been solved with the help of the BNDSCO program. Activity of constraints and the form of the optimal design for the various boundary conditions have been show.

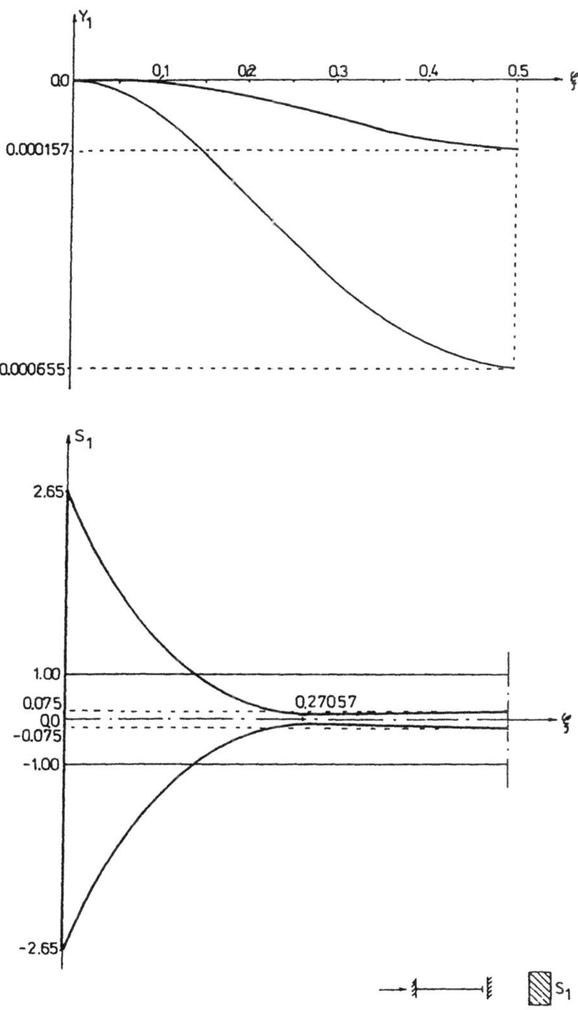

Fig.1 Optimal bar with the strength constraints

References

[1] R. Bulirsch: *Die Mehrzielmethode zur numerischen Lösung von nichtlinearen Randwertoroblemen und Aufgaben der optimalen Steuerung*, Report, Cral-Cranz Gesellschaft (1971)

[2] A. Gajewski and M. Życzkowski: *Optymalne kształtowanie ustrojów prętowych przy warunkach stateczności*, Wybrane zagadnienia stateczności konstrukcji. ISBN 83-04-02408X 133-226 (1987)

[3] H.J. Oberle and W. Grimm: *BNDSCO-A program for the numerical solution of optimal control problems*, DLR IB 515-89/22

[4] H.J. Pesch: *Real-time computation of feedback controls for constrained optimal control problems P.1*, Optimal Control Applications & Methods Vol.10,129-145 (1989)

[5] H. Maurer and H.D. Mittelmann: *The non-linear beam via optimal control with bounded state variables*, Optimal Control Applications & Methods Vol.12,19-31 (1991)

[6] L. Mikulski: *Die Anwendung der Mehrzielmethode zur Optimierung von elastischen Stäben*, Operations Research Proceedings DGOR Vorträge der 19. Jahrestagung ISBN 3-540-55081-X Springer-Verlag 1992.

Author's address

Leszek Mikulski, Institut für Baumechanik der TU Krakau
Warszawska 24, PL-31-155 KRAKÓW, Polen

Combining Direct and Indirect Methods in Optimal Control: Range Maximization of a Hang Glider

R. Bulirsch, E. Nerz, H. J. Pesch, O. von Stryk

Abstract. When solving optimal control problems, indirect methods such as multiple shooting suffer from difficulties in finding an appropriate initial guess for the adjoint variables. For, this initial estimate must be provided for the iterative solution of the multipoint boundary-value problems arising from the necessary conditions of optimal control theory. Direct methods such as direct collocation do not suffer from this problem, but they generally yield results of lower accuracy and their iteration may even terminate with a non-optimal solution. Therefore, both methods are combined in such a way that the direct collocation method is at first applied to a simplified optimal control problem where all inequality constraints are neglected as long as the resulting problem is still well-defined. Because of the larger domain of convergence of the direct method, an approximation of the optimal solution of this problem can be obtained easier. The fusion between direct and indirect methods is then based on a relationship between the Lagrange multipliers of the underlying nonlinear programming problem to be solved by the direct method and the adjoint variables appearing in the necessary conditions which form the boundary-value problem to be solved by the indirect method. Hence, the adjoint variables, too, can be estimated from the approximation obtained by the direct method. This first step then facilitates the subsequent extension and completition of the model by homotopy techniques and the solution of the arising boundary-value problems by the indirect multiple shooting method. Proceeding in this way, the high accuracy and reliability of the multiple shooting method, especially the precise computation of the switching structure and the possibility to verify many necessary conditions, is preserved while disadvantages caused by the sensitive dependence on an appropriate estimate of the solution are considerably cut down. This procedure is described in detail for the nu-

---

merical solution of the maximum-range trajectory optimization problem of a hang glider in an upwind which provides an example for a control problem where appropriate initial estimates for the adjoint variables are hard to find.

1. Introduction and Survey of Numerical Methods

Complex optimal control problems, such as those origin from applications in aeronautics, astronautics, and robotics, can today be solved by sophisticated numerical methods. If the accuracy of the solution and the judgement of its optimality holds the spotlight, the multiple shooting method (see [3], [32], [12], [13], [28], [14], [21], and [17]) seems to be superior over other methods. This can be attested also by the complexity of the problems which have been successfully treated in the references [6], [7] (maximum payload ascent of a two-stage-to-orbit vehicle), [4], [5], [10] (maximum payload missions to planetoids (Vesta, Flora) or to the planet Neptune), [8], [9] (abort landing of an airplane in a windshear), [28] (optimal heating and cooling by solar energy), [29] (time optimal control of a robot), and [30] (singular controls in trajectory optimization problems), to cite only a few of the many papers. However, the multiple shooting method is often assessed by users as difficult to handle because not only a deep knowledge of the calculus of variations is required, but the user has to have also a deep insight into the physical nature of the problem in order to get around the obstacle of finding an appropriate initial guess for starting the iteration process. These numerical difficulties are caused by the relatively small domain of convergence of the Newton method which is built-in in the multiple shooting method, and augmented by the lack of information about the adjoint variables which one has to deal with when using an indirect method. Moreover, the switching structure, i.e., the partition of the optimal trajectory into different subarcs such as bang-bang or singular subarcs and unconstrained or constrained subarcs, can be obtained only by applying homotopy techniques (see, e.g., [9]). Within such a homotopy chain, that is a family of subproblems where the solution of one problem serves as an initial guess for a neighboring problem, the computation of often some hundred boundary-value problems is required.

These difficulties are typical for those indirect methods which solve the boundary-value problem obtained via the elimination of the control variables by means of the minimum principle. Other indirect methods, so-called gradient methods such as described in [22], [2], [11], [15], [35], and [26], use the minimum principle directly.

In contrast to this, the optimal control problem can be transformed into a nonlinear programming problem for the direct approach by parameterizing the control variables. The methods described, e.g., in [23], [24], [1], [18], and [20] use explicit numerical integration of the equations of motion, while the control variables are chosen from a finite dimensional space. This explicit integration can be avoided if the state variables are also parameterized or, in other words, if they are also chosen from a finite dimensional space. The equations of motion are then satisfied only pointwise by prescribing so-called collocation conditions. A description of methods belonging to this class can be found, e.g., in [31], [16], [34], [33], and [19].

Among the indirect methods, the multiple shooting method has several advantages, for example, its outstanding accuracy and the possibility to verify many necessary conditions. In addition, inequality constraints and interior point constraints can be treated, too, and, which is of increasing interest, the method is qualified for an application on vector or parallel computers (see [21]). Among the direct methods, direct collocation has the advantage that no explicit integration must be carried through. Thus, this method is very efficient.

Recently, a so-called hybrid approach was suggested (see [34] and [33]) where just those two methods, direct collocation and multiple shooting, are combined in the following way: The numerical approximation of the adjoint variables of the Lagrangian of the associated nonlinear programming problem is used to approximate the adjoint variables of the optimal control problem; see [34] for details. This idea amalgamates the two classes of methods in order to benefit from their advantages without taking into account their disadvantages.

The present paper describes the numerical procedure in solving an optimal control problem from real-life applications and discusses the benefits of this approach. The problem solved here describes the range maximization of a hang glider in an upwind. Many of the numerical difficulties appearing during the process of solution go with the known sensitivity of this kind of flight vehicle.

2. Optimal Control Problem: Maximum Range Flight of a Hang Glider

The maximum range flight of a hang glider through a given thermal can be modelled by the following optimal control problem: The vehicle is approximately described as a point mass subject to its weight W, a lift force L perpendicular to the velocity v_r relative to the air, and a drag force D opposite to v_r. The relative velocity vector v_r is at an angle η relative to the horizontal plane. The motion of the hang glider is restricted to a vertical plane. Thus we have four state variables: the horizontal distance x, the altitude y, the horizontal absolute velocity component v_x, and the vertical absolute velocity component v_y; see Fig. 1. The given thermal is assumed to have a distribution with respect to the horizontal distance x as given by the upward wind velocity $u_a(x)$,

$$u_a(x) = u_{a\,\max} \exp\left(-\left(\frac{x}{R} - 2.5\right)^2\right)\left(1 - \left(\frac{x}{R} - 2.5\right)^2\right) \tag{1}$$

where $5\,R$ denotes the horizontal extend of the thermal (here $R = 100$ [m]) and $u_{a\,\max}$ gives the maximal upwind velocity (here $u_{a\,\max} = 2.5$ [m s^{-1}]). A similar problem is described in [25] for the minimum time flight of a sailplane

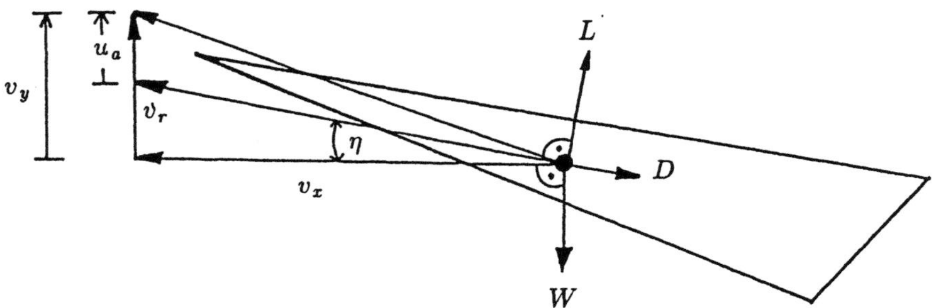

Fig. 1. Forces and velocity components.

through a thermal of the type (1).

Thus we have the following equations of motion,

$$\dot{x} = v_x \,, \qquad \dot{v}_x = \frac{1}{m}\left(-L\,\sin\eta - D\,\cos\eta\right)\,,$$

$$\dot{y} = v_y \,, \qquad \dot{v}_y = \frac{1}{m}\left(L\,\cos\eta - D\,\sin\eta - W\right) \tag{2}$$

with

$$\eta = \arctan\left(\frac{v_y - u_a(x)}{v_x}\right)\,, \qquad v_r = \sqrt{v_x^2 + (v_y - u_a(x))^2}\,,$$

$$L = c_L\,\frac{1}{2}\,\rho\,S\,v_r^2\,, \qquad D = c_D(c_L)\,\frac{1}{2}\,\rho\,S\,v_r^2\,, \qquad W = m\,g\,.$$

The hang glider is controlled via the lift coefficient c_L. The drag coefficient c_D is assumed to be a quadratic function of the lift coefficient. Based on data for a high performance hang glider of the type *Saphir 17* (see [36]), this leads to the quadratic polar

$$c_D(c_L) = c_{D_0} + k\,c_L^2 \tag{3}$$

with values $c_{D_0} = 0.034$ and $k = 0.069662$. In addition, the lift coefficient is constrained,

$$c_L \le c_{L\max} := 1.4\,. \tag{4}$$

Further constants are $m = 100$ [kg] (mass of vehicle and pilot), $S = 14$ [m^2] (wing area), $\rho = 1.13$ [kg m^{-3}] (air density corresponding to standard pressure and temperature at a height of about 1000 m above sea level), and $g = 9.81$ [m s^{-2}] (gravitational acceleration).

The model is completed by the following boundary conditions where the direct starting and landing phase is excluded because of the difficulties in modelling them appropriately,

$$
\begin{aligned}
x(0) &= 0 \ [\text{m}] \ , & x(t_f) &\overset{!}{=} \max \ , \\
y(0) &= 1000 \ [\text{m}] \ , & y(t_f) &= 900 \ [\text{m}] \ , \\
v_x(0) &= v_{x\,\text{McC}} := 13.23 \ [\text{m/s}] \ , & v_x(t_f) &= v_{x\,\text{McC}} \ [\text{m/s}] \ , \\
v_y(0) &= v_{y\,\text{McC}} := -1.288 \ [\text{m/s}] \ , & v_y(t_f) &= v_{y\,\text{McC}} \ [\text{m/s}] \ .
\end{aligned}
\tag{5}
$$

A given difference between initial and terminal altitude is to be used to maximize the range with initial and terminal velocity prescribed. Here, $v_{x\,\text{McC}}$ and $v_{y\,\text{McC}}$ denote the components of the so-called McCready velocity, which is associated with the velocity of best gliding.

By means of the minimum principle the optimal control function can be eliminated in terms of the state and the adjoint variables; cf., e.g., [2]. Hence, we have

$$
c_L =
\begin{cases}
c_L^{\text{free}} := -\dfrac{1}{2k} \dfrac{\lambda_{v_x} \sin\eta - \lambda_{v_y} \cos\eta}{\lambda_{v_x} \cos\eta + \lambda_{v_y} \sin\eta} & \text{if } c_L^{\text{free}} < c_{L\,\max} \\[2ex]
c_{L\,\max} & \text{if } c_L^{\text{free}} \geq c_{L\,\max}
\end{cases}
\tag{6}
$$

where the adjoint variables satisfy the differential equations

$$
\dot{\lambda}_\diamond = -\frac{\partial H}{\partial \diamond} \ , \quad \diamond \in \{x, y, v_x, v_y\}
\tag{7}
$$

with the Hamiltonian defined by

$$
H = \lambda_x \dot{x} + \lambda_y \dot{y} + \lambda_{v_x} \dot{v}_x + \lambda_{v_y} \dot{v}_y \ .
\tag{8}
$$

To give an impression of the complexity of the adjoint variables, one of the equations is presented here,

$$
\begin{aligned}
\dot{\lambda}_x = \frac{1}{m} \Bigg[&\lambda_{v_x} \left(-c_L \, \rho \, S \, (v_y - u_a(x)) \sin\eta - L \, \frac{v_x^2}{\sqrt{v_x^2 + (v_y - u_a(x))^2}^3} \right. \\
&\left. - c_D(c_L) \, \rho \, S \, (v_y - u_a(x)) \cos\eta + D \, \frac{v_x(v_y - u_a(x))}{\sqrt{v_x^2 + (v_y - u_a(x))^2}^3} \right) \\
&+ \lambda_{v_y} \left(c_L \, \rho \, S \, (v_y - u_a(x)) \cos\eta - L \, \frac{v_x(v_y - u_a(x))}{\sqrt{v_x^2 + (v_y - u_a(x))^2}^3} \right. \\
&\left. - c_D(c_L) \, \rho \, S \, (v_y - u_a(x)) \sin\eta - D \, \frac{v_x^2}{\sqrt{v_x^2 + (v_y - u_a(x))^2}^3} \right) \Bigg]
\end{aligned}
$$

$$\cdot \left[u_a(x) + u_{a\,\max} \exp\left(-\left(\frac{x}{R} - 2.5\right)^2 \right) \left(-\frac{2}{R}\left(\frac{x}{R} - 2.5\right) \right) \right] \ .$$

The boundary-value problem is completed by the transversality conditions

$$\lambda_x(t_f) = -1 \ , \quad H\,|_{t=t_f} = 0 \ . \tag{9}$$

After the transformation $\tau := t/t_f$ of the interval $[0, t_f]$ onto $[0, 1]$, the equations (2), (7), (5), and (9) describe a two-point boundary-value problem for 9 unknowns. Note that the final time t_f then is an additional dependent variable introduced by that transformation. The right-hand side of the system of differential equations depends via (6) on the sign of a so-called switching function,

$$S := c_L^{\text{free}} - c_{L\,\max} \ . \tag{10}$$

Thus, we have a so-called two-point boundary-value problem with switching function. Alternatively, we can formulate a multipoint boundary-value problem which is based on a hypothesis of the switching structure. For example, if the optimal trajectory is assumed to have one interior constrained subarc, a multipoint boundary-value problem can be stated having one additional interior boundary condition at both the entry and the exit point of that constrained subarc. Because of the continuity of the control function, the interior boundary conditions are

$$\begin{aligned}
c_L^{\text{free}}\,|_{t=t_{\text{entry}}} &= c_{L\,\max}\,|_{t=t_{\text{entry}}} \ , \\
c_L^{\text{free}}\,|_{t=t_{\text{exit}}} &= c_{L\,\max}\,|_{t=t_{\text{exit}}} \ .
\end{aligned} \tag{11}$$

With respect to the convergence behaviour of the multiple shooting method, the latter formulation is more advantageous than the formulation using switching functions; see [28]. Note that it is, in this case, important to examine the solution of the multipoint boundary-value problem whether the sign of the switching function (10) and the control law according to (6) correspond with the control law based on the hypothesis. See [8, 9] for techniques how to reveal and adapt the switching structure for problems with multiple subarcs.

Herewith all information is provided to treat the problem by an indirect method; the above analysis can be omitted when applying a direct method.

3. Numerical Procedure: Combination of Direct and Indirect Methods

3.1 Attempt of the construction of a starting trajectory using multiple shooting. Using the indirect approach, the most promising way to obtain a candidate for an optimal solution of a given problem is to embed this problem into a family of subproblems. By homotopy techniques the solution of one problem out of that family then serves as an initial guess for the solution of a neighboring problem. Starting with a simplified problem, the given optimal control problem can be solved via the solution of a whole chain of boundary-value problems.

For the problem under consideration, we first omit the control constraint (4), and we also neglect the upwind by setting the parameter $u_{a\,\mathrm{max}} = 0$. So, the maximum lift coefficient $c_{L\,\mathrm{max}}$ as well as $u_{a\,\mathrm{max}}$ will play the role of homotopy parameters. Then we have the following information about the adjoint variables λ_x and λ_y,

$$\lambda_x(t) = \mathrm{const} = -1 \, , \quad \lambda_y(t) = \mathrm{const} \, .$$

However, no information about λ_{v_x} and λ_{v_y} is available. This poor knowledge of the adjoint variables causes the numerical integration to fail for both backward and forward integration unless the adjoint variables are properly guessed. Usually many attemps must be undertaken to obtain a trajectory which at least has some relevancy. This trajectory or may be a part of it then would provide the first boundary-value problem of the aforementioned family from which we could start the homotopy.

3.2 Construction of a starting trajectory using direct collocation. Applying the direct collocation method [33], convergence cannot be obtained for the full model directly. We have to apply homotopy techniques, too. For lower initial velocity components, here $v_x(0) = 11$ $[\mathrm{m\,s}^{-1}]$ and $v_y(0) = -1.1$ $[\mathrm{m\,s}^{-1}]$, and for the simplified model where both the upwind and the constraint of the lift coefficient are neglected, a solution can be obtained by the direct collocation method even when starting the iteration with the following simple initial guess. This initial estimate is constituted by the linear polynomial which interpolates the boundary values for the state variables and by $c_L \equiv 1$ for the control function. The McCready velocity components and the upwind are then introduced step by step. For the upwind the parameter $u_{a\,\mathrm{max}}$ is increased to $u_{a\,\mathrm{max}} = 2$ $[\mathrm{m\,s}^{-1}]$ in steps of 0.5 $[\mathrm{m\,s}^{-1}]$. A grid of 21 equidistant points is used for the discretization of the time interval throughout the whole homotopy.

Thereafter, the approximation is improved by a grid refinement; 37 non-equidistant grid points are chosen so that the error function $d(\tau) := \max_i \delta_i \, |f_i(p, u, \tau) - p_i'(\tau)|$ with appropriate scaling factors $\delta_i > 0$ is approximately equally distributed over the interval $[0, 1]$. Here, p denotes the piecewise cubic vector polynomial interpolating the state vector and its derivatives at the grid points. The variable u denotes the control function, and the f_i's are the components of the right-hand side. The variable $\tau := t/t_f$ is the normalized time. We finally end up with an approximate solution provided by the collocation method from which an approximation of the adjoint variables can be obtained according to [34] with an accuracy sufficient to yield convergence by the multiple shooting software package [28].

Figures 2–6 show the solution obtained by the direct collocation method (dashed line) and the improved solution obtained by the multiple shooting method (solid line). The differences for the horizontal distance x and the altitude y are below the drawing accuracy. The approximation for the velocity component v_y shows the largest differences; see Fig. 5. The values for the maximum range are $x(t_f) =$ 1201.65 [m] with $t_f = 96.444$ [s] obtained by the collocation method and $x(t_f) =$ 1201.63 [m] with $t_f = 96.438$ [s] obtained by the multiple shooting method. In Figure 3 the grid points are marked which have been used for the collocation method. Figures 7–9 show the accuracy of the initial guess of the adjoint variables based on their relationship, according to [34], to the multipliers associated with the nonlinear programming problem. Instead of the graph of the constant adjoint variable λ_y, its approximations are given here: we obtain $\lambda_y \approx -10.275$ from the collocation method and $\lambda_y \approx -10.274$ from the multiple shooting method.

The difficulties in obtaining the numerical solution of the problem are caused by the high sensitivity of the solution with respect to its initial values. A numerical integration of the initial-value problem associated with the solution of the boundary-value problem fails if the integration is carried through over the entire flight time interval at one stroke. However, the numerical integration of the initial-value problem can be carried through if, as in the multiple shooting algorithm, a series of initial-value problems is solved over smaller subintervals, where the initial values are always redefined at the grid points of the discretization using the approximation obtained by the multiple shooting method. The different pieces of the trajectory then match with an accuracy of at least 5 digits. That sensitivity also explains why such a relatively large number of grid points is to be used when going over from the collocation method to the multiple shooting method. The higher number of grid points provides a better estimate of the adjoint variables. As a rule of thumb, the adjoint variables must be approximated to an accuracy of at least 2 digits to provide convergence of the multiple shooting iteration if the problem to be solved is as sensitive as the hang glider problem. During the subsequent homotopy steps with the multiple shooting method, the number of multiple shooting nodes can then be decreased again.

The question now arises when the transition from the collocation to the multiple shooting method should be done. Generally speaking, the transition should be done preferably for a simplier model. For example, the transition fails when the control variable inequality constraint, too, is taken into account for the solution using the collocation method. On the other hand, if the transition is made for a too simple version of the problem as in [27] where the difference between initial and terminal altitude is reduced to 10 [m] and where the upwind as well as the constraint of the lift coefficient are also neglected at the beginning, a higher amount of computation is needed because of the smaller domain of convergence of the multiple shooting method. This is caused by the smaller homotopy step sizes. Following this way, the step size for the first homotopy where the difference between initial and ter-

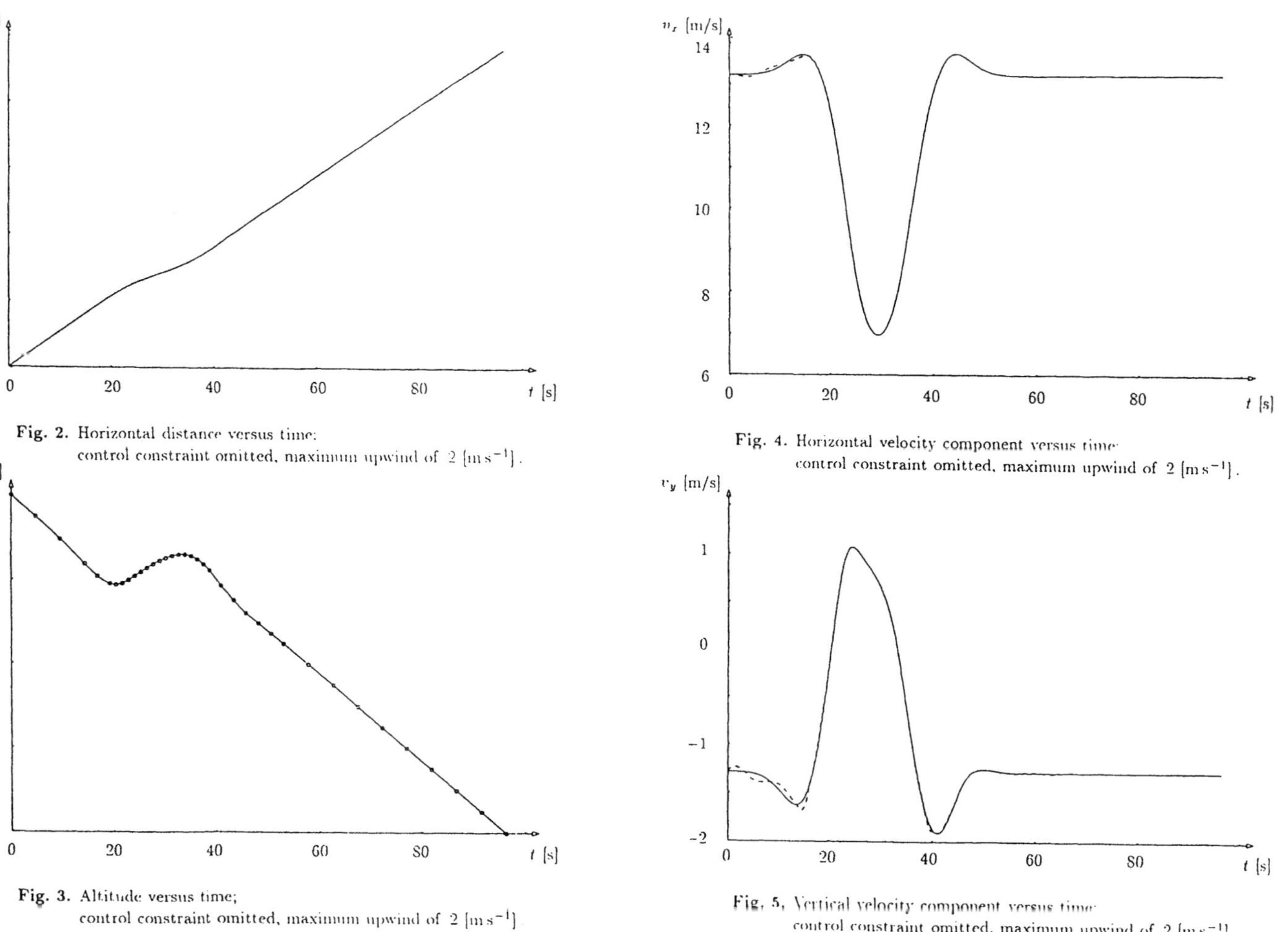

Fig. 2. Horizontal distance versus time: control constraint omitted, maximum upwind of 2 [m s⁻¹].

Fig. 3. Altitude versus time; control constraint omitted, maximum upwind of 2 [m s⁻¹].

Fig. 4. Horizontal velocity component versus time: control constraint omitted, maximum upwind of 2 [m s⁻¹].

Fig. 5. Vertical velocity component versus time: control constraint omitted, maximum upwind of 2 [m s⁻¹].

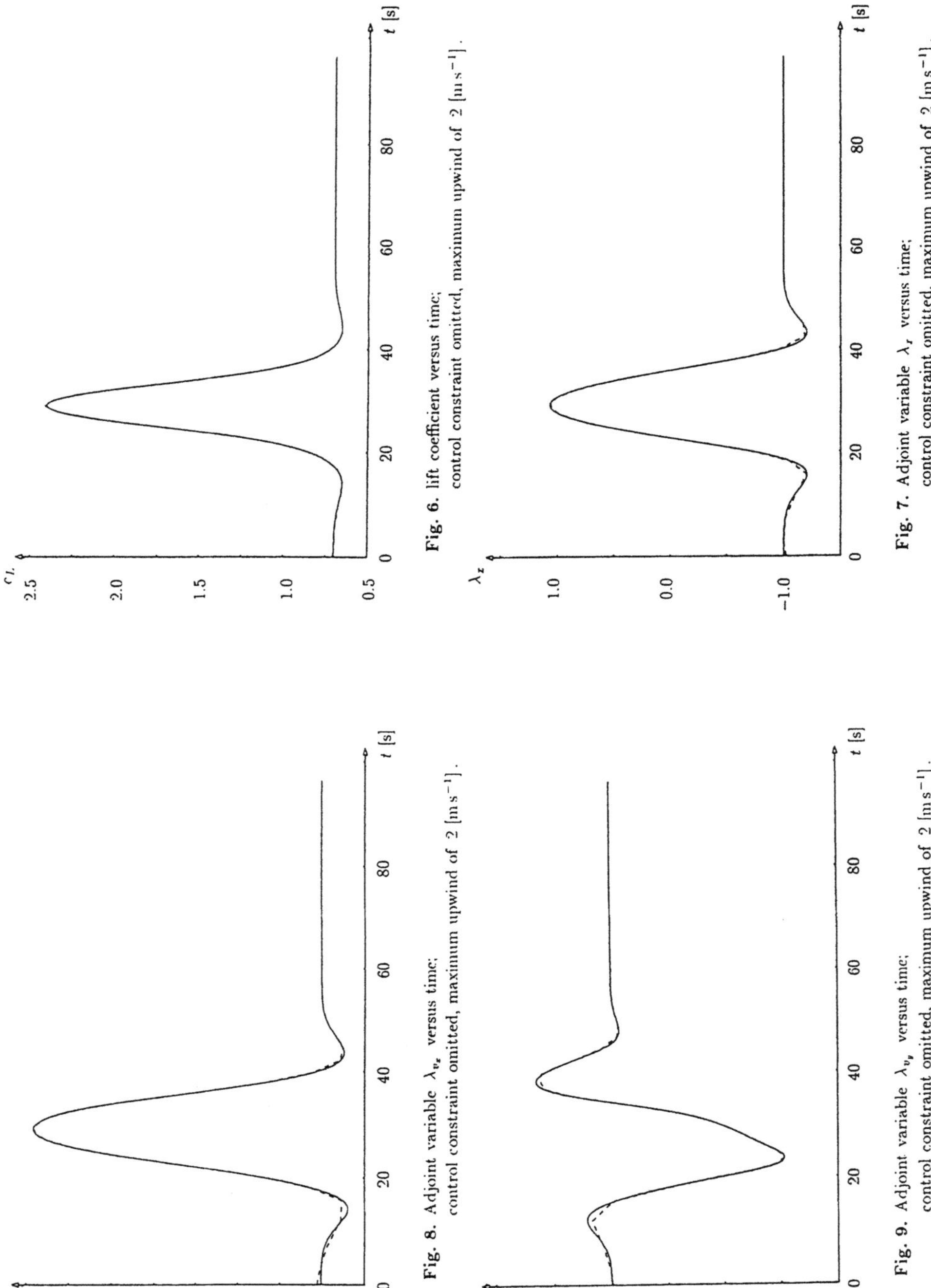

Fig. 6. lift coefficient versus time; control constraint omitted, maximum upwind of 2 $[\mathrm{m\,s^{-1}}]$.

Fig. 7. Adjoint variable λ_z versus time; control constraint omitted, maximum upwind of 2 $[\mathrm{m\,s^{-1}}]$.

Fig. 8. Adjoint variable λ_{v_x} versus time; control constraint omitted, maximum upwind of 2 $[\mathrm{m\,s^{-1}}]$.

Fig. 9. Adjoint variable λ_{v_y} versus time; control constraint omitted, maximum upwind of 2 $[\mathrm{m\,s^{-1}}]$.

minal altitude must be increased varies between about 10^{-3} [m] to about 2 [m] when using the multiple shooting method. In a second homotopy, the effect of the thermal must be then brought into the game by increasing the parameter $u_{a\,max}$. Thereby, the minimum homotopy step size is 10^{-2} [ms^{-1}]. Recall the homotopy step size of 0.5 [ms^{-1}] for $u_{a\,max}$ when using the direct collocation method.

3.3 Introducing the control variable inequality constraint using homotopy and multiple shooting. From Fig. 6, we easily obtain a hypothesis of the switching structure: there will be only one constrained subarc when introducing the control constraint via the parameter $c_{L\,max}$ moderately. Some of the results for this homotopy are given in Figs. 10–12. The solid lines indicate the extremal values $c_{L\,max} = 2.38$ (start of the homotopy) and $c_{L\,max} = 1.4$ (end of the homotopy); compare Figs. 4 and 5, too. The intermediate values $c_{L\,max} = 2.0$ and $c_{L\,max} = 1.7$ are given by the dashed and the dashed-dotted lines, respectively.

4. Numerical Results: The optimal trajectory

To complete the solution, a very last homotopy step must be performed to achieve the desired maximum upwind of $u_{a\,max} = 2.5$ ms^{-1}. Figures 13–17 show the optimal trajectory obtained by the multiple shooting method. The maximum range is $x(t_f) = 1247.60$ [m], the final time is $t_f = 98.380$ [s], and the switching times are $t_{entry} = 23.301$ [s] and $t_{exit} = 33.250$ [s]. The two switching points are indicated in the figures by the vertical dashed lines.

The results indicate the gain of range caused by the upwind. To increase the potential energy, the altitude has to be increased. To stay as long as possible in the upwind, the horizontal velocity component has to be decreased. Comparing the results for the maximum range trajectory of the hang glider presented here with the minimum time trajectory of a sailplane presented in [25], we see that the two-dimensional model still gives meaningful results for upwind velocities considered here. In the sailplane problem of [25] strong upwind velocities cause a breakdown of the vertical plane model. The optimal trajectory there shows a horizontal velocity component which is negative in the upwind and indicates that the pilot should gain altitude by flying circles in the thermal. This point of a model breakdown is, however, not reached here.

5. Conclusions

Despite the superiority of the multiple shooting method with respect to accuracy and reliability, which is hardly obtainable by any other method for the solution of optimal control problems, its use is often difficult and laborious since an appropriate guess of initial data, in particular, of the adjoint variables as well as of the switching points must be provided. In this paper it is shown how to overcome this obstacle when solving a real-life problem. By using a direct collocation method the

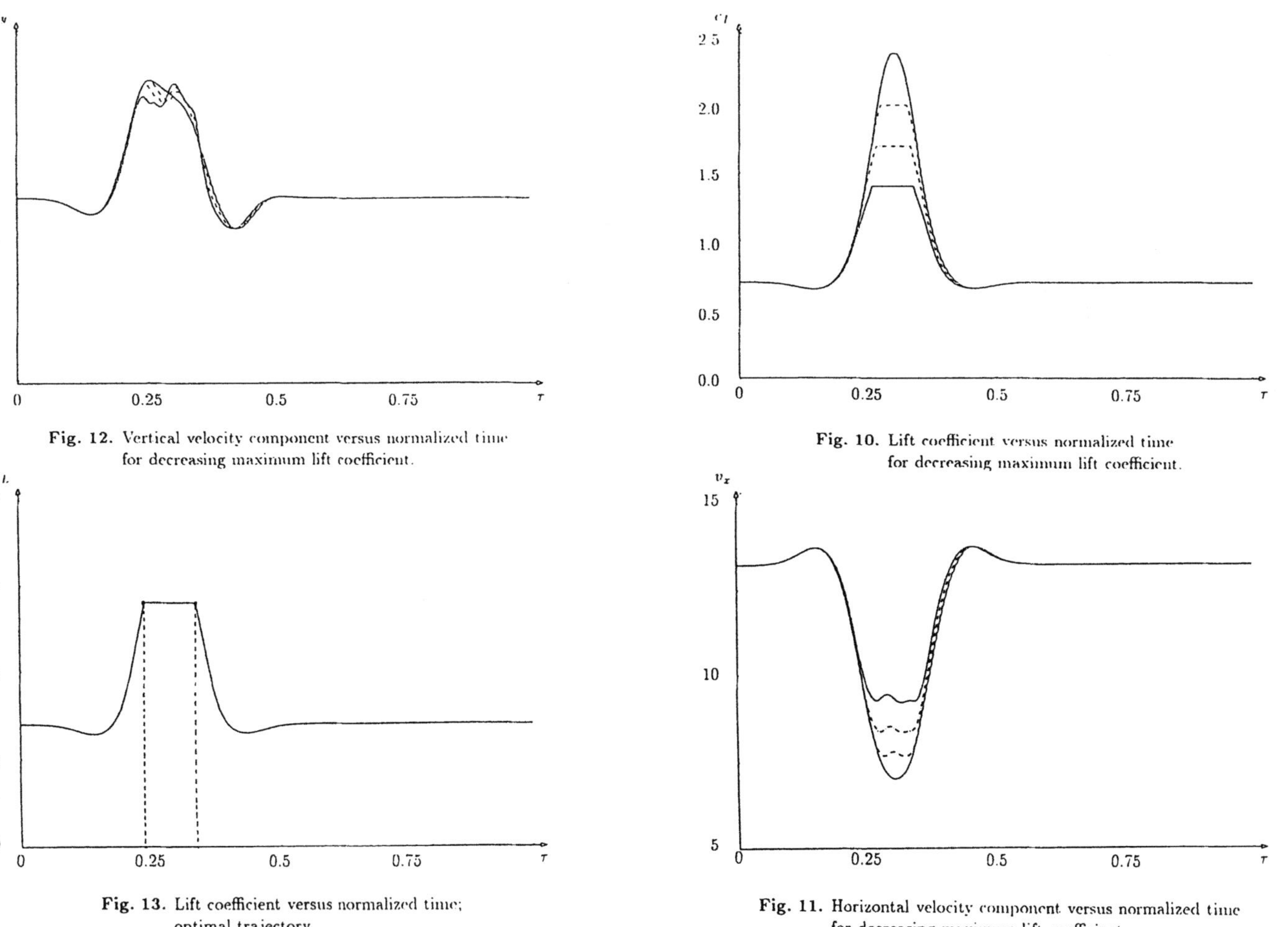

Fig. 12. Vertical velocity component versus normalized time for decreasing maximum lift coefficient.

Fig. 10. Lift coefficient versus normalized time for decreasing maximum lift coefficient.

Fig. 13. Lift coefficient versus normalized time; optimal trajectory.

Fig. 11. Horizontal velocity component versus normalized time for decreasing maximum lift coefficient.

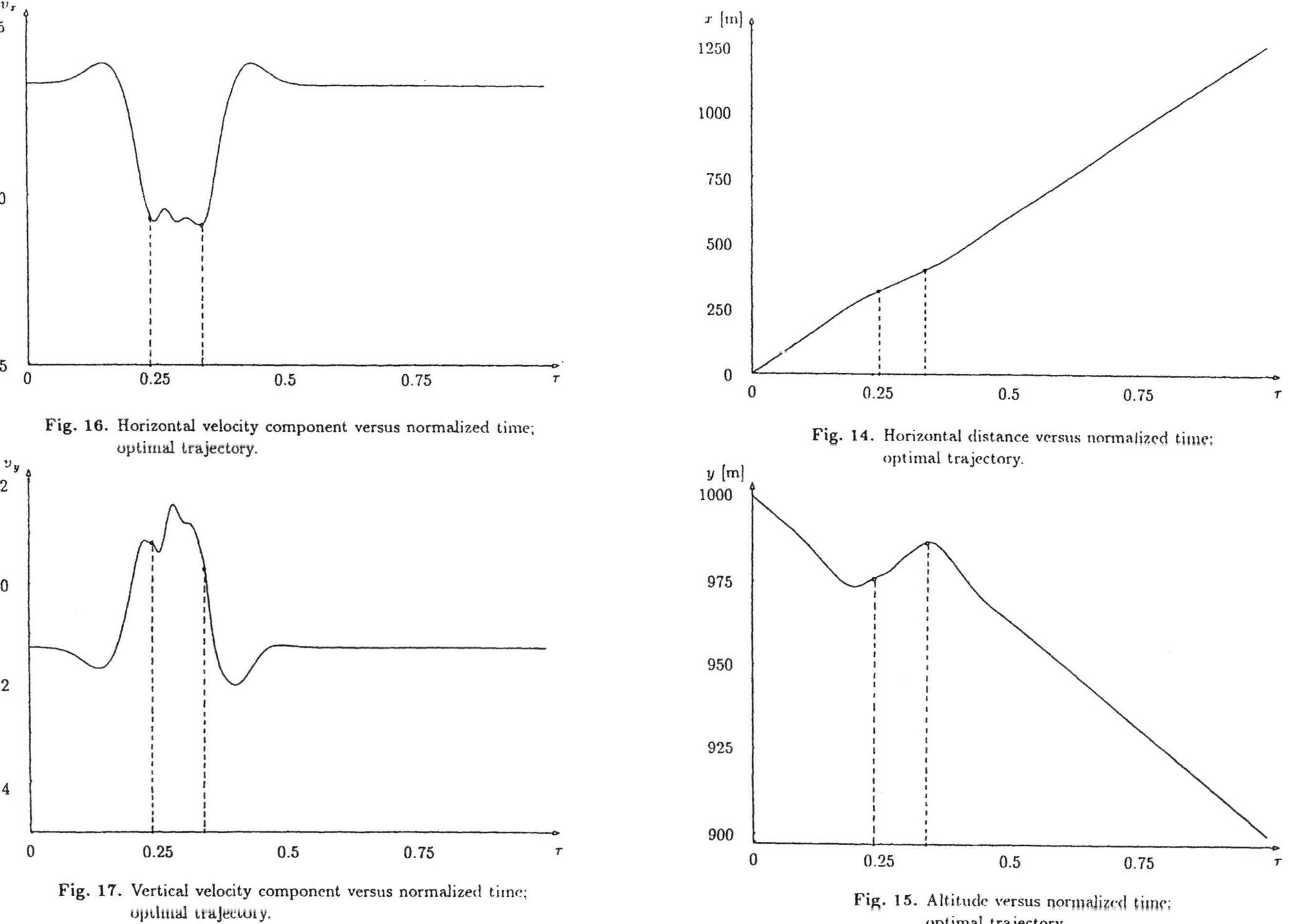

Fig. 16. Horizontal velocity component versus normalized time; optimal trajectory.

Fig. 17. Vertical velocity component versus normalized time; optimal trajectory.

Fig. 14. Horizontal distance versus normalized time; optimal trajectory.

Fig. 15. Altitude versus normalized time; optimal trajectory.

adjoint variables can be estimated from the Lagrange parameters of the underlying nonlinear programming problem. For problems of moderate degree of complexity, the approximations of both the state and the adjoint variables provided by the direct collocation method are accurate enough to yield convergence with the multiple shooting method. At this point of investigation homotopy techniques still must be used to introduce inequality constraints imposed on the model. Future investigations will try to fill this gap to obtain also inequality-constrained optimal solutions by multiple shooting directly using a pre-computation with an improved direct collocation method.

Acknowledgements

This research was supported by the German National Science Foundation (Deutsche Forschungsgemeinschaft) through the Sonderforschungsbereich 255 (Transatmosphärische Flugsysteme)

References

[1] Bock, H. G. and Plitt, K. J.: *A Multiple Shooting Algorithm for Direct Solution of Optimal Control Problems*, Proceedings of the 9th IFAC Worldcongress, Budapest, 1984, Vol. IX, Colloquia 14.2, 09.2, 1984.

[2] Bryson, A. E. and Ho, Y. C.: *Applied Optimal Control*, New York: Hemisphere (Rev. Printing), 1975.

[3] Bulirsch, R.: *Die Mehrzielmethode zur numerischen Lösung von nichtlinearen Randwertproblemen und Aufgaben der optimalen Steuerung*, Carl-Cranz Gesellschaft, Oberpfaffenhofen, Report der Carl-Cranz Gesellschaft, 1971; Munich University of Technology, Department of Mathematics, Munich, Reprint, 1985.

[4] Bulirsch, R. and Callies, R.: *Optimal Trajectories for an Ion Driven Spacecraft from Earth to the Planetoid Vesta*, Proc. of the AIAA Guidance, Navigation and Control Conference, New Orleans, 1991, AIAA Paper No. 91-2683, 1991.

[5] Bulirsch, R. and Callies, R.: *Optimal Trajectories for a Multiple Rendezvous Mission to Asteroids*, 42nd International Astronautical Congress, Montreal, 1991, IAF-Paper No. IAF-91-342, 1991.

[6] Bulirsch, R., Chudej, K., and Reinsch, K. D.: *Optimal Ascent and Staging of a Two-Stage Space Vehicle System*, Jahrestagung der Deutschen Gesellschaft für Luft- und Raumfahrt, Friedrichshafen, 1990, DGLR-Jahrbuch 1990, Vol. 1, 243–249, 1990.

[7] Bulirsch, R. and Chudej, K.: *Ascent Optimization of an Airbreathing Space Vehicle*, Proc. of the AIAA Guidance, Navigation and Control Conference, New Orleans, 1991, AIAA Paper No. 91-2656, 1991.

[8] Bulirsch, R., Montrone, F., and Pesch, H. J.: *Abort Landing in the Presence of a Windshear as a Minimax Optimal Control Problem, Part 1: Necessary Conditions*, J. of Optimization Theory and Applications **70**, 1–23, 1991.

[9] Bulirsch, R., Montrone, F., and Pesch, H. J.: *Abort Landing in the Presence of a Windshear as a Minimax Optimal Control Problem, Part 2: Multiple Shooting and Homotopy*, J. of Optimization Theory and Applications **70**, 221–252, 1991.

[10] Callies, R.: *Optimal Design of a Mission to Neptune*, in: Bulirsch, R., Miele, A., Stoer, J., and Well, K. H. (eds): Optimal Control, Proc. of the Conf. in Optimal Control and Variational Calculus, Oberwolfach, 1991, Lecture Notes in Control and Information Sciences, Berlin, Heidelberg, New York, London, Paris, Tokyo: Springer, this issue.

[11] Chernousko, F. L. and Lyubushin, A. A.: *Method of Successive Approximation for Solution of Optimal Control Problems*, Optimal Control Applications and Methods **3**, 101–114, 1982.

[12] Deuflhard, P.: *A Relaxation Strategy for the Modified Newton Method*, in: Bulirsch R., Oettli, W., and Stoer, J. (eds.), Optimization and Optimal Control, Proceedings of a Conference Held at Oberwolfach, 1974, Lecture Notes in Mathematics **477** Berlin, Heidelberg, New York: Springer, 59–73, 1975.

[13] Deuflhard, P.: *A Modified Newton Method for the Solution of Ill-conditioned Systems of Nonlinear Equations with Application to Multiple Shooting*, Numerische Mathematik **22**, 289–315, 1974.

[14] Deuflhard, P. and Bader, G.: *Multiple Shooting Techniques Revisited*, in: Deuflhard P. and Hairer, E. (eds.), Numerical Treatment of Inverse Problems in Differential and Integral Equations, Proceedings of an International Workshop, Heidelberg, 1982 Progress in Scientific Computing **2**, Boston: Birkhäuser, 74–94, 1983.

[15] Gottlieb, R. G.: *Rapid Convergence to Optimum Solutions Using a Min-H Strategy* AIAA J. **5**, 322–329, 1967.

[16] Hargraves, C. R. and Paris, S. W.: *Direct Trajectory Optimization Using Nonlinear Programming and Collocation*, AIAA Journal of Guidance and Control **10**, 338–342, 1987.

[17] Hiltmann, P.: *Numerische Lösung von Mehrpunkt-Randwertproblemen und Aufgaben der optimalen Steuerung mit Steuerfunktionen über endlichdimensionalen Räumen*, Munich University of Technology, Department of Mathematics, Doctoral Thesis, 1990.

[18] Horn, K.: *Solution of the Optimal Control Problem Using the Software Package STOMP*, to appear in: Bernhard, P. and Bourdache-Siguerdidjane, H. (eds.), Proc. of the 8th IFAC Workshop on Control Applications of Nonlinear Programming and Optimization, Paris, 1989, Oxford: IFAC Publications, 1991.

[19] Jänsch, C. and Paus, M.: *Aircraft Trajectory Optimization with Direct Collocation Using Movable Gridpoints*, in: Proceedings of the American Control Conference, San Diego, 262–267, 1990.

[20] Jänsch, C., Schnepper, K., and Well, K. H.: *Ascent and Descent Trajectory Optimization of Ariane V/Hermes*, in: AGARD Conf. Proc. No. 489 on Space Vehicle Flight Mechanics, 75th Symp. of the AGARD Flight Mechanics Panel, Luxembourg, 1989.

[21] Kiehl, M: *Vectorizing the Multiple-Shooting Method for the Solution of Boundary-Value Problems and Optimal-Control Problems*, in: Dongarra, J., Duff, I., Gaffney, P., and McKee, S. (eds.), Proceedings of the 2nd International Conference on Vector and Parallel Computing Issues in Applied Research and Development, Tromsø, 1988, London: Ellis Horwood, 179–188, 1989.

[22] Kelley, H. J., Kopp, R. E., and Moyer, H. G.: *Successive Approximation Techniques*

for Trajectory Optimization, Proc. Symp. on Vehicle System Optimization, New York, 1961.

[23] Kraft, D.: *FORTRAN Computer Programs for Solving Optimal Control Problems*, Deutsche Forschungs- und Versuchsanstalt für Luft- und Raumfahrt, Oberpfaffenhofen, Report 80-03, 1980.

[24] Kraft, D.: *On Converting Optimal Control Problems into Nonlinear Programming Codes*, in: Schittkowski, K. (ed.) Computational Mathematical Programming, Berlin: Springer (NATO ASI Series **15**), 261–280, 1985.

[25] Lorenz, J.: *Numerical Solution of the Minimum-Time Flight of a Glider Through a Thermal by Use of Multiple Shooting Methods*, Optimal Control Applications and Methods **6**, 125–140, 1985.

[26] Miele, A.: *Gradient Algorithms for the Optimization of Dynamic Systems*, in: Leondes, C. T., Control and Dynamic Systems **16**, New York: Academic Press, 1–52, 1980.

[27] Nerz, E.: *Optimale Steuerung eines Hängegleiters*, Munich University of Technology, Department of Mathematics, Diploma Thesis, 1990.

[28] Oberle, H. J.: *Numerische Berechnung optimaler Steuerungen von Heizung und Kühlung für ein realistisches Sonnenhausmodell*, Habilitationsschrift, Munich University of Technology, Munich, Germany, 1982.

[29] Oberle, H. J.: *Numerical Computation of Singular Control Functions for a Two-Link Robot Arm*, in: Bulirsch, R., Miele, A., Stoer, J., and Well, K. H. (eds): Optimal Control, Proc. of the Conf. in Optimal Control and Variational Calculus, Oberwolfach, 1986, Lecture Notes in Control and Information Sciences **95**, Berlin, Heidelberg, New York, London, Paris, Tokyo: Springer, 244–253, 1987.

[30] Oberle, H. J.: *Numerical Computation of Singular Functions in Trajectory Optimization Problems*, J. Guidance and Control **13**, 153–159, 1990.

[31] Renes, J. J.: *On the Use of Splines and Collocation in a Trajectory Optimization Algorithm Based on Mathematical Programming*, National Aerospace Laboratory, Amsterdam, Report No. NLR-TR-78016 U, 1978.

[32] Stoer, J. and Bulirsch, R.: *Introduction to Numerical Analysis*, New York: Springer, 1980.

[33] von Stryk, O.: *Numerical Solution of Optimal Control Problems by Direct Collocation*, in: Bulirsch, R., Miele, A., Stoer, J., and Well, K. H. (eds): Optimal Control, Proc. of the Conf. in Optimal Control and Variational Calculus, Oberwolfach, 1991, Lecture Notes in Control and Information Sciences, Berlin, Heidelberg, New York, London, Paris, Tokyo: Springer, this issue.

[34] von Stryk, O. and Bulirsch, R.: *Direct and Indirect Methods for Trajectory Optimization*, to appear in Annals of Operations Research, 1991.

[35] Tolle, H.: *Optimierungsverfahren*, Berlin: Springer, 1971.

[36] *Drachenfliegermagazin*, München: Ringier Verlag, issue 7, 1988.

Prof. Dr. Roland Bulirsch, Dipl. Math. Edda Nerz, Priv.-Doz. Dr. Hans Josef Pesch, Dipl. Math. Oskar von Stryk, Mathematisches Institut, Technische Universität München, Postfach 20 24 20, D–8000 München 2

Periodic Optimal Trajectories with Singular Control for Aircraft with High Aerodynamic Efficiency

G. Sachs, K. Lesch, H.G. Bock, M. Steinbach

Abstract. Fuel minimum range cruise of aircraft with high aerodynamic efficiency is considered as an optimal periodic control problem. Optimality conditions for trajectories with singular arcs and state variable constraints are derived.

Computation of periodic optimal trajectories in the case addressed presents strong requirements on the numerical algorithm. Computational difficulties become larger when aerodynamic efficiency is increased and wing loading is decreased. The specific nature of the numerical problems encountered and the means used to overcome them are described.

1   Nomenclature

C_D	drag coefficient
C_L	lift coefficient
C_L^*	lift coefficient at maximum lift/drag ratio
c	normalized period length
D	drag
E	aerodynamic efficiency, $E = (C_L/C_D)_{max}$
g	acceleration due to gravity
H	Hamiltonian
h	altitude
J	performance criterion
K	factor according to lift dependent drag, $C_D = C_{D0} + KC_L^2$
L	lift

m	mass
m_V	exponent denoting the effect of speed on fuel consumption
P	propulsive power, $P = TV$
S	reference area
$\overline{S}$	switching function
T	thrust
V	speed
V_0^*	speed for best glide ratio at sea level ($h = 0$)
x	horizontal coordinate
y	state variable vector
γ	flight path angle
δ	throttle setting
ε	glide ratio
λ	Lagrange multiplier
ϱ	atmospheric density
σ	fuel consumption factor
ξ	independent variable
$-$	a bar denotes a normalized quantity, e.g. $\overline{V}$

2   Introduction

The cruise of aircraft for maximizing range classically consists of a steady-state flight at constant speed and/or altitude with a basically straight trajectory. The control and state variables are chosen such that the fuel consumption is minimized. However, it has been shown that this type of cruise is not generally optimal, and results have been presented for optimal aircraft cruise which shows a periodic behavior (Refs. 1-10). The trajectory of such a cruise basically consists of periodically repeated cycles each of which shows a climb followed by a descent, with thrust alternately operated at high and low throttle settings.

In some cases, periodic optimal cruise shows singular arcs. This means that the throttle setting takes on intermediate values between the maximum and minimum bounds during a non-zero portion of the flight time. It has been shown that the dependence of thrust specific fuel consumption on speed has a significant effect as regards the existence of singular arcs (Refs. 9, 10).

It is the purpose of this paper to provide more insight into periodic optimal cruise of aircraft with singular arcs. Emphasis is placed on aircraft with high aerodynamic efficiency like modern motorgliders. Accordingly, the computational difficulties which exist in determining periodic optimal trajectories for such aircraft are of particular interest. Special numerical techniques are applied concerning the formulation of the multipoint boundary value problem and the solution of the linearized system in the iterative procedure used.

3 Problem Formulation

The optimal control problem consists of minimizing the fuel consumption per range of aircraft. This can be formulated by introducing the performance criterion

$$(3.1) \quad J = \frac{m_f(x_{\mathrm{cyc}})}{x_{\mathrm{cyc}}}$$

where m_f is the consumed fuel and x_{cyc} is the (horizontal) length of one period of the trajectory. A period may be considered as a basic element of the whole trajectory.

The performance criterion is subject to the equations of motion. For a non-dimensional form, the following normalization is introduced:

$$(3.2) \quad \overline{V} = \frac{V}{V_0^*}, \quad \overline{h} = \frac{hg}{V_0^{*2}}, \quad \overline{m}_f = \frac{m_f}{m}, \quad \xi = \frac{x}{x_{\mathrm{cyc}}}, \quad c = \frac{x_{\mathrm{cyc}}g}{V_0^{*2}}$$

$$\overline{C}_{\mathrm{L}} = \frac{C_{\mathrm{L}}}{C_{\mathrm{L}}^*}, \quad \overline{C}_{\mathrm{D}} = \frac{C_{\mathrm{D}}}{C_{\mathrm{L}}^*}, \quad \overline{T} = \frac{T}{mg}, \quad \overline{J} = \frac{\overline{m}_f}{c}$$

Consequently, the equations of motion may be written as

$$\frac{\mathrm{d}\overline{V}}{\mathrm{d}\xi} = c\frac{1}{\overline{V}}\left[\frac{\overline{T}(\overline{V},\overline{h};\delta) - \overline{\varrho}\,\overline{V}^2\overline{C}_{\mathrm{D}}}{\cos\gamma} - \tan\gamma\right]$$

$$(3.3) \quad \frac{\mathrm{d}\gamma}{\mathrm{d}\xi} = c\left[\frac{\overline{\varrho}\,\overline{C}_{\mathrm{L}}}{\cos\gamma} - \frac{1}{\overline{V}^2}\right]$$

$$\frac{\mathrm{d}\overline{h}}{\mathrm{d}\xi} = c\tan\gamma$$

$$\frac{\mathrm{d}\overline{m}_f}{\mathrm{d}\xi} = c\frac{V_0^*}{g}\frac{\dot{\overline{m}}_f}{\overline{V}\cos\gamma}$$

The aerodynamic drag polar is modelled as

$$(3.4) \quad \overline{C}_D = \frac{\varepsilon_{\min}}{2}(1 + \overline{C}_L^2)$$

The thrust model accounts for the effect of speed, altitude and throttle setting. It may be expressed as

$$(3.5) \quad \overline{T}(\overline{V}, \overline{h}; \delta) = \delta \overline{T}_{\max}(\overline{V}, \overline{h})$$

Fuel consumption characteristics are described by the relation

$$(3.6) \quad \dot{\overline{m}}_f = \dot{\overline{m}}_{f_0}(\overline{h}) + \delta g \sigma(\overline{V}) \overline{T}_{\max}(\overline{V}, \overline{h})$$

The mass of the aircraft can be considered constant for one period, since the fuel consumed during such a time interval is small as compared with the total mass, i.e.,

$$(3.7) \quad \overline{m}_f(1) - \overline{m}_f(0) \ll 1$$

Periodicity of the flight path implies the boundary conditions

$$(3.8) \quad \overline{V}(0) = \overline{V}(1), \qquad \gamma(0) = \gamma(1), \qquad \overline{h}(0) = \overline{h}(1)$$

The initial condition for the fuel mass can be written as

$$(3.9) \quad \overline{m}_f(0) = 0$$

Control variables are the lift coefficient $\overline{C}_L$ and the throttle setting δ which are subject to the inequality constraints

$$(3.9) \quad \overline{C}_{L_{\min}} \le \overline{C}_L \le \overline{C}_{L_{\max}}, \qquad 0 \le \delta \le 1$$

The atmospheric model which is used for describing air density and thrust dependence on altitude agrees with the ICAO Standard Atmosphere (Ref. 11).

The periodic control problem is to find the control histories $\overline{C}_L$ and δ, the initial states $(\overline{V}(0), \gamma(0), \overline{h}(0))$ and the periodic cycle length c which minimize the performance criterion $\overline{J} = \overline{m}_f(1)/c$ subject to the dynamic system Eq. (3.3), the boundary conditions Eqs. (3.8, 3.9), and the inequality constraints for the control variables Eq. (3.10).

4 Conditions for Optimality and Singular Control

Necessary conditions for optimality can be determined by applying the minimum principle. For this purpose, the Hamiltonian is defined as

$$(4.1) \qquad H = \lambda_V \frac{\mathrm{d}\overline{V}}{\mathrm{d}\xi} + \lambda_\gamma \frac{\mathrm{d}\gamma}{\mathrm{d}\xi} + \lambda_h \frac{\mathrm{d}\overline{h}}{\mathrm{d}\xi} + \lambda_f \frac{\mathrm{d}\overline{m}_f}{\mathrm{d}\xi} ,$$

where the Lagrange multipliers $\lambda = (\lambda_V, \lambda_\gamma, \lambda_h, \lambda_f)^T$ have been adjoined to the dynamic system of Eq. (3.3). The Lagrange multipliers are determined by

$$(4.2) \qquad \frac{\mathrm{d}\lambda_V}{\mathrm{d}\xi} = -\frac{\partial H}{\partial \overline{V}} , \quad \frac{\mathrm{d}\lambda_\gamma}{\mathrm{d}\xi} = -\frac{\partial H}{\partial \gamma} , \quad \frac{\mathrm{d}\lambda_h}{\mathrm{d}\xi} = -\frac{\partial H}{\partial \overline{h}} , \quad \frac{\mathrm{d}\lambda_f}{\mathrm{d}\xi} = 0 ,$$

with boundary conditions

$$(4.3) \qquad \lambda_V(0) = \lambda_V(1) , \ \ \lambda_\gamma(0) = \lambda_\gamma(1) , \ \ \lambda_h(0) = \lambda_h(1) , \ \ \lambda_f(1) = 1/c$$

The optimal controls $\overline{C}_L$ and δ are such that H is minimized pointwise along the optimal trajectory. For this reason, $\overline{C}_L$ is determined either by

$$(4.4) \qquad \partial H/\partial \overline{C}_L = 0$$

or by the constraining bounds of Eq. (3.10).

With regard to throttle setting, H is linear in δ. Therefore, a bang-bang type of control and/or singular arcs can exist. Evaluation of the switching function

$$(4.5) \qquad \overline{S}(y, \lambda) = \frac{\partial}{\partial \delta} H(y, \lambda; \overline{C}_L, \delta)$$

leads to

$$(4.6) \qquad \begin{aligned} \delta &= 0 \quad \text{if} \ \ \overline{S}(y, \lambda) > 0 , \\ \delta &= 1 \quad \text{if} \ \ \overline{S}(y, \lambda) < 0 , \\ \delta &: \text{singular} \quad \text{if} \ \ \overline{S}(y, \lambda) = 0 \ \text{on a finite interval of} \, \xi , \end{aligned}$$

where $y = (\overline{V}, \gamma, \overline{h}, \overline{m}_f)^T$.

The order of the singular arc is $p = 1$, because

$$(4.7) \qquad \frac{\partial}{\partial \delta} \overline{S}^{(1)} = 0 , \quad \frac{\partial}{\partial \delta} \overline{S}^{(2)} \neq 0$$

where

$$(4.8) \quad \overline{S}^{(i)}(y, \lambda; \overline{C}_\mathrm{L}, \delta) = \frac{d^i}{d\xi^i} \overline{S}(y, \lambda; \overline{C}_\mathrm{L}), \quad i = 1, 2$$

The relation $\overline{S}^{(2)}(y, \lambda; \overline{C}_\mathrm{L}, \delta) = 0$ can be used to find the singular control

$$(4.9) \quad \delta = \delta_\mathrm{sing}(y, \lambda; \overline{C}_\mathrm{L})$$

The switching structure of the optimal control is considered to be known, and the conditions described above have to be checked and evaluated along the trajectory.

The system described by Eq. (3.3) is autonomous, so that the Hamiltonian H is constant. Since furthermore the cycle lenght c is considered free, the following relation holds:

$$(4.10) \quad H = \frac{\overline{m}_\mathrm{f}(1)}{c}$$

5   Further Considerations

Additional conditions exist for constraint arcs where the altitude range is limited. This concerns a minimum admissible altitude and, in some cases, an additional maximum altitude limit

$$(5.1) \quad G_1(\overline{h}) = -\overline{h} \leq 0, \qquad G_2(\overline{h}) = \overline{h} - \overline{h}_\mathrm{max} \leq 0$$

The conditions for dealing with the constraints are formulated according to Ref. 13.

There are additional necessary optimality conditions which have been evaluated in the investigation. They may be briefly addressed as

a) Legendre Clebsch condition (Ref. 12),

b) Kelley condition (Ref. 12),

c) Goh condition (Ref. 12),

d) Robbins equality condition (Ref. 12).

6 Periodic Optimum Trajectories with Singular Control

As an illustration for the type of trajectories of interest, some examples are considered first. They are presented in Figs. 1-3.

In Fig. 1, state and control variables of optimal trajectories for minimum fuel consumption per range are shown. A periodic optimal trajectory can be characterized as consisting of a climbing flight phase at maximum thrust setting followed by a sinking phase at zero thrust. The singular thrust arcs may be considered as elements connecting these phases. The throttle control shows a pronounced singular behavior for both transitions during which it changes from one boundary value to the other. The lift coefficient which is the other control appears to be activated primarily during the singular arcs.

The trajectory belonging to a low aerodynamic efficiency shows changes more gradual. By increasing aerodynamic efficiency, the relative length of the singular arcs decreases. Accordingly, the slopes of the singular thrust arcs become steeper and the lift coefficient shows less changes.

More insight into the behavior of the singular arcs is presented in Figs. 2 and 3 which show the switching functions and their derivatives for the cases considered in Fig. 1. As may be seen, similar characteristics develop when changing the aerodynamic efficiency from a low to a high value. This particularly concerns the slopes of the functions in the neighbouring regions of the singular arcs. A pronounced example is presented in the lowest part of Fig. 3 which shows a peak type behavior of a function which, in fact is smooth.

7 Numerical Problems and Improvements

The examples presented in Figs. 1-3 provide an indication of the numerical problems which exist when aerodynamic efficiency is increased. Another effect which produces numerical difficulties is due to a decrease of wing loading m/S. This is illustrated in Fig. 4 which shows homotopy paths concerning solutions when changing both factors addressed, with $V_0^* = \sqrt{m/S}\sqrt{2g/(C_L^* \varrho_0)}$ used as a measure for wing loading. The break-down points due to a convergence failure are indicated. They concern an increase of aerodynamic efficiency as well as a decrease of wing loading.

An improvement is possible by normalizing state and adjoint variables such that they are of about equal magnitude. Another measure applied for improving convergence is a different time scaling of each shooting interval. In addition, an alternative formulation of the singular control and the switching conditions can be applied where the adjoint variables λ_V and λ_γ are eliminated. With this formulation, it is possible to stay on the singular surface during the integration. The standard formulation where λ_V and λ_γ are not eliminated leaves the singular surface during integration because of the numerical deficiencies concerning $\overline{S} = 0$ and $\overline{S}^{(1)} = 0$. A final measure which may be applied concerns a refinement of the multiple shooting grid up to 60 nodes.

With the measures described a homotopy continuation could be achieved. This was applied to a homotopy path showing a combination of aerodynamic efficiency and wing loading as indicated in Fig. 4. A convergence failure again occured before regions of interest for aerodynamic efficiency and wing loading were reached.

In order to find out more about the nature of the difficulties, the last solution that could be obtained before the homotopy break-down was analyzed. A new generalized version of the Multiple Shooting code PARFIT (Ref. 14) was applied for a detailed investigation of the numerical multi-point boundary value problem (MPBVP). Even though this code can handle over- and underdetermined systems (see below), the numerical MPBVP was always formulated as a regular (nonlinear) system with fixed "phase" of the periodic trajectory. In the following part, the results of our analysis are summarized and the strategy used for overcoming the difficulties is outlined.

1) The norms of the sensitivity matrices on the shooting intervals $I_j = [\tau_j, \tau_{j+1}]$,

$$(7.1) \qquad G_j = \frac{\partial}{\partial s_j} z(\tau_{j+1}; \tau_j, s_j), \qquad z(\tau_j; \tau_j, s_j) = s_j$$

are extremely large. Fig. 5 shows the error propagation factors

$$(7.2) \qquad \gamma_j = \|G_j\|_\infty = \max_i \left\{ \sum_k |(G_j)_{ik}| \right\},$$

which indicate that an initial error δs_j at τ_j may be blown up to a magnitude $\gamma_j \|\delta s_j\|_\infty$ at τ_{j+1}. In other words, $\kappa_j = \log_{10} \gamma_j$ decimal digits are lost during numerical integration on I_j. Obviously, the differential equations are very unstable.

The instability does not only influence the computation of the Newton step in each iteration, there is also an important side effect. All the interior point conditions are not taken at shooting nodes. Hence, the computation of the corresponding derivative matrices

$$(7.3) \qquad D_j = \frac{\partial}{\partial s_j} r\left(z(0), \ldots, z(1)\right), \qquad z(\xi) = z(\xi; \tau_j, s_j) \text{ on } I_j$$

involves multiplication by certain transition matrices. Thus, they are also large and loss of decimals at fractions of subintervals decreases their accuracy.

2) The minimum principle approach may always cause stability problems to some extent because of the symplectic structure of the canonical system. But even with unstable differential equations the MPBVP may be well-posed, provided there is some dichotomy and the boundary conditions are well-posed. The latter point was investigated in more detail using sophisticated linear algebra methods. A condensing algorithm of Ref. 14 was used to generate an orthonormal base for the linearized problem such that the solution space of the matching conditions and the solution space of the multi-point boundary conditions are spanned as orthogonal complements. This allows elimination of the variables representing the discretization of the ordinary differential equations system in an optimal manner, i.e., in a way that corresponds to the dichotomy properties of the differential equations. Therefore the condition number of the special condensed system generated this way describes the actual well-posedness of the boundary conditions.

The condition number of the condensed system was found to be extremely large with very small singular values present; estimates are shown in Fig. 6. Hence, the multi-point boundary conditions were in fact ill-posed. As a consequence, no reliable Newton increments could be obtained as solution of the linearized system. This turned out to be the most serious problem at the break-down point of the homotopy.

To overcome the difficulties, the round-off errors had to be reduced and the MPBVP had to be reformulated. In the following, the techniques used successfully to achieve these goals are briefly described.

1a) The error propagation factors on the intervals are drastically reduced by refining the mesh. (Taking a fraction α of subinterval I_j approximately reduces γ_j to γ_j^α.) It may be noted that refining the mesh improves the numerical behavior only as long as the reduction in γ_j is not counter-balanced by increasing roundoff errors in linear algebra operations.

During continuation of the homotopy, error propagation became the dominant difficulty, and the mesh was subsequently refined using 32, 64, 96 and 192 subintervals. Along with this, the distribution of the nodes was adapted to the local stability properties.

1b) Conventional (external) numerical differentiation approximates the transition matrices with low accuracy because of nonlinear coupling and instability. To get more accurate derivative matrices, the "Internal Numerical Differentiation" technique described in Ref. 14 was applied. In the most difficult cases it was still necessary to use the highest available integration accuracy (error per step $\leq 10^{-13}$). The CPU time spent for integrating and generating derivative matrices was in the range of 90 to 99.5% of the total CPU time required by PARFIT.

2) To reduce the condition number, the MPBVP was reformulated as described in the next section, and stable linear algebra methods were applied. After condensing the system by the aforementioned special algorithm, the condensed system is treated by a stable solver for constrained least-squares problems, which is capable of solving over- and underdetermined problems and includes pivoting and a rank and condition analysis. The combination of both algorithms provides a powerful tool not only for analyzing the MPBVP but also for developing an appropriate numerical formulation, thus leading to considerable insight.

The crucial point in solving a difficult optimal control problem is the numerical formulation of the associated MPBVP. Usually several equivalent formulations exist leading to more or less well-posed boundary conditions.

Since the precise switching structure is considered to be known in the problem dealt with, a "variable mesh" technique as described in Ref. 13 was applied. Thus placing the switching points on shooting nodes reduces round-off errors and makes the numerical MPBVP more stable. This technique also prevents switching points from traversing nodes during iteration, a very important advantage especially for fine meshes.

After choosing this problem transformation still a large number of equivalent sets of necessary conditions exist, especially as regards the choice of entry and exit conditions of singular arcs. The periodicity of the autonomous control problem adds an additional "phase shift" degree of freedom, hence the set of necessary conditions derived by applying the minimum principle is degenerate. Since switch and jump conditions must be satisfied, the phase should be fixed, and the singularity calls for a normalization condition. (Alternatively, one could simply solve the singular system using a pseudo-inverse, at least in the absence of discontinuities.)

In our problem physical knowledge and mathematical experience were used to choose an appropriate numerical MPBVP. It should be pointed out, however, that any number of necessary conditions could be included in the MPBVP (in a least-squares sense), and the linear solver would automatically pick an appropriate combination by its pivoting algorithm. This allows for an easy and safe treatment of the problem without investigating the details of its nature.

8   Fuel Savings

An evaluation concerning the reduction of fuel consumption per range is presented in Fig. 7 which shows a comparison between periodic optimal cruise and the best steady-state cruise. The result is that the gain due to periodic optimal control is comparatively small. The gains are reduced when aerodynamic efficiency is increased and maximum engine power available is decreased.

The fuel consumption model used for the results presented in Fig. 7 shows a linear dependence on throttle setting, with zero fuel flow at idling ($\overline{\dot{m}}_{f_0} = 0$). This model may be considered as an idealized characteristic. A model more realistic accounts for a non-zero fuel consumption at idling $\overline{\dot{m}}_{f_0} \neq 0$. This is illustrated in Fig. 8 which shows both models addressed. For the more realistic model, the gains achievable with optimal periodic cruise can reach a level of practical significance when the engine is switched off during phases for which $\delta = 0$ is required by the optimality conditions and when the maximum altitude is constrained such as for aircraft with non-pressurized cabins (Ref. 15). An evaluation is presented in Fig. 9 which provides a quantitative insight into the gains achievable for a wide range of relevant aircraft performance parameters (Ref. 9).

9 Conclusions

For minimizing fuel consumption in aircraft range cruise, it is shown that periodic optimal trajectory solutions with singular control exist. Singular arcs are favored when thrust specific fuel consumption is linearly dependent on speed. Such a characteristic is of practical significance because it is existent with propeller driven aircraft.

The improvements due to periodic optimal control are also considered. For a wide range of relevant aircraft performance parameters, it is shown which are the reductions in fuel consumption as compared with the best steady-state cruise. The improvements appear to be comparatively small when an idealized fuel consumption model is considered. There may be a significant improvement for a more realistic model accounting for a non-zero fuel flow at idling.

To generate optimal trajectories when changing aerodynamic efficiency and wing loading, a homotopy technique is used. In the progress of the homotopy, convergence problems result when aerodynamic efficiency is increased and wing loading is decreased. To reach a domain valid for modern motorgliders, special numerical improvements are required. After analyzing the problem, reasons for the numerical difficulties were found in extremely large sensitivity matrices on the shooting intervals and very ill-posed multipoint boundary conditions. Successful measures to decrease the sensitivity matrices are the refinement of the multiple shooting mesh and the use of the "Internal Numerical Differentiation" technique together with a very high integration accuracy. The multiple point boundary value problem was reformulated and a stable condensing algorithm was applied.

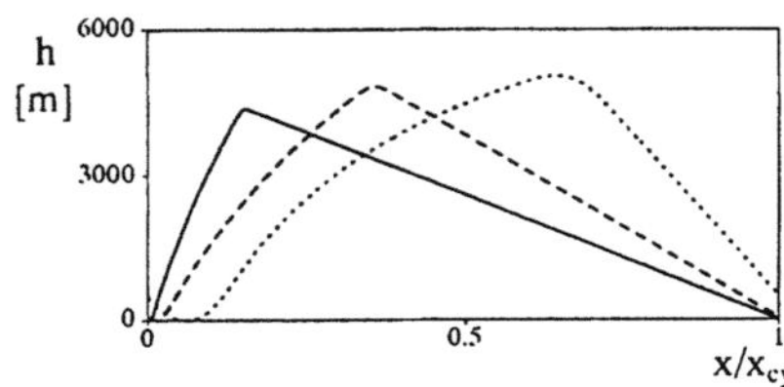
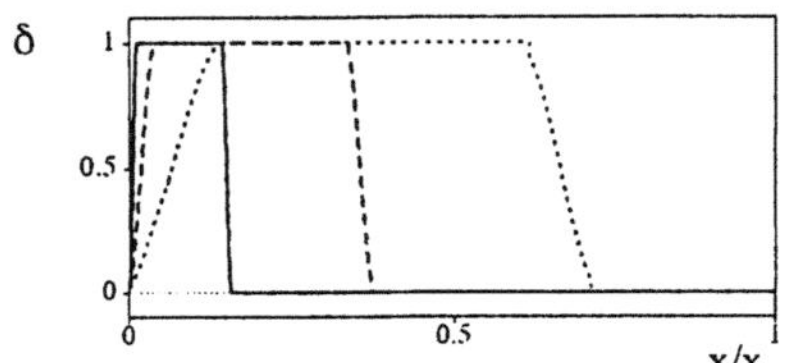
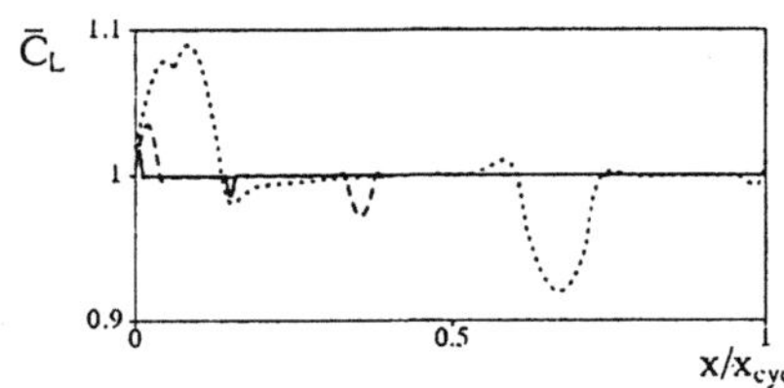

I $\cdots\cdots$ $E = 3.4,\ m/S = 345\,\mathrm{kg/m^2},$
$P_{\max}/(D_{\min}V_0^*) = 2.3,$
$x_{\mathrm{cyc}} = 51.4\,\mathrm{km}$

II $----$ $E = 15,\ m/S = 214\,\mathrm{kg/m^2},$
$P_{\max}/(D_{\min}V_0^*) = 4.0,$
$x_{\mathrm{cyc}} = 116\,\mathrm{km}$

III ——— $E = 33,\quad m/S = 45\,\mathrm{kg/m^2},$
$P_{\max}/(D_{\min}V_0^*) = 8.9,$
$x_{\mathrm{cyc}} = 174\,\mathrm{km}$

Fig. 1 Optimal periods with singular arcs, altitude and control variables

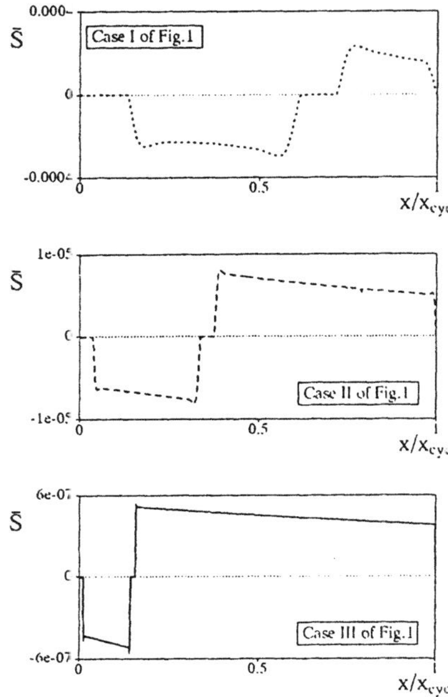

Fig. 2   Optimal periods with singular arcs, switching functions

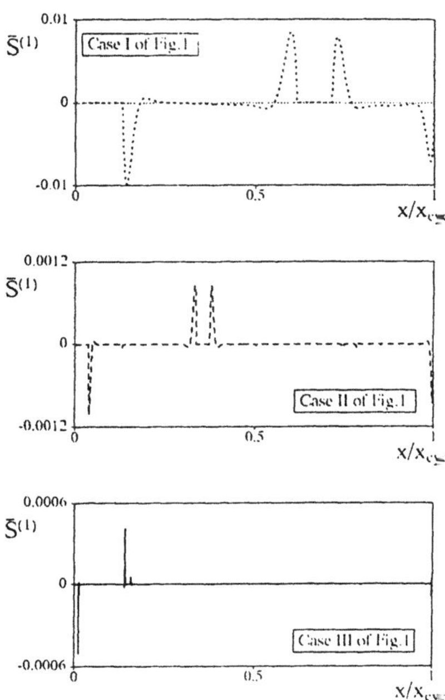

Fig. 3   Optimal periods with singular arcs, time derivatives of switching functions

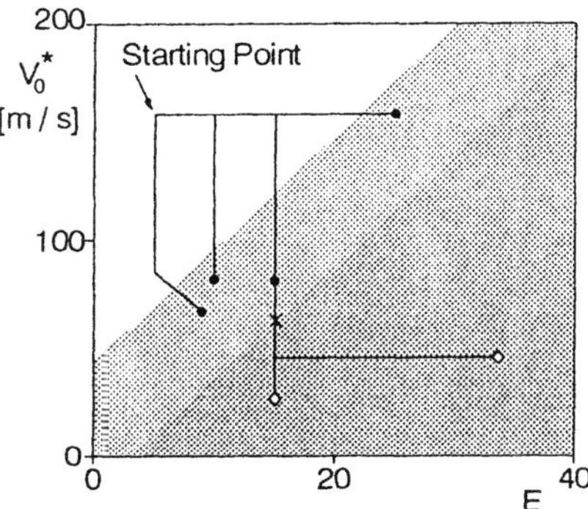

Fig. 4   Homotopy path
- ● Break down due to convergence failure, initial problem formulation
- × Break down due to convergence failure after normalization and time scaling
- ◇ Parameter values reached after numerical improvements

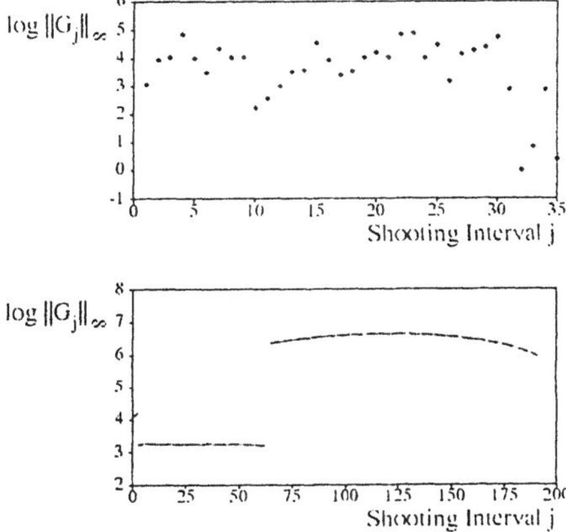

Fig. 5   Norms of transition matrices for solution at break-down point (32 intervals) and at end point of homotopy (192 intervals) $E/(m/S) = 0.76$ kg/m^2, $C_L^* = 0.78$,

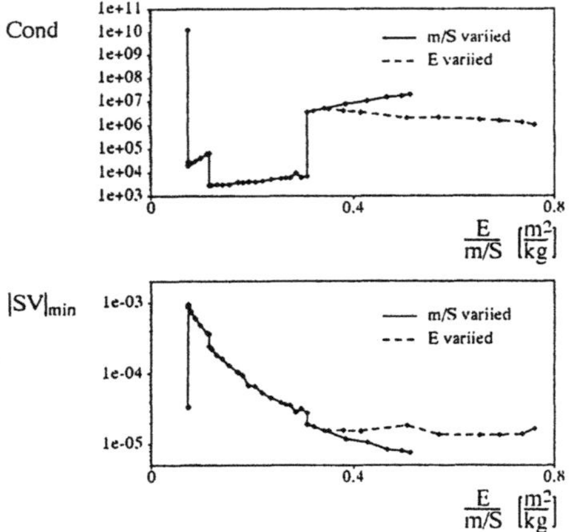

Fig. 6 Sub-condition numbers *Cond* of multipoint boundary conditions along homotopy branches and estimate of corresponding minimum singular values *SV*

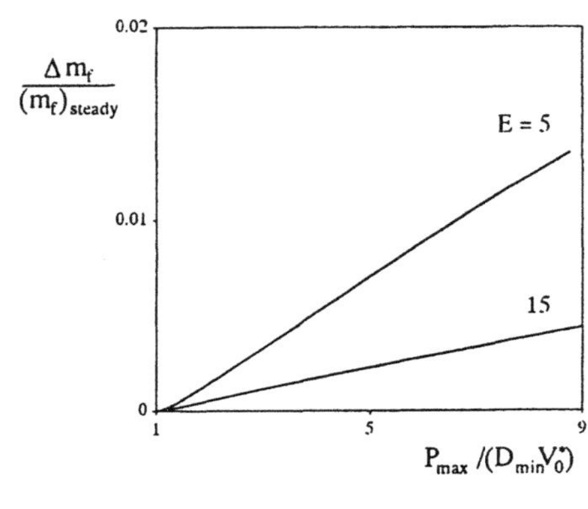

Fig. 7 Fuel saving Δm_f due to periodic optimal cruise with singular control $(m_f)_{steady}$: best steady-state cruise

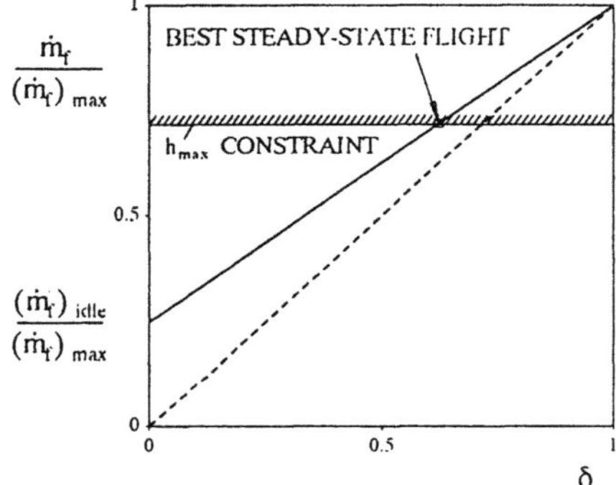

Fig. 8 Model with non-zero fuel consumption at idle thrust setting

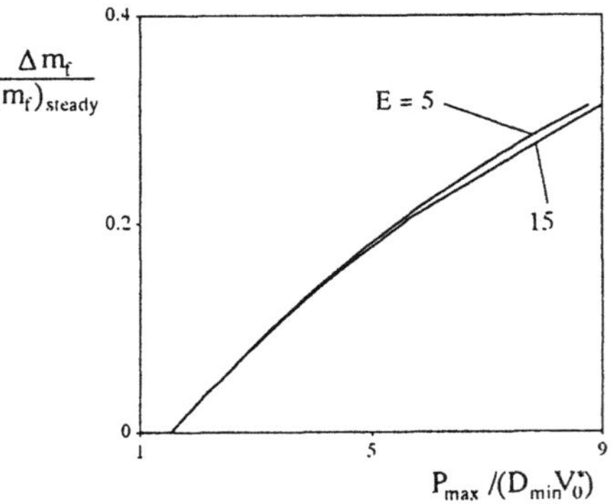

Fig. 9 Fuel saving Δm_f due to periodic optimal cruise $(m_f)_{steady}$: best steady-state cruise with a 3.5 km altitude limit $(\dot{m}_f)_{idle}/(\dot{m}_f)_{max} = 0.1$

References

1. Speyer, J.L., "Nonoptimality of the Steady-State Cruise of Aircraft," AIAA Journal, Vol. 14, 1976, pp. 1604-1610.

2. Speyer, J.L., Dannemiller, D., Walker, D.,"Periodic Optimal Cruise of an Atmospheric Vehicle," Journal of Guidance, Control, and Dynamics, Vol. 8, 1985, pp. 31-38.

3. Gilbert, E.G., "Vehicle Cruise: Improved Fuel Economy by Periodic Control," Automatica, Vol. 12, 1976, pp. 159-166.

4. Breakwell, J.V., Shoee, H., "Minimum Fuel Flight Paths for Given Range," AIAA Paper, 1980, No. 80-1660.

5. Grimm, W., Well, K.-H., Oberle, H.J., "Periodic Control for Minimum Fuel Aircraft Trajectories", Journal of Guidance, Control, and Dynamics, Vol. 9, 1986, pp. 169-174.

6. Menon, P.K.A., "Study of Aircraft Cruise," Journal of Guidance, Control, and Dynamics, Vol. 12, 1989, pp. 631-639.

7. Sachs, G., "Verringerung des Treibstoffverbrauchs durch periodische Optimalflugbahnen," DGLR Jahrbuch, 1984, 090-1 - 09-17.

8. Sachs, G., Christodoulou, T., "Reducing Fuel Consumption of Subsonic Aircraft by Optimal Cyclic Cruise," Journal of Aircraft, Vol. 24, 1987, pp. 616-622.

9. Sachs, G., Lesch K., "Fuel Savings by Optimal Aircraft Cruise with Singular and Chattering Control," Lecture Notes in Control and Information Sciences, Springer-Verlag, (1990), pp. 590-599.

10. Sachs, G., Lesch K., "Optimal Periodic Cruise with Singular Control," AIAA Guidance, Navigation and Control Conference Proceedings, 1990, pp. 1586-1594.

11. ICAO Standard Atmosphere, International Civil Aviation Organization, Montreal, 1964.

12. Kumar, R., Kelley, H.J., "Singular Optimal Atmospheric Rocket Trajectories," Journal of Guidance, Control, and Dynamics, Vol. 11, 1988, pp. 305-312.

13. Bock , H.G., "Numerische Behandlung von Zustandsbeschränkten und Chebychef-Steuerungsproblemen," Course DR 3.10 of the Carl-Cranz-Gesellschaft, Oberpfaffenhofen, Germany, 1981.

14. Bock, H.G., "Randwertproblemmethoden zur Parameteridentifizierung in Systemen nichtlinearer Differentialgleichungen", Dissertation, Universität Bonn, 1985, Bonner Mathematische Schriften 183, Bonn, 1987.

15. ICAO Standard and Recommended Practices, Operation of Aircraft, Annex 6, 1986.

Authors' Addresses

Prof. Dr.-Ing. Gottfried Sachs and
Dipl.-Ing. Klaus Lesch

Institute of Flight Mechanics
and Flight Control
Technische Universität München
Arcisstr. 21
8000 München 2
Germany

Prof. Dr. Hans Georg Bock and
Dipl.-Math. Marc Steinbach

Interdisciplinary Center
of Scientific Computing
Universität Heidelberg
Im Neuenheimer Feld 368
6900 Heidelberg
Germany

Optimal Flight Paths with Constrained Dynamic Pressure

Werner Grimm

Abstract. This paper presents optimal aircraft motion subject to constrained dynamic pressure. The aircraft is modeled as a point mass in the three-dimensional space. Two optimal control problems are solved: 1) range maximization in fixed time, 2) minimum time intercept of a moving target. At first all combinations of active state and control constraints are considered to determine the possible types of optimal control. They are composed to give the optimal switching structures of 1) and 2). Extremals for various boundary conditions are obtained with multiple shooting. Along the dynamic pressure boundary a) full throttle, b) partial throttle and c) minimum thrust subarcs occur. b) is a singular control, c) is a chattering control.

Nomenclature

A	longitudinal acceleration dV/dt	n	load factor
a	speed of sound, $a = a(h)$	q	dynamic pressure, $q = \rho(h)V^2/2$
C_D	drag coefficient, $C_D = C_D(M, C_L)$	R	distance of interceptor and target
C_L	lift coefficient	S	wing reference area
D	drag, $D = D(h, V, n)$	s	switching function for the
d	final distance to the target		power setting, $s = \lambda_V + \lambda_q \rho V$
E	specific energy, $E = h + V^2/(2g)$	T	thrust
g	gravitational acceleration,	t	time
	$g = 9.80665\ m/s^2$	U	$U = \sqrt{\lambda_\gamma^2 + (\lambda_\chi/\cos\gamma)^2}$
H	Hamiltonian	V	speed
h	altitude	W	aircraft weight
M	Mach number, $M = V/a(h)$	x	x-coordinate
m	artificial control in the extended	y	y-coordinate
	dynamical model	z	state vector, $z = (x, y, h, V, \gamma, \chi)^T$

γ	flight path angle	ξ	power setting
λ	Lagrange multiplier	ρ	air density
μ	bank angle	ϕ	performance index
ν	multiplier to adjoin the terminal	χ	heading angle
	constraint to the performance index	ψ	terminal constraint function

Subscripts

0	initial value	min	minimum value
f	final value	max	maximum value
T	target variable		

1   Introduction

In terms of optimal control pursuit and escape maneuvers of fighter aircraft are often
formulated as minimum time intercept or maximum range trajectories. In many cases
the extremals reach the sea level, enter the dynamic pressure boundary or are constrained by the maximum Mach number. Then different types of optimal control can
occur, regular and singular ones with respect to the linearly appearing thrust control.
Finding the optimal sequence of subarcs is a mathmatical challenge in each case. The
necessary conditions of optimality lead to ill-conditioned multipoint boundary value
problems (MPBVP's), which can efficiently be solved by multiple shooting (Bulirsch
1971). The corresponding computer code BNDSCO (Oberle et al. 1985) was used by the
authors mentioned below. Shinar et al. (1986) present minimum time intercept trajectories. However, the boundary conditions are such that altitude and speed constraints
are avoided. Seywald (1984) computed maximum range trajectories with unspecified
terminal state. The results are composed of two arcs. An unconstrained arc is followed
by a subarc on the q_{max}-boundary. Maximum thrust is optimal throughout. Seywald
continued the study in his dissertation (1990), where he specified the terminal state for a
trajectory coming from the dash-point. The result is a complicated sequence of singular
and chattering arcs, giving an idea of the complexity of the problem. In this paper two
optimal control problems are considered:
1) range maximization in fixed time
2) minimum time intercept of a moving target
The dynamic pressure boundary is the relevant state constraint. The planar model used
by Seywald (1984, 1990) is extended to three dimensions. The different types of optimal control are summarized and the switching structures for the particular boundary
conditions are determined.

2 The Dynamical Model and its "Relaxation"

The following equations of motion are an extension of the usual point mass model for aircraft motion in the way that the *hodograph* (the set of all admissible tangent vectors $(\dot{x}, \dot{y}, \dot{h}, \dot{V}, \dot{\gamma}, \dot{\chi})^T$) is convex.

$$
\begin{aligned}
(2.1) \qquad \dot{x} &= V \cos\gamma \cos\chi \\
(2.2) \qquad \dot{y} &= V \cos\gamma \sin\chi \\
(2.3) \qquad \dot{h} &= V \sin\gamma \\
(2.4) \qquad \dot{V} &= g\left(\frac{T(h,V,\xi) - D(h,V,n)}{W} - \sin\gamma\right) = A(h,V,\gamma,n,\xi) \\
(2.5) \qquad \dot{\gamma} &= \frac{g}{V}(m\cos\mu - \cos\gamma) \\
(2.6) \qquad \dot{\chi} &= \frac{g}{V}\frac{m\sin\mu}{\cos\gamma}
\end{aligned}
$$

The controls are the load factor n, the power setting ξ, the bank angle μ and the artificial control m, which are subject to the control constraints

$$
\begin{aligned}
(2.7) \qquad 0 &\leq \xi \leq 1, \\
(2.8) \qquad n &\leq n_{max}(= 5), \\
(2.9) \qquad n &\leq n_S(h,V) = q(h,V)SC_{L,max}(M)/W, \\
(2.10) \qquad 0 &\leq m \leq n,
\end{aligned}
$$

where $q(h,V) = \rho(h)V^2/2$ denotes dynamic pressure and $C_{L,max}$ the maximum lift coefficient. (2.1)–(2.6) is an extension of the original point mass model, which is the special case

$$
(2.11) \quad m = n.
$$

Under restriction (2.11) the hodograph is not convex. Its projection into the $(\dot{V}, \dot{\gamma})$-plane is depicted in Fig.1. To properly include the "chattering controls" occurring in the original model the hodograph is extended to its convex hull (see Fig.1). The fourth control variable m constrained by (2.10) accomplishes this extension.

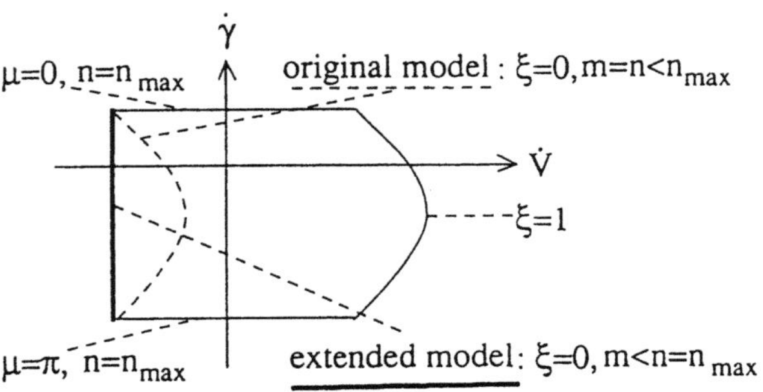

*F*igure 1: Set of admissible pairs $(\dot{V}, \dot{\gamma})$ in the original and extended dynamical model.

Thrust and drag are modelled as follows:

$$T(h, V, \xi) = T_{min}(h, V) + \xi\,(T_{max}(h, V) - T_{min}(h, V)),$$
$$D(h, V, n) = q(h, V)\, S\, C_D\,(M(h, V),\, C_L(h, V, n))$$

with $C_L(h, V, n) = \frac{nW}{q(h,V)S}$ and $M(h, V) = \frac{V}{a(h)}$.

For each altitude h the "corner speed" $V^*(h)$ is defined by

$$(2.12) \quad n_{max} = n_S(h, V^*(h)).$$

For $V > V^*(h)$ (2.8) is the more stringent constraint, for $V < V^*(h)$ $n_s(h, V)$ is the smaller bound for n. In Fig.3 the corner velocity locus is depicted together with the "loft ceiling", the set of all pairs (h, V) satisfying $n_S(h, V) = 1$.
The relevant state constraint of the model is the dynamic pressure boundary:

$$(2.13) \quad q \leq q_{max}(= 83.85\,kN/m^2)$$

T_{min}, T_{max}, C_D and $C_{L,max}$ are analytic functions which are carefully adapted to table data of a high performance aircraft. A detailed model description is given by Grimm (1992). Strictly speaking, (2.1)–(2.6) are only valid for small angles of attack. Further simplifying assumptions are flat earth, constant weight W and the absence of side force and wind.

3   Optimal Control Problems and Boundary Conditions

The problem is to find the optimal controls $n(t), \mu(t), \xi(t)$ and $m(t)$, which minimize the performance index $\phi(x_f, y_f, h_f, V_f)$ subject to the terminal condition $\psi(x_f, y_f, h_f, V_f) = 0$. Two examples for ϕ and ψ are considered:

(3.1)   range maximization in fixed time: $\phi = -x_f, \psi = t_f - t_{f1}$

(3.2)   minimum time intercept: $\phi = t_f, \psi = R_f^2 - d^2$

In (3.2) $R(t)$ is the distance to a nonmaneuvering target:

$$(3.3) \quad R = \sqrt{(x - x_T)^2 + (y - y_T)^2 + (h - h_T)^2},$$

$$(x_T, y_T, h_T) = (x_{T0}, y_{T0}, h_{T0}) + t \cdot (\dot{x}_T, \dot{y}_T, 0).$$

$\dot{x}_T$ and $\dot{y}_T$ are constant velocity components of the target. The constants in (3.1) and (3.2) are $t_{f1} = 100\ sec$ and $d = 1\ km$ (final distance to the target). Six representative initial pairs (h_0, V_0) are chosen in the flight envelope of the aircraft (see Table 1). Each pair is denoted by a code consisting of a cipher between 1 and 3 and a letter (A or B). There are always two points which belong to the same energy level as indicated by the leading cipher. One of them is characterizied by $V_0 > V^*(h_0)$ (indicated by the letter A), the reverse relation holds for the other one (the B-case). The initial values (h_0, V_0) can be seen in Fig.3.

example	1A	1B	2A	2B	3A	3B
E_0 [km]	3.8	3.8	14.55	14.55	25.3	25.3
h_0 [km]	1	3	6	12	10	16
h_{T0} [km]	0	0	0	3	5	8

Table 1: Initial values of aircraft and target

The inital state of the aircraft is given by $x_0 = y_0 = 0$, h_0 from Table 1,
$V_0 = \sqrt{2g(E_0 - h_0)}$ with E_0 and h_0 taken from Table 1, $\gamma_0 = 0, \chi_0 = 120°$.

For the minimum time intercept the target motion is given by $x_{T0} = 30\ km$ ($x_{T0} = 20\ km$ for example 1B), $y_{T0} = 0$, $\dot{x}_T = 200\ m/s$, $\dot{y}_T = 0$, h_{T0} from Table 1.

4   Necessary Conditions of Optimality

Let $z = (x, y, h, V, \gamma, \chi)^T$ denote the state vector. For an optimal solution of problems (3.1), (3.2) there exists an adjoint vector $\lambda_z = (\lambda_x, \lambda_y, \lambda_h, \lambda_V, \lambda_\gamma, \lambda_\chi)^T$ with the following properties: The optimal control minimizes the Hamiltonian

$$(4.1) \qquad H = \lambda_z^T \dot{z} + \lambda_q \dot{q} - \lambda_n (n_S(h, V) - n).$$

$\dot{z}$ in (4.1) must be replaced by the expressions in (2.1)–(2.6). $\dot{q}$ must be substituted by Eq. (4.2) below. The q_{max}-constraint (2.13) is a first order state constraint, i.e. the controls explicitly occur in the first derivative

$$(4.2) \qquad -\dot{q} = -\rho V (\frac{1}{2} \frac{\rho_h}{\rho} V^2 \sin\gamma + A(h, V, \gamma, n, \xi))$$

of the constraint function $q_{max} - q$. $-dq/dt$ and the $C_{L,max}$-constraint (2.9) are adjoined to the Hamiltonian with multipliers λ_q and λ_n, respectively. λ_q and λ_n are nonnegative, as long as the corresponding constraints are active, and zero else:

$$(4.3) \qquad \lambda_q \geq 0, \qquad \text{if } q = q_{max}, \qquad \lambda_q = 0 \text{ else,}$$
$$(4.4) \qquad \lambda_n \geq 0, \qquad \text{if } n = n_S(h, V), \qquad \lambda_n = 0 \text{ else.}$$

If the q_{max}-constraint ($C_{L,max}$ -constraint) is active, $\lambda_q(\lambda_n)$ are determined from $0 = \partial H/\partial n$. The other control constraints (2.7),(2.8) and (2.10) do not depend on state variables. Thus, they do not affect the structure of the adjoint equations and therefore are not adjoined to the Hamiltonian. H is linear in ξ; the coefficient of ξ is

$$(4.5) \qquad s = \lambda_V + \lambda_q \rho V.$$

s acts as a "switching function" for the power setting:

$$(4.6) \qquad \begin{aligned} s > 0 &\implies \xi = 0, \\ s < 0 &\implies \xi = 1, \\ 0 < \xi < 1 &\implies s = 0. \end{aligned}$$

$\dot{\lambda}_z$ satisfies the adjoint equation $\dot{\lambda}_z = -H_z^T$. $\lambda_z(t_f)$ is given by the transversality conditions: There exists a scalar ν such that

$$(4.7) \qquad \lambda_z^T(t_f) = \frac{\partial}{\partial z_f}(\phi(z_f, t_f) + \nu \cdot \psi(z_f, t_f)),$$

$$(4.8) \qquad H|_{t=t_f} = -\frac{\partial}{\partial t_f}(\phi(z_f, t_f) + \nu \cdot \psi(z_f, t_f)).$$

5 Different Types of Optimal Control

Depending on the set of active constraints the application of the Minimum Principle (Eq. (4.1)) leads to different types of optimal control. The whole set of different types divides into two groups, where

$$(5.1) \quad U = \sqrt{\lambda_\gamma^2 + (\lambda_\chi/\cos\gamma)^2}$$

is positive for one group and zero for the other one. The two theorems in this section are valid under the following assumptions:

$$(5.2) \quad \lambda_x^2 + \lambda_y^2 > 0$$

(5.2) is satisfied for the optimal control problems (3.1),(3.2) .

$$(5.3) \quad C_{D,C_L} = 0 \quad \text{for} \quad C_L = 0, \quad C_{D,C_L} > 0 \quad \text{for} \quad C_L > 0,$$

$$C_{D,C_L C_L} > 0$$

(5.3) holds for a symmetric and convex drag polar as used in this paper. The following theorems are reported without proofs, which are given by Grimm (1992). They are formulated for the case $n_{max} < n_S(h, V)$. For the reversal analogous results hold.

Theorem 5.1: Let (m, n, μ, ξ) be the optimal control vector under assumptions (5.2),(5.3), $|\gamma| < \pi/2$ and $U > 0$. Then the following statements hold:

$$(5.4) \quad m = n > 0$$

$$(5.5) \quad \cos\mu = -\frac{\lambda_\gamma}{U}, \quad \sin\mu = -\frac{\lambda_\chi}{U\cos\gamma}$$

The following types of optimal control are the only possible ones:

I.1: $n < n_{max}$, $\xi = 1$. $s \leq 0$ is additional optimality condition.

  I.1.1: $q < q_{max}$. n is the solution of

$$(5.6) \quad Ug - sVA_n = 0.$$

I.1.2: $q = q_{max}$. n is the solution of $\dot{q} = 0$ (with $\xi = 1$). Additional optimality condition is $\lambda_q \geq 0$ with λ_q determined from (5.6).

I.2: $n = n_{max}$.

I.2.1: $q < q_{max}$, $\xi = 1$. Additional optimality conditions are $s \leq 0$ and $Ug - sVA_n \geq 0$.

I.2.2: $q < q_{max}$, $\xi = 0$. Additional optimality condition is $s \geq 0$.

I.2.3: $q < q_{max}$, $s \equiv 0$. ξ is determined from $0 = dH_V/dt$ (and must satisfy $0 \leq \xi \leq 1$).

I.2.4: $q = q_{max}$, $s \equiv 0$. ξ is determined from $0 = dq/dt$. Additional optimality conditions are $0 \leq \xi \leq 1$ and $\lambda_V \leq 0$.

Theorem 5.2: Let (m, n, μ, ξ) be the optimal control vector along a subarc with $U \equiv 0$. Assumptions (5.2), (5.3) and $|\gamma| < \pi/2$ shall be satisfied. Then the following statements hold:

$$(5.7) \quad \chi = const. \quad (\Longrightarrow \mu \in \{0, \pi\})$$

$$(5.8) \quad \lambda_\gamma, \lambda_\chi \equiv 0$$

The following types of optimal control are the only possible ones.

II.1: $s > 0$, $\xi = 0$, $n = n_{max}$.

II.1.1: $q < q_{max}$. $m \cos \mu$ is determined from $0 = dH_\gamma/dt$.

II.1.2: $q = q_{max}$. $m \cos \mu$ is determined from $0 = d^2q/dt^2$. According to (2.10) II.1.1 and II.1.2 are only valid for $m \leq n_{max}$.

II.2: $s \equiv 0$, $q = q_{max}$. $m \cos \mu$ is determined from $0 = dH_\gamma/dt$.

As long as the q_{max}-boundary is not active Theorems 5.1 and 5.2 yield exactly one type of optimal control on each edge and surface of the boundary of the hodograph (see Fig. 2). Along the dynamic pressure boundary only such control values are admissible which satisfy $\dot{q} \leq 0$ and thus would not cause a constraint violation. According to (4.2) $\dot{q} \leq 0$ is equivalent to $\dot{V} \leq -V^2 \sin \gamma \cdot \rho_h/(2\rho)$. As can be seen from this inequality $\dot{q} = 0$ is a line perpendicular to the $\dot{V}$-axis in the $(\dot{V}, \dot{\gamma})$-space and the admissible control set is the part of the hodograph on the left side of the line in Fig. 2. In the case $q = q_{max}$ the optimal control always is a point on the intersection of the hodograph and the line

$\dot{q} = 0$. There is a singular type of optimal control in the interior of the intersection and regular type on the boundary.

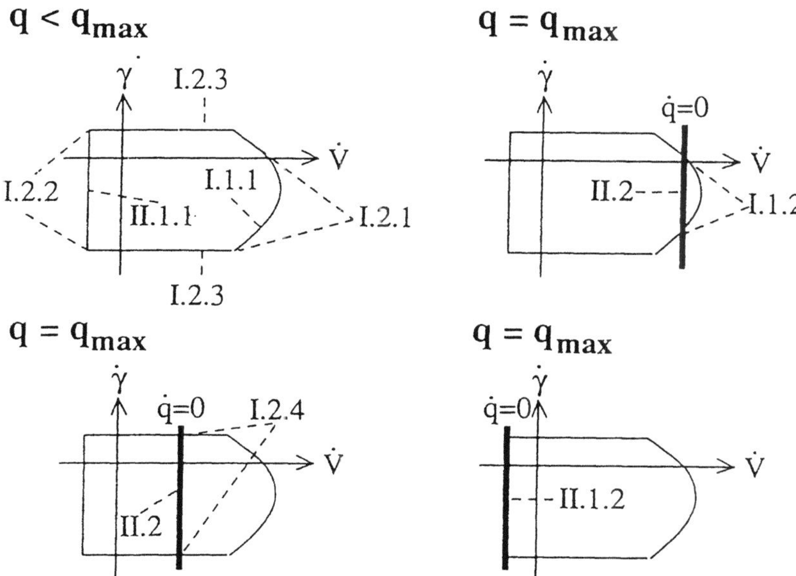

Figure 2: Different types of optimal control on the boundary of the hodograph.

6   Numerical Results

In the examples defined in section 3 control types I.1.1, I.1.2, I.2.1, II.1.2 and II.2 occur. In all cases one of the following structures is taken (see Table 2):

1: I.2.1 – I.1.1
2: I.2.1 – I.1.1 – I.1.2
3: I.2.1 – I.1.1 – I.1.2 – II.2
4: I.2.1 – I.1.1 – I.1.2 – II.2 – II.1.2

In any case there is an initial turning maneuver with maximum load factor (control type I.2.1). If $V_0 > V^*(h_0)$ (initial states 1A, 2A, 3A) the load factor is constrained by n_{max}. In the B-cases characterizied by $V_0 < V^*(h_0)$ the first subarc is a flight along the $C_{L,max}$-limit (2.9). Turning is necessary because of the initial value $\chi_0 = 120°$. The aircraft must perform a heading change of about 120° to maximize range in x-direction or to intercept the target, which always flies along the x-axis.

example	1A	1B	2A	2B	3A	3B
max. range	3	3	2	2	2	2
intercept	4	3	4	4	3	1

Table 2: Optimal switching structures

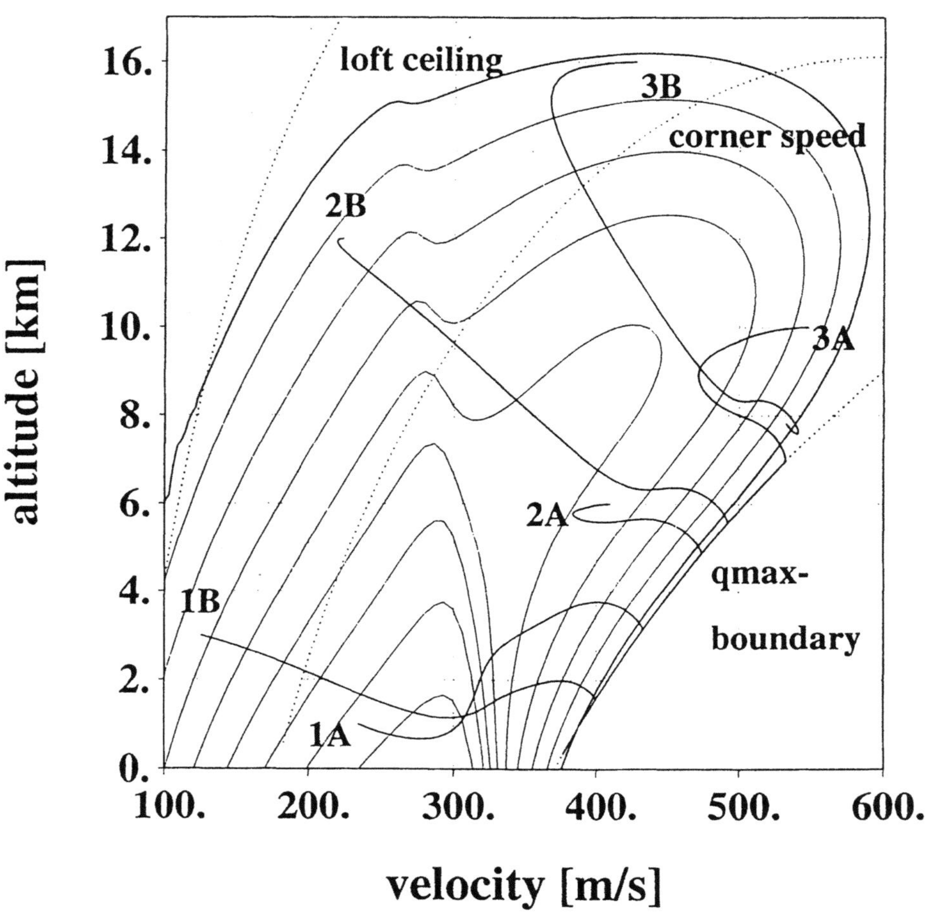

Figure 3: Minimum time intercept of a low flying target. Final altitude of the constrained portion on the q_{max}-boundary: $1A : h_f = 283\,m, 1B : h_f = 382\,m, 2A : h_f = 700\,m, 2B : h_f = 3482\,m, 3A : h_f = 5403\,m, 3B : h_f = 7819\,m$.

7   Conclusions

In this paper optimal flight paths are presented, which maximize range in specified time or minimize the final time on intercepting a moving target. The analysis focusses on the optimal control structure along the dynamic pressure boundary. The following conclusions can be drawn from the results: Maximum range trajectories keep full throttle along the q_{max}-boundary. Only if the final state on the constraint is inside the flight envelope a singular arc at the end occurs. Target altitude determines the optimal control structure of a minimum time intercept maneuver. On the transition from a high flying to a low flying target following cases occur.

a) The constraint is not touched at all.

b) The constraint becomes active but thrust is maximal throughout.

c) The constraint is entered with full throttle and terminates with a singular arc.

d) Sequence c) plus a "chattering" arc with minimum thrust and maximum drag.

Types of optimal control other than a)–d) can occur for different cost functions and boundary conditions. Again, new phenomena must be expected if the coordinates in the horizontal plane do not affect the optimal control problem or the drag polar does not satisfy the assumptions made in this paper.

References

Bulirsch, R., Die Mehrzielmethode zur numerischen Lösung von nichtlinearen Randwertproblemen und Aufgaben der optimalen Steuerung. Report of Carl-Cranz-Gesellschaft e.V., Oberpfaffenhofen, 1971.

Grimm, W., Lenkung von Hochleistungsflugzeugen - Vergleich von optimaler Steuerung und fastoptimaler Echtzeitsteuerung. Dissertation, Department of Mathematics, Munich University of Technology, 1992.

Oberle, H.J., Grimm, W., Berger, E.G., BNDSCO - Rechenprogramm zur Lösung beschränkter optimaler Steuerungsprobleme - Benutzeranleitung. TUM-M8509, Department of Mathematics, Munich University of Technology, 1985.

Seywald, J., Reichweitenoptimale Flugbahnen für ein Überschallflugzeug. Diploma thesis, Department of Mathematics, Munich University of Technology, 1984.

Seywald, J., Optimal Control Problems with Switching Points. Ph. D. Thesis, Virginia Polytechnic Institute and State University, Blacksburg, Va., 1990.

Shinar, J., Well, K.H., Järmark, B.S.A., Near-Optimal Feedback Control for Three-Dimensional Interceptions. Proceedings of the 15th Congress of ICAS, London, Sept. 1986.

Author's Address

Dr. rer. nat. W. Grimm
Institute of Flight Mechanics and Control
University of Stuttgart
Forststr. 86
D–7000 Stuttgart
Germany

Optimal Ascent of a Hypersonic Space Vehicle

Kurt Chudej

Abstract: An ascent optimization of a hypersonic Sänger type lower stage is presented. The optimal control problem is reduced to a multipoint boundary-value problem by calculus of variations and solved by the multiple shooting method. A good first estimate of state and especially adjoint variables to start the multiple shooting algorithm is computed via a special direct collocation method. Accurate solutions including the switching structure are obtained through this hybrid approach due to Bulirsch and von Stryk.

1. Introduction

The space vehicle considered here is a two-stage Sänger type vehicle, with a turbo and ramjet engine in the lower stage and a conventionally rocket-propelled upper stage (c.p. e.g. [12, 14, 7, 23, 24]). The whole system will be launched horizontally and will be able to deliver into orbit either a manned shuttle type upper stage or an unmanned cargo unit. The lower stage is winged and capable of performing cruising flights due to its airbreathing engines. The optimization in this paper is focused on the ascent of the first stage, including the optimization of the switching from turbo to ramjet acceleration. Due to its importance on the resulting flight path a dynamic pressure constraint is considered in the problem formulation.

The model used here is a refinement of a formerly investigated model of a two-stage rocket-propelled space transporter, where simultaneously an ascent and mass ratio optimization of both stages was performed using a two-dimensional flight scenario (Shau [25], [19]) and a three-dimensional rotating Earth model (Bulirsch and Chudej [5], [6]), respectively. Some ideas for such a space transportation system have already been developed by Eugen Sänger in the 30ies.

The investigated problem includes several control constraints and a state constraint of order one. The transformation of the state constraint by differentiation yields an additional control constraint which affects simultaneously three of the four controls. This yields very complicated case dependent control laws. Moreover the Hamiltonian is not regular and the control function not scalar, so the theoretical results concerning the existence of boundary subarcs and contact points of state constraints in dependence of their order apply only in a slightly modified sense [13].

A hybrid approach consisting of a special direct collocation method and the multiple shooting method is used to calculate an accurate numerical solution (von Stryk and Bulirsch [28]).

2. Problem Formulation

In the following we give, in short, the mathematical description of the model:

State Variables		Control Functions	
v	velocity	C_L	lift coefficient
γ	path inclination	μ	bank angle
χ	azimuth inclination	δ	mass flow
h	altitude	ε	thrust angle
Λ	geographical latitude		
θ	geographical longitude		
m	mass		
x	$= (v, \gamma, \cdots, m)^T$	u	$= (C_L, \mu, \delta, \varepsilon)^T$

Other Important Quantities

a	speed of sound	Ma	mach number
D, L	drag and lift force	Q	switching function
C_D, C_L	drag and lift coefficient	q	dynamic pressure
f	right hand side of o.d.e.	r_0	Earth's radius, $R = r_0 + h$
F	reference area	T	thrust force (turbo/ramjet)
g	gravitational acceleration	t	time
H	Hamiltonian	t_e	switching time of engines
I_{sp}	specific impuls (turbo/ramjet)	t_f	separation time of stages
J	performance index	ϱ	atmospheric density
λ	adjoint variables	ω	angular velocity

The objective of the optimization is to minimize the fuel consumption of the lower stage, i.e.

$$J[u] = -m(t_f) \overset{!}{=} \min$$

s.t. the following constraints: The equations of motion of a point mass over a spherical and rotating Earth with no wind in the atmosphere (see e.g. Miele [18] Chap. 4-5[4], Vinh

---

[4] Side force $\tilde{Q} = 0$ is assumed. Calculus of variations yields optimal thrust sideslip angle $\nu = 0$, which is therefore already eliminated in the formulation of equations of motion used here [5].

et. al. [30]) are used to describe the position and velocity of the spacecraft.

$$\dot{v} = \Big[T(v,h)\,\delta\,\cos\varepsilon - D(v,h;C_L)\Big]\frac{1}{m} - g(h)\sin\gamma +$$
$$+ \;\omega^2 R\cos\Lambda(\sin\gamma\cos\Lambda - \cos\gamma\sin\chi\sin\Lambda)$$

$$\dot{\gamma} = \Big[T(v,h)\,\delta\,\sin\varepsilon + L(v,h;C_L)\Big]\frac{\cos\mu}{mv} - \Big[\frac{g(h)}{v} - \frac{v}{R}\Big]\cos\gamma +$$
$$+ \;2\,\omega\cos\chi\cos\Lambda + \omega^2\cos\Lambda\,(\sin\gamma\sin\chi\sin\Lambda + \cos\gamma\cos\Lambda)\,\frac{R}{v}$$

$$\dot{\chi} = \Big[T(v,h)\,\delta\,\sin\varepsilon + L(v,h;C_L)\Big]\frac{\sin\mu}{mv\cos\gamma} - \cos\gamma\cos\chi\tan\Lambda\frac{v}{R} +$$
$$+ \;2\,\omega\,(\sin\chi\cos\Lambda\tan\gamma - \sin\Lambda) - \omega^2\cos\Lambda\sin\Lambda\cos\chi\,\frac{R}{v\cos\gamma}$$

$$\dot{h} = v\sin\gamma$$
$$\dot{\Lambda} = \cos\gamma\sin\chi\frac{v}{R}$$
$$\dot{\theta} = \cos\gamma\cos\chi\frac{v}{R\cos\Lambda}$$
$$\dot{m} = -\frac{T(v,h)}{g_0\,I_{sp}(v,h)}\,\delta$$

The following abbreviations are used:

$$\varrho(h) = \varrho_0\,\exp(-h/h_{scal})$$
$$g(h) = g_0\,(r_0/R)^2$$
$$Ma = v/a(h)$$
$$a(h) = \sum_{i=0}^{6}\tilde{a}_i\,h^i$$

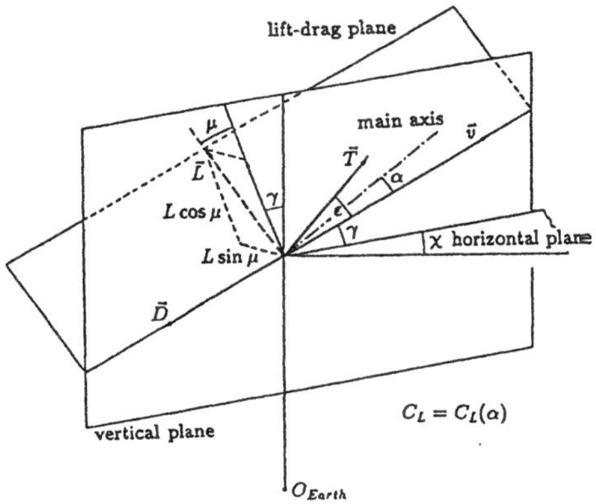

Fig. 1: State and Control Variables

The thrust model is based on tabular data of the maximum thrust $T(v,h)$ and the specific impuls $I_{sp}(v,h)$ of a turbo and ramjet engine [8], which are approximated by C^∞-functions through a nonlinear least squares approach. The used drag and lift model consists of a Mach-dependent quadratic polar [8].

$$L(v,h;C_L) = q(v,h)\,F\,C_L$$
$$D(v,h;C_L) = q(v,h)\,F\,C_D(Ma;C_L)$$
$$C_D(Ma;C_L) = C_{D_0}(Ma) + k(Ma)\,C_L^2$$
$$q(v,h) = \varrho(h)\,v^2/2$$

The following control constraints on the linear control throttle setting δ and the non-linear control C_L hold:

$$0 \leq \delta \leq 1 \quad ; \quad |C_L| \leq C_{L,max}$$

To avoid overloading of the spacecraft structure the dynamic pressure q is constrained (state constraint of order one):

$$S(x) := q(v, h) - q_{max} \leq 0$$

The space transportation system is launched horizontally at (nearly) sea level at prescribed geographical latitudes (Equator, Kourou (French-Guayana), Istres (France), ...). The optimization includes only the first phase of the ascent from launch until stage separation of the two stages, therefore the state variables at the time of stage separation t_f are prescribed here. The time of switching the engines t_e is optimized due to given values of altitude h_e and velocity v_e.

$$
\begin{array}{lllllllll}
v(0) & = & v_0 & \quad v(t_e) & = & v_e & \quad v(t_f) & = & v_f \\
\gamma(0) & = & 0 & & & & \quad \gamma(t_f) & = & \gamma_f \\
\chi(0) & = & \chi_0 & & & & \quad \chi(t_f) & = & \chi_f \\
h(0) & = & 0 & \quad h(t_e) & = & h_e & \quad h(t_f) & = & h_f \\
\Lambda(0) & = & \Lambda_0 & & & & \quad \Lambda(t_f) & = & \Lambda_f \\
\theta(0) & = & 0 & & & & \\
m(0) & = & m_0 & & & &
\end{array}
$$

3. Necessary Conditions of Calculus of Variations

A scetch of the necessary conditions of optimal control theory used is given (see e.g. [2]), which yield the multipoint boundary-value problem used for the numerical treatment. The necessary conditions for state constrained optimal control problems of Bryson et.al. [1] and Jacobson et. al. [13] treat only scalar controls, therefore, the more general conditions of Maurer [15, 16]) are used. The state constraint is of first order since $\frac{dS}{dt}$ explicitly depends on some components of the control vector.

$$
\begin{aligned}
S(x) & = q(v, h) - q_{max} \\
\frac{dS}{dt} = S^1(x, u) & = \frac{\partial q}{\partial v} \dot{v}(x, u) + \frac{\partial q}{\partial h} \dot{h}(x).
\end{aligned}
$$

Therefore, the modified Hamiltonian $H(x, u, \lambda, \eta) = \lambda^T f(x, u) + \eta S^1(x, u)$ is introduced. The adjoint variables λ satisfy the Euler-Lagrange-Equation

$$-\frac{d\lambda^T}{dt} = \frac{\partial H(x, u, \lambda, \eta)}{\partial x}.$$

At the entry point t_{entry} and the exit point t_{exit} of a boundary arc, there hold:

$$\lambda^T(t_{entry}^+) = \lambda^T(t_{entry}^-) - \kappa S_x(t_{entry}), \kappa \geq 0 \; ; \; \lambda^T(t_{exit}^+) = \lambda^T(t_{exit}^-)$$

At a contact point $t_{contact}$ of a boundary arc the following condition holds:

$$\lambda^T(t_{contact}^+) = \lambda^T(t_{contact}^-) - \kappa S_x(t_{contact}), \kappa \geq 0.$$

Moreover, η satisfies $\eta(t)\, S(x(t)) \equiv 0$ on $[0, t_f]$. The following sign conditions hold:

$$\eta \geq 0 \; ; \; \dot\eta \leq 0$$

Finally the optimal control u satisfies the minimum principle for all admissible controls u^* with respect to the augmented Hamiltonian

$$u = \arg\min_{u^*} H(x, u, \lambda, \eta) \quad \text{and} \quad H(x, u, \lambda, \eta) \equiv const \text{ on } t \in [0, t_f].$$

The formulas for the treatment of singular arcs are omitted here due to the lack of space (see Oberle [21], [4]). The treatment of the interior point conditions (e.g. at t_e) can be found in [2] if the state constraint is inactive, a slight generalization of Lemma 4.3 in Maurer [17] is applied if the state constraint is active.

4. Multipoint Boundary-Value Problem

In summary, the set of all necessary conditions leads to a multipoint boundary-value problem with jump conditions of the following type:

Find the n-dimensional vector function $z(t) = (x, \lambda, \kappa)^T$ and the parameters $\tau_1, \ldots, \tau_s, t_f$ satisfying

$$\dot z(t) = F(t, z(t)) = \begin{cases} F_0(t, z(t)) & \text{if } 0 < t < \tau_1 \\ \vdots \\ F_s(t, z(t)) & \text{if } \tau_s < t < t_f \end{cases}$$

$$r_i(t_f, z(0), z(t_f)) = 0 \qquad 1 \leq i \leq \bar n$$
$$r_i(\tau_{j_i}, z(\tau_{j_i}^-), z(\tau_{j_i}^+)) = 0 \quad \bar n + 1 \leq i \leq (n+1)(s+1), j_i \in \{1, \ldots, s\}$$

where F is a combination of the right hand sides of the state and adjoint equations and some so-called trivial equations of type $\dot\kappa = 0$ for each jump parameter κ.

This multipoint boundary-value problem with jump conditions is solved by the multiple shooting method (Bulirsch [3], [26], [9], [20], [22], [11]). The GBS-extrapolation method or high order Runge-Kutta-Fehlberg methods with step size control are used as initial value problem solvers [26, 10]. After a solution is obtained an a posteriori check of the additional sign conditions is made.

In order to start the multiple shooting method, a rather good first estimate of state and adjoint variables is necessary.

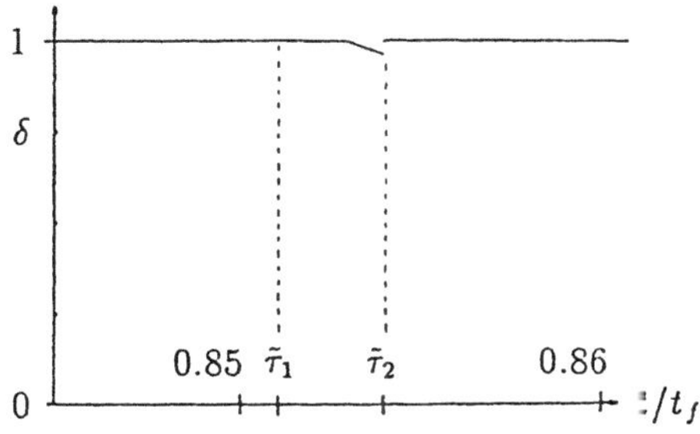

Fig. 2: Boundary Arc (Detail)

The original optimal control problem is therefore converted into a nonlinear programming problem via collocation. The solution is computed by a special direct collocation method [27]. For more details of this hybrid approach see von Stryk and Bulirsch [28] and von Stryk [29].

5. Numerical Results

Some diagrams of the optimized ascent trajectory of the lower stage of this particular Sänger type space vehicle are presented. By a detailed investigation and application of

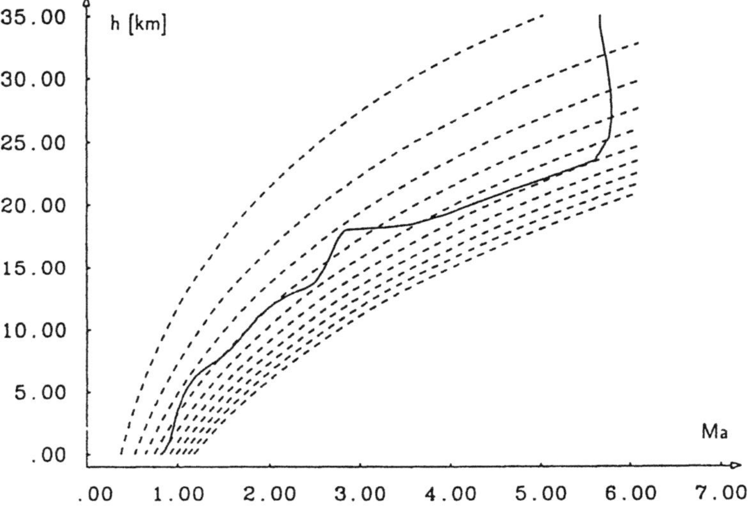

Fig. 3: Velocity-Altitude Diagram

the above mentioned minimum principle one can derive a whole set of different control laws on the boundary arcs of the dynamic pressure constraint, depending on the activity or inactivity of the several control constraints. These different control laws are then competing for the global minima of the Hamiltonian. These competing control laws can be separated into two main groups, either with maximum thrust or with singular thrust. A detailed picture of a boundary arc, which is divided into subarcs with optimal controls of both main groups is given in Fig. 2, see also [4]. Figure 3 illustrates the altitude-velocity diagram with lines of constant dynamic pressure. Due to the dynamic pressure constraint the altitude is a monotonously increasing function for the turbo powered part of the flight path. The ramjet powered part of the trajectory is dominated by the influence of the dynamic pressure constraint manifold. At the end of the flight time the two stage space vehicle system maneuvers to reach the here prescribed stage separation conditions. Further investigations which will include the change of some model details as well as some upgrading of the existing theory to the here encountered complex structure of the optimal control problem will be published in future.

Acknowledgement

This work was partly supported by the Deutsche Forschungsgemeinschaft (DFG) through the SFB 255 Transatmosphärische Flugsysteme. The author is indebted to Prof. Bulirsch for his great encouragement and support. He thanks also the numerical analysis and optimal control group of Prof. Bulirsch for helpful comments and discussions.

References

[1] Bryson, A.E.; Denham, W.F.; Dreyfus, S.E.: *Optimal Programming Problems with Inequality Constraints I: Necessary Conditions for Extremal Solutions.* AIAA Journal 1 (1963) 2544-2550.

[2] Bryson, A.E.; Ho, Y.-C.: *Applied Optimal Control.* Hemisphere Publishing Corp., Washington, D.C., 1975^2.

[3] Bulirsch, R.: *Die Mehrzielmethode zur numerischen Lösung von nichtlinearen Randwertproblemen und Aufgaben der optimalen Steuerung.* Report der Carl-Cranz-Gesellschaft e.V., Oberpfaffenhofen, Germany (1971). Nachdruck: Mathematisches Institut, TU München, Germany (1985).

[4] Bulirsch, R.; Chudej, K.: *Ascent Optimization of an Airbreathing Space Vehicle.* Proc. of the AIAA Guidance, Navigation and Control Conference, New Orleans, Louisiana, Paper-No. AIAA-91-2656 (1991) 520-528.

[5] Bulirsch, R.; Chudej, K.: *Staging and Ascent Optimization of a Dual-Stage Space Transporter.* Z. Flugwiss. Weltraumf. 16 (1992) 143-151.

[6] Chudej, K.; Bulirsch, R.; Reinsch, K.-D.: *Optimal Ascent and Staging of a Two-Stage Space Vehicle System.* Deutscher Luft- und Raumfahrtkongreß, Friedrichshafen, Germany, Paper-No. DGLR-90-137 (1990) 243-249.

[7] Cliff, E.; Schnepper, K.; Well, K.H.: *Performance Analysis of a Transatmospheric Vehicle.* AIAA 2nd International Aeroplane Conference, Orlando, Florida, Paper-No. AIAA-90-5257 (1990).

[8] Data of the Sonderforschungsbereich 255: TU München, Germany (1991).

[9] Deuflhard, P.: *A Relaxation Strategy for the Modified Newton Method.* In: Optimization and Optimal Control (R. Bulirsch, W. Oettli, J. Stoer eds.) Lecture Notes in Mathematics 477, Springer, Berlin, 1975.

[10] Hairer, E.; Nørsett, S.P.; Wanner, G.: *Solving Ordinary Differential Equations I, Nonstiff Problems.* Springer, Berlin, 1987.

[11] Hiltmann, P.: *Numerische Lösung von Mehrpunkt-Randwertproblemen und Aufgaben der optimalen Steuerung mit Steuerfunktionen über endlichdimensionalen Räumen.* Dissertation, Mathematisches Institut, TU München, Germany (1990).

[12] Högenauer, E.: *Raumtransporter.* Z. Flugwiss. Weltraumf. 11 (1987) 309-316.

[13] Jacobson, D.H.; Lele, M.M.; Speyer, J.L.: *New Necessary Conditions of Optimality for Control Problems with State-Variable Inequality Constraints.* J. Math. Anal. Appl. 35 (1971) 255-284.

[14] Koelle, D.E.; Kuczera, H.: *Sänger Space Transportation System.* 41st IAF-Congress, Dresden, Germany, Paper-No. IAF-90-175 (1990).

[15] Maurer, H.: *Optimale Steuerprozesse mit Zustandsbeschränkungen.* Habilitaticn, Mathematisches Institut, Universität Würzburg, Germany (1976).

[16] Maurer, H.: *Differential Stability in Optimal Control Problems.* Appl. Math. Optim. 5 (1979) 283-295.

[17] Maurer, H.: *On the Minimum Principle for Optimal Control Problems with State Constraints.* Report No. 41, Rechenzentrum der Universität Münster, Germany (1979).

[18] Miele, A.: *Flight Mechanics I, Theory of Flight Paths.* Addison-Wesley, Reading, Massachusetts, 1962.

[19] Oberle, H.J.: *Numerical Computation of Minimum-Fuel Space-Travel Problems by Multiple Shooting.* Report M-7635, Institut für Mathematik, TU München, Germany (1976).

[20] Oberle, H.J.: *Numerische Berechnung optimaler Steuerungen von Heizung und Kühlung für ein realistisches Sonnenhausmodell.* Habilitation, Report M-8310, Institut für Mathematik, TU München, Germany (1982).

[21] Oberle, H.J.: *Numerical Computation of Singular Functions in Trajectory Optimization Problems.* J. Guidance Control Dynamics 13 (1990) 153-159.

[22] Oberle, H.J.; Grimm, W.: *BNDSCO – A Program for the Numerical Solution of Optimal Control Problems, User Guide.* Report IB/515-89/22, DLR, Oberpfaffenhofen, Germany (1989).

[23] Sachs, G.; Schoder, W.: *Optimal Separation of Lifting Vehicles in Hypersonic Flight.* Proc. of the AIAA Guidance, Navigation and Control Conference, New Orleans, Louisiana, Paper-No. AIAA-91-2657 (1991) 529-536.

[24] Schöttle, U.M.; Hillesheimer, M.: *Performance Optimization of an Airbreathing Launch Vehicle by a Sequential Trajectory Optimization and Vehicle Design Scheme.* AIAA Guidance, Navigation and Control Conference, New Orleans, Louisiana, Paper-No. AIAA-91-2655 (1991).

[25] Shau, G.-C.: *Der Einfluß flugmechanischer Parameter auf die Aufstiegsbahn von horizontalstartenden Raumtransportern bei gleichzeitiger Bahn- und Stufungsoptimierung.* Dissertation, Fakultät für Maschinenbau und Elektrotechnik, TU Braunschweig, Germany (1973).

[26] Stoer, J.; Bulirsch, R.: *Numerische Mathematik 2.* Springer, Berlin, 1990^3.

[27] von Stryk, O.: *Ein direktes Verfahren zur näherungsweisen Bahnoptimierung von Luft- und Raumfahrzeugen unter Berücksichtigung von Beschränkungen.* Z. angew. Math. Mech. 71, 6 (1991) T705-706.

[28] von Stryk, O.; Bulirsch, R.: *Direct and Indirect Methods for Trajectory Optimization.* Annals of Operations Research (1992) 357-373.

[29] von Stryk, O.: *Numerical Solution of Optimal Control Problems by Direct Collocation*. In: Optimal Control (R. Bulirsch, A. Miele, J. Stoer, K.H. Well eds.) Birkhäuser, Basel, 1992.

[30] Vinh, N.X.; Busemann, A.; Culp, R.D.: *Hypersonic and Planetary Entry Flight Mechanics*. The University of Michigan Press, Ann Arbor, 1980.

Author's address

Dipl.-Math. Kurt Chudej
Mathematisches Institut, TU München,
Postfach 20 24 20, D-8000 München 2, Germany.

International Series of Numerical Mathematics, Vol. 111, ©1993 Birkhäuser Verlag Basel

Controllability Investigations of a Two-Stage-To-Orbit Vehicle

Bernd Kugelmann and Hans Josef Pesch

Abstract. For the real-time computation of closed-loop controls, guidance methods have been recently developed by the authors which are based on the theory of neighboring extremals and/or the multiple shooting algorithm. Because of their close relationship, these methods have a comparable domain of controllability, i.e., the set of all deviations from the nominal optimal trajectory which can be compensated in real time. Therefore, the multiple shooting method can be used to get an estimation for the size of these domains. In this way, the domain of controllability has been investigated for the payload maximum ascent of a two-stage rocket-propelled space vehicle from Earth into a circular target orbit. Here, the staging and the mass ratio of the two stages are optimized simultaneously.

1. Introduction

One of the key problems for a practical application of solutions of optimal control problems is the availability of feedback schemes that can be applied in real-time. Recently, two related numerical methods for the fast approximation of closed-loop controls were published; see Kugelmann and Pesch 1990a, 1990b and Pesch 1989a, 1989b. With respect to their theoretical part, these methods are based on the theory of neighboring extremals and, with respect to their numerical part, they are based on the multiple shooting algorithm (see Bulirsch 1971, Stoer and Bulirsch 1980 and Oberle 1982). These guidance methods provide neighboring optimal feedback schemes and do not track the nominal optimal solution in the sense of automatic control. More precisely, these feedback schemes approximate the actual optimal solution to the first order because of the underlying linearization around the nominal solution.

This paper is in final form and no version of it will be submitted for publication elsewhere.

The two neighboring guidance methods differ with respect to the necessary amount of online computation, since they are designed for different safety demands for the feedback solutions. This will be briefly summarized:

The feedback method of Kugelmann and Pesch (1990a, 1990b) leads to a linear guidance law of the form

$$\delta u(t) = \Lambda(t)\,\delta x(t) \tag{1}$$

where δx and δu denote the deviations of the state and the control vector; Λ is the precomputed gain matrix. Thus, we have only matrix-times-vector operations to be performed during the process at each time instant where the reference trajectory is to be corrected. Because of the simplicity of the scheme, the frequency of the corrections of the nominal control can be adjusted to the flow of data provided by the measurement devices. From the theory one knows that optimality conditions and constraints are satisfied to the first order.

We now turn to the second method described in Pesch 1989a and 1989b. The observance of the constraints can be checked explicitly by this method. The price to be payed for this, is a single numerical integration of the equations of motion during which the feedback scheme (1) is continuously applied. Although this method is more costly, it can still be used in real time. For more details see the papers mentioned above.

On the other hand, the multiple shooting method itself can be applied in real time if special features are implemented to reduce the amount of computation needed for the numerical solution of the large number of initial-value problems (see Bulirsch and Branca 1974 and Pesch 1990). The multiple shooting method is, by its construction, a method which is especially well suited for an implementation on parallel computers. If the multiple shooting method is used for guidance purposes, one can benefit from the fact that the reference trajectory and the associated Jacobian can be up-dated periodically, i.e., solutions according to previous disturbances serve as new reference trajectories. Therefore, a parallel multiple shooting method will be a strong competitor for the neighboring optimal guidance schemes in the future.

2. The Basic Idea

Those feedback methods and the multiple shooting method, too, have in common that information of and along the nominal trajectory must be provided. We first consider the multiple shooting method. Since this method essentially is a Newton method for the solution of large systems of nonlinear equations, a good initial guess of the solution must be available. In case the multiple shooting method is used for guidance purposes, the initial guess, of course, is given by the nominal trajectory. The Jacobian can then be precomputed and up-dated in each iteration by Broyden approximations.

We now consider the aforementioned neighboring optimal feedback schemes again. These schemes are based on the linearization of the actual trajectory around the nominal one. Note that the Newton method is based on linearization, too. This explains why the domain of controllability of these methods approximates the domain of convergence of the multiple shooting method. However, this is true only if the ordinary Newton method, i.e., the Newton method without any relaxation technique, is used. See Pesch 1990 for more details.

For a first investigation of the controllability of a new problem, one therefore can avoid the laborious and tedious analytical work to be done before the neighboring optimal feedback schemes can be applied. It suffices just to compute the domain of convergence of the multiple shooting method because this domain is qualitatively similar to the controllability domains of the feedback schemes.

In the present paper, we apply this idea to investigate the controllability of a two-stage-to-orbit vehicle of Sänger type (see Shau 1973 and Bulirsch, Chudej and Reinsch 1990) during its ascent to an Earth orbit.

3. Solving Optimal Control Problems by Multiple Shooting

The multiple shooting method and the aforementioned feedback methods, too, can be applied to the following class of optimal control problems:

Performance index:

$$I[u] = \varphi(x(t_f), t_f) + \bar{\varphi}(x(\tau_b^-), x(\tau_b^+)) + \int_0^{t_f} L(x(t), u(t))\, dt \tag{2}$$

where

$$\varphi: \mathbb{R}^n \times \mathbb{R}_+ \to \mathbb{R}\ , \quad \bar{\varphi}: \mathbb{R}^{2n} \to \mathbb{R}\ , \quad \text{and } L: \mathbb{R}^{n+k} \to \mathbb{R}\ .$$

Constraints:

$$\dot{x}(t) = f(x(t), u(t))\ , \qquad f: \mathbb{R}^{n+k} \to \mathbb{R}^n\ ,$$

$$x(0) = x_0\ , \qquad x_0 \in \mathbb{R}^n \text{ given}\ ,$$

$$\psi(x(t_f), t_f) = 0\ , \qquad \psi: \mathbb{R}^n \times \mathbb{R}_+ \to \mathbb{R}^q, \quad q \leq n\ ,$$

$$C(x(t), u(t)) \leq 0\ , \qquad C: \mathbb{R}^{n+k} \to \mathbb{R}^l\ , \tag{3}$$

$$S(x(t)) \leq 0\ , \qquad S: \mathbb{R}^n \to \mathbb{R}^{\bar{l}}\ ,$$

$$N(x(\tau_a)) = 0\ , \qquad N: \mathbb{R}^n \to \mathbb{R}^{\bar{q}}$$

$$\bar{\psi}(x(\tau_b^-), x(\tau_b^+)) = 0\ , \qquad \bar{\psi}: \mathbb{R}^{2n} \to \mathbb{R}^n\ .$$

Notations: x state vector, u control vector, t_f terminal time, φ, $\bar{\varphi}$, L given cost functions, x_0 initial conditions, ψ terminal conditions, C control constraints.

S state constraints, N interior point conditions, τ_a interior point, $\bar{\psi}$ discontinuity conditions, τ_b discontinuity point.

It is known that the necessary conditions of optimal control theory lead to a multipoint boundary-value problem with jump conditions of the form

$$\dot{y}(t) = f(t, y(t)) = \begin{cases} f_0(t, y(t)) & \text{for } 0 \le t \le \tau_1, \\ \vdots \\ f_s(t, y(t)) & \text{for } \tau_s \le t \le \tau_{s+1}, \end{cases} \tag{4}$$

$$y(\tau_k^+) = \sigma_k(\tau_k, y(\tau_k^-)) , \quad 1 \le k \le s ,$$
$$r_i(y(0), y(t_f)) = 0 , \quad 1 \le i \le N_1 , \tag{5}$$
$$r_i(\tau_{k_i}, y(\tau_{k_i}^-)) = 0 , \quad N_1 + 1 \le i \le N + s .$$

Equation (4) is a piecewise defined system of differential equations for the function $y: [0, t_f] \to \mathbb{R}^N$. The vector function is composed of the state variables, the adjoint variables and, in general, of additional parameters describing possible discontinuities of the adjoint variables. The switching points τ_k for $1 \le k \le s$ with

$$0 =: \tau_0 < \tau_1 < \ldots < \tau_s < \tau_{s+1} := t_f$$

are to be determined as part of the solution to the problem. The above formulation postulates that the switching structure and its associated control laws are known in advance. That means the control function is given as a function of the state and the adjoint variables on each switching segment $\tau_k \le t \le \tau_{k+1}$. The paper of Bulirsch et. al. (1991a, 1991b) shows how to treat such complex optimal control problems from the analysis of the necessary conditions to the numerical solution using the multiple shooting method in connection with homotopy techniques. Special emphasis is laid on how to get the correct switching structure in case of multiple subarcs which may be caused by state constraints and singular controls.

4. Computation of the Controllability Regions

By means of the multiple shooting idea this multipoint boundary-value problem is reduced by discretization to a high-dimensional system of nonlinear equations,

$$F(s) = 0 \tag{6}$$

which is solved by a modified Newton method

$$s^{(i+1)} = s^{(i)} - \lambda \, DF(s^{(i)})^{-1} \, F(s^{(i)}) . \tag{7}$$

λ is a relaxation parameter.

Comparing the multiple shooting algorithm with the neighboring optimal feedback schemes one can see that the operations *discretization* and *linearization* are interchanged. While for the feedback schemes the linearization can be almost completely accomplished analytically the linearization step in the multiple shooting algorithm must be done numerically which requires the approximation of the Jacobian $DF(s^{(i)})$. Therefore, the system of linear equations to be solved in each iteration step of the Newton method (7) corresponds to the system of linear equations which is solved for each correction step of the neighboring feedback schemes. This relation explains the qualitative coincidence of the controllability domains of the feedback schemes with the domain of convergence for the unrelaxed ($\lambda = 1$) Newton method.

Therefore, a controllability analysis can be performed parallel to the process of developing complex optimal control models. We only have to compute the domain of convergence of the unrelaxed multiple shooting method. The advantage is evident: no analytical derivatives must be computed for a preliminary model and the same software package, with slight modifications, can be used for the computation of an optimal nominal solution and its corresponding controllability region.

5. The Optimal Control Problem for the Sänger Ascent

The Sänger II project of a European space transportation system is, at present, in the focus of industrial and scientific discussion and development. In the future, a completely reusable space transportation system will be necessary to maintain and service cost-efficiently the planned international space station.

The mathematical model for the space vehicle considered here goes back to a model presented by Shau (1973) in his thesis in 1973. Some ideas for such a space transportation system have been already developed by Eugen Sänger in the 30ies. His investigations were published in 1962. Recently, Shau's model was further improved by Bulirsch et. al. (1990).

The two-stage-to-orbit vehicle is designed to launch horizontally and to deliver either a manned shuttle or an unmanned cargo unit. The first stage is equipped with wings and airbreathing engines. It is capable of performing cruising flights. Since, at present, realistic data for airbreathing engines are still in the process of development the model considered here has a rocket propulsion. So, it can be considered as a first step on the way of development towards a realistic Sänger II model. Here, we briefly summarize the model of Bulirsch et. al. (1990) and say how the model has to be modified for the investigation of its controllability capacity.

The payload of the TSTO vehicle is to be maximized by means of the control variables, the staging time and the mass ratio of the two stages:

$$m_{\text{payload}} = m(t_f) - \Psi_I(m(t_s^+) - m(t_f)) \overset{!}{=} \max \tag{8}$$

with the stage separation condition

$$m(t_s^+) = m(t_s^-) - \Psi_{II}(m_0 - m(t_s^-)) \ . \tag{9}$$

Here, t_f and t_s denote the unspecified terminal time and the time of stage separation, respectively. The functions Ψ_I and Ψ_{II} describe the structural mass consisting of the engines and the fuel tank in dependence of the fuel used for the two stages. For details see Bulirsch et. al. (1990).

The equations of motion over a spherical Earth with no wind in the atmosphere are (see Vinh et. al. 1980)

$$\dot{v} = \frac{1}{m}\left[T(v,h;b)\cos\epsilon - D(v,h;u)\right] - g(h)\sin\gamma$$

$$+ \omega^2\left(r_0 + h\right)\cos\Lambda\left(\sin\gamma\cos\Lambda - \cos\gamma\sin\chi\sin\Lambda\right) \ ,$$

$$\dot{\gamma} = \frac{1}{mv}\left[T(v,h;b)\sin\epsilon + L(v,h;u)\right]\cos\mu$$

$$- \left[\frac{g(h)}{v} - \frac{v}{r_0 + h}\right]\cos\gamma + 2\omega\cos\chi\cos\Lambda$$

$$+ \omega^2\frac{r_0 + h}{v}\cos\Lambda\left(\sin\gamma\sin\chi\sin\Lambda + \cos\gamma\cos\Lambda\right) \ ,$$

$$\dot{\chi} = \frac{1}{mv\cos\gamma}\left[T(v,h;b)\sin\epsilon + L(v,h;u)\right]\sin\mu$$

$$- \frac{v}{r_0 + h}\cos\gamma\cos\chi\tan\Lambda \tag{10}$$

$$+ 2\omega\left(\sin\chi\cos\Lambda\tan\gamma - \sin\Lambda\right)$$

$$- \omega^2\frac{r_0 + h}{v\cos\gamma}\cos\Lambda\sin\Lambda\cos\chi \ ,$$

$$\dot{h} = v\sin\gamma \ ,$$

$$\dot{\Lambda} = \frac{v}{r_0 + h}\cos\gamma\sin\chi \ ,$$

$$\dot{\Theta} = \frac{v}{(r_0 + h)\cos\Lambda}\cos\gamma\cos\chi \ ,$$

$$\dot{m} = -b \ .$$

Notations: state variables: v velocity, γ path inclination, χ azimuth inclination, h altitude, Λ geographical latitude, Θ geographical longitude, m mass; control variables: u angle of attack, μ lateral inclination angle, b mass flow, ϵ thrust angle; constants: r_0 Earth's radius, ω angular velocity, g_0 gravitational constant.

The formulas for the thrust T, the drag D, the lift L, and the gravitational acceleration $g(h)$ can be found in Bulirsch et. al. (1990). Moreover, the mass flow for both stages is subject to the constraints

$$0 \leq b \leq b_{I,\max} \quad \text{for} \quad 0 < t < t_s ,$$
$$0 \leq b \leq b_{II,\max} \quad \text{for} \quad t_s < t < t_f . \tag{11}$$

The boundary conditions are

$$v(0) = 275\,[\text{m/s}] , \qquad \psi_1(v(t_f), \chi(t_f), h(t_f), \Lambda(t_f)) = 0 ,$$
$$\gamma(0) = 0\,[\text{rad}] , \qquad \gamma(t_f) = 0\,[\text{rad}] ,$$
$$h(0) = 0\,[\text{m}] , \qquad h(t_f) = 500,000\,[\text{m}] ,$$
$$m(0) = m_0\,[\text{kg}] , \qquad \psi_2(\chi(t_f), \Lambda(t_f)) = 0 , \tag{12}$$
$$\chi(0) = \text{ free} ,$$
$$\Theta(0) = 0\,[\text{rad}] ,$$

$\Lambda(0)$ given; different prescribed launch latitudes

with

$$\psi_1 = [v(t_f) \cos\chi(t_f) + \omega\,(h(t_f) + r_0)\,\cos\Lambda(t_f)]^2$$
$$+ [v(t_f)\sin\chi(t_f)]^2 - \frac{g_0\,r_0^2}{h(t_f) + r_0} ,$$
$$\psi_2 = \cos\zeta - \cos\chi(t_f)\cos\Lambda(t_f)$$

where ζ denotes the prescribed inclination angle of the target orbit.

A more realistic model which includes data for a turbo and ramjet propulsion can be found in Bulirsch and Chudej (1991).

6. The Guidance Problem for the Sänger Ascent

For the guidance problem which has to be solved during the flight, it is no longer meaningful to regard the mass ratio as a variable which is to be optimized. Therefore, the performance index and the stage separation condition have to be reformulated: Again, the payload is to be maximized,

$$m_{\text{payload}} = m(t_f) \overset{!}{=} \max \tag{13}$$

subject to the stage separation condition, now given as interior point conditions,

$$m(t_s^+) = \text{optimal value of the open-loop solution at } t_s^+ ,$$
$$m(t_s^-) = \text{optimal value of the open-loop solution at } t_s^- . \tag{14}$$

In addition, the initial values have to be changed according to the measured state variables. The end conditions may also be altered if a neighboring target orbit is to be headed for.

7. The Nominal Solution

The nominal optimal control problem was solved for various launch latitudes by Bulirsch et. al. (1990). The solution for the latitude $\Lambda(0) \approx 48$ deg (Munich) is given in the following altitude-velocity diagram; see Fig. 1.

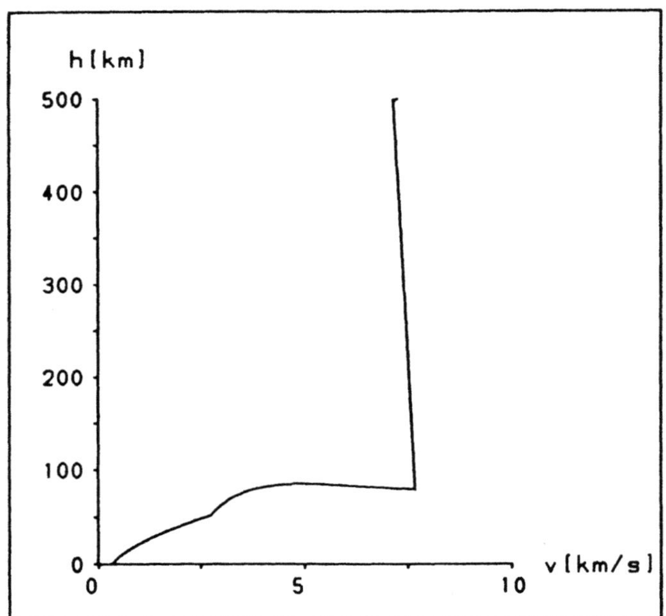

Fig. 1: altitude versus velocity

The optimal switching structure is given by

$$b = b_{I,\max} \mid_{t_s} \quad b = b_{II,\max} \mid_{t_{\text{off}}} \quad b = 0 \mid_{t_{\text{on}}} \quad b = b_{II,\max} \;, \tag{15}$$

$$t_s \approx 135.8 \text{ s} \;, \quad t_{\text{off}} \approx 369.4 \text{ s} \;, \quad t_{\text{on}} \approx 3167.1 \text{ s} \;,$$

$$t_f \approx 3170.0 \text{ s}$$

where t_s denotes the stage separation time. At this time the fuel of the first stage is assumed to be totally consumed. The times t_{off} and t_{on} denote where the engines of the second stage are switched off and switched on, respectively.

Below an altitude of 90 km, the flight path is strongly influenced by the atmosphere. Beyond this point, the trajectory turns out to be a Hohmann transfer. For more details see Bulirsch et. al. (1990).

8. Controllability Domains

As explained before, the controllability domains of the guidance methods of Kugelmann, Pesch (1990a, 1990b) and Pesch (1989a, 1989b) can be estimated by the domain of convergence of the multiple shooting method when using the unrelaxed Newton method. This domain of convergence can be only computed by try and error: For a given correction time t_0 and a prescribed deviation $\delta x(t_0)$ from the reference trajectory, the multiple shooting method is applied to this neighboring boundary-value problem. The initial trajectory is always given by the optimal solution presented above. According to the switching segment, the correction point $t_\bullet$ belongs to, the switching structure of the multipoint boundary-value problem has to be adjusted, i.e., the multipoint boundary conditions have to be reformulated. For example, if deviations are measured shortly after the stage separation the switching structure reduces to

$$b = b_{II,\max} \quad \big|_{t_{\text{off}}} \quad b = 0 \quad \big|_{t_{\text{on}}} \quad b = b_{II,\max} \; .$$

If the multiple shooting method converges for the perturbed problem associated with the pair $(t_0, \delta x(t_0))$, this pair is assumed to belong to the n-dimensional controllability tube around the reference flight path. The size of this multidimensional tube characterizes the performance of a guidance method. Instead of this tube, we describe here the controllability regions. They indicate the extremal deviations—the maximal positive and the minimal negative deviations—of one state variable at a time t_0 which can be successfully compensated if all other state variables are assumed to be undisturbed. These controllability regions correspond to the cross-sections of the multidimensional tube with the n orthogonal cylinder surfaces parallel to the coordinate planes.

These controllability regions are given in Tables 1 and 2.

These results clearly indicate that the size of the controllability regions is large enough for practical purposes for that part of the trajectory which is within the Earth's atmosphere. Especially at take-off, all state variables can be corrected within a wide range. Note that the obtained negative deviations of the inclination angle and the altitude at the take-off are results of the mathematical optimization; they are not to be interpreted physically.

Table 1 Controllable Deviations
from the Reference Trajectory
for the Velocity and the Path Inclination

t_0 [s]	$m^{1)}$	v [m/s] opt	δv [m/s] min	max	γ [deg] opt	$\delta\gamma$ [deg] min	max
0	3	275	-77	192	0.0	-22.9	58.1
67	3	909	-671	843	28.2	-37.0	45.8
155	2	2957	-1073	690	8.6	-5.3	1.8
414	1	7668	0	0	-0.1	0.0	0.0
1315	1	7550	0	0	1.6	0.0	0.0
2104	1	7321	0	0	1.8	0.0	0.0
2795	1	7177	0	0	0.8	0.0	0.0

[1)] number of switching points

Table 2 Controllable Deviations
from the Reference Trajectory
for the Azimuth and the Altitude

t_0 [s]	m	χ [deg] opt	$\delta\chi$ [deg] min	max	h [km] opt	δh [km] min	max
0	3	-1.1	-157.9	167.3	0.0	-2.3	11.7
67	3	-0.8	-15.6	14.1	19.6	-6.9	25.0
155	2	-2.7	-15.6	12.5	62.1	-21.3	20.6
414	1	-16.9	0.0	0.0	79.1	0.0	0.0
1315	1	-50.0	0.0	0.1	178.5	-0.9	0.9
2104	1	-42.1	0.0	0.1	364.5	-0.4	0.4
2795	1	-5.6	0.0	0.0	481.6	0.0	0.0

Outside the atmosphere, for correction times t_0 with $h(t_0) > 70$ km , the vehicle can be controlled only by its thrust. Since the thrust is only active at the end of the flight for a very short time (about 2.9 s) it is obvious that there is almost no maneuverability if the nominal switching structure is maintained. However, the switching structure can be changed only by introducing an additional interval where the thrust is active. This requires that the interior boundary conditions have to be modified which can be hardly done automatically. This is a consequence of the fact that a change of the switching structure results in a point of nondifferentiability of the solution with respect to the perturbations, i.e., the initial values of the state variables. Since for the aforementioned neighboring optimum guidance methods the switching structure of the nominal trajectory must also coincide with that of the actual trajectory—differentiability with respect of the perturbations must be also assumed for the theory of neighboring extremals—we did not proceed beyond this point. Note that, independent of the guidance method used, there are severe mathematical reasons which prevent that the actual flight path can be successfully controlled according to a switching structure different from the nominal one.

To complete the numerical results, the controllability with respect to the mass and the latitude are discussed. Note that the longitude is decoupled in the equations of motion. Since deviations of the nominal mass influence only the optimality criterion—there is no end condition on this variable—the size of the controllable deviations within the atmosphere is considerably larger than the size of those deviations which may be expected in reality. In contrast, there is almost no possiblity to compensate perturbances of the latitude as long as the inclination of the target orbit is prescribed. If this end condition is omitted, realistic deviations from the nominal latitude can be also compensated.

Note that, by the procedure described in this paper, the performance of the neighboring optimum guidance methods with respect to disturbances of the air density can also be approximated by the corresponding domain of convergence of the multiple shooting method. More detailed investigation for disturbances of the air density will be done as soon as the Sänger ascent can be modelled more realistically.

9. Conclusions

Because of the common theoretical background of guidance methods which are based on the theory of neighboring extremals and the multiple shooting method, the domain of controllability for these guidance methods can be approximated by the domain of convergence of the multiple shooting method when the unrelaxed Newton method is used in the multiple shooting algorithm to solve the resulting system of nonlinear equations. This procedure enables a fast and convenient computation of the controllability performance of the neighboring optimum guidance methods without the laborious implemention of the highly efficient guidance scheme. Especially, during the phase of development of complex models the controllability can be easily investigated parallel to the computation of optimal trajectories of different models. There is only one software package to be used, namely an implementation of the multiple shooting algorithm. Numerical results are presented for the payload maximum ascent of a two-stage-to-orbit vehicle of the Sänger type, a problem for which the mathematical model is permanently reformulated. The size of the controllability regions is large enough for practical purposes as long as the atmosphere can be used to control the vehicle.

Acknowledgements

This research was supported by the German National Science Foundation (Deutsche Forschungsgemeinschaft) through the Sonderforschungsbereich 255 (Transatmosphärische Flugsysteme).

References

Bulirsch, R. (1971), Die Mehrzielmethode zur numerischen Lösung von nichtlinearen Randwertproblemen und Aufgaben der optimalen Steuerung. Deutsche Forschungs- und Versuchsanstalt für Luft- und Raumfahrt, Report of the Carl-Cranz Gesellschaft, Oberpfaffenhofen, Germany.

Bulirsch, R. and H.-W. Branca (1974), Computation of Real-Time-Control in Aerospace Applications by Multiple Shooting Procedures. In: R. A. Willoughby (Ed.), Stiff Differential Systems, Proc. of a Conference held at Wildbad, Germany, 1973, Plenum Press, New York, New York, pp. 49–50.

Bulirsch, R., K. Chudej and K. D. Reinsch (1990), Optimal Ascent and Staging of a Two-stage Space Vehicle System. Jahrestagung der Deutschen Gesellschaft für Luft- und Raumfahrt, Friedrichshafen, 1990, DGLR-Jahrbuch 1990, Vol. 1, 243–249.

Bulirsch, R. and K. Chudej (1991), Ascent Optimization of an Airbreathing Space Vehicle. AIAA Guidance, Navigation and Control Conference, New Orleans, Louisiana, 1991, AIAA Paper No. 91-2656.

Bulirsch, R., F. Montrone and H. J. Pesch (1991a), Abort Landing in the Presence of a Windshear as a Minimax Optimal Control Problem, Part 1: Necessary Conditions. J. of Optimization Theory and Applications, Vol. 70, 1–23.

Bulirsch, R., F. Montrone and H. J. Pesch (1991b), Abort Landing in the Presence of a Windshear as a Minimax Optimal Control Problem, Part 2: Multiple Shooting and Homotopy. J. of Optimization Theory and Applications, Vol. 70, 221–252.

Kugelmann, B. and H. J. Pesch (1990a), New General Guidance Method in Constrained Optimal Control, Part 1: Numerical Method. J. of Optimization Theory and Applications, Vol. 67, 421–435.

Kugelmann, B. and H. J. Pesch (1990b), New General Guidance Method in Constrained Optimal Control, Part 2: Application to Space Shuttle Guidance. J. of Optimization Theory and Applications, Vol. 67, 437–446.

Oberle, H. J. (1982), Numerische Berechnung optimaler Steuerungen von Heizung und Kühlung für ein realistisches Sonnenhausmodell. Habilitationsschrift, Munich University of Technology, Munich, Germany.

Pesch, H. J. (1989a), Real-time Computation of Feedback Controls for Constrained Optimal Control Problems, Part 1: Neighbouring extremals. Optimal Control Applications & Methods, Vol. 10, 129–145.

Pesch, H. J. (1989b), Real-time Computation of Feedback Controls for Constrained Optimal Control Problems, Part 2: A Correction Method Based on Multiple Shooting. Optimal Control Applications & Methods, Vol. 10, 147–171.

Pesch, H. J. (1990), Optimal and Nearly Optimal Guidance by Multiple Shooting. In: Centre National d'Etudes Spatiales (Ed.), Proc. of the Inter. Symp. Mécanique Spatiale - Space Dynamics, Toulouse, France, 1989, Cepadues Editions, Toulouse, France, 761–771.

Sänger, E. (1962), Raumfahrt — gestern, heute und morgen. Astronautica Acta VIII 5, 323–343.

Shau, G.-C. (1973), Der Einfluß flugmechanischer Parameter auf die Aufstiegsbahn von horizontal startenden Raumtransportern bei gleichzeitiger Bahn- und Stufungsoptimierung. Dissertation, Department of Mechanical and Electrical Engineering, University of Technology, Braunschweig, Germany.

Stoer, J. and R. Bulirsch (1980), Introduction to Numerical Analysis. Springer, New York, New York.

Vinh, N. X., A. Busemann and R. D. Culp (1980), Hypersonic and Planetary Entry Flight Mechanics. University of Michigan Press, Ann Arbor, Michigan.

Author's address

Dr. Bernd Kugelmann, Priv.-Doz. Dr. Hans Josef Pesch, Mathematisches Institut, Technische Universität München, Postfach 20 24 20, D–8000 München 2

Optimal Design of a Mission to Neptune

Rainer Callies

Abstract. For an interplanetary spacecraft from the Earth to the planet of Neptune trajectory optimization and the optimization of spacecraft design are tightly coupled to increase overall system performance.

For each set of boundary conditions and constraints the problem of trajectory optimization is transformed into a multi-point boundary value problem. The spacecraft is described by a very accurate model. Model parameters and model functions are integrated into the framework of differential equations. The solution of the resulting problem is by the multiple shooting method.

For a journey to Neptune with a launch in the year 2001 and without any swing-bys at other planets or moons the optimal spacecraft needs 18 years and has an initial mass of about 1.7 tons. The launch window covers the full year of 2001.

1. Introduction

In late August 1989, nearing the end of its epic 12-year journey across the solar system, *Voyager 2* paid a visit to the distant blue planet Neptune. Its average distance from Sun is 30.1 AU and its sidereal period 165.5 tropical years. The scientific results of this encounter are exciting: Pictures from Neptune with the Great Dark Spot, the discovery of the neptunian rings, measurements of the magnetosphere – to mention only a few. And as a grand finale, Voyager transmitted images to Earth of Triton, this frozen world of rock and ice with geysers of liquid and gaseous nitrogen.

But, as has been the case with all other planetary encounters, many more questions have been opened up than answered. Without a return to Neptune – not only with a short flyby, but with orbiting missions – most of these questions will remain unanswered. Thus NASA is looking into the possibility of sending a spacecraft to orbit Neptune, carrying better instruments than did *Voyager* and returning data for more than four years [1]. However, the situation is not very encouraging: A rendezvous mission is more difficult to perform than a simple flyby. The necessary velocity increment for an orbiting mission is much higher, and so is the flight time: clearly a challenge to spacecraft technology as well as to components' lifetime. In addition, *Voyager 2* could increase its velocity by swing-bys at Jupiter, Saturn and Uranus; this technique is mostly thought to be inevitable for such a mission. But it will last many decades until there are again such favourable positions of several outer planets relative to the Earth.

In the present paper we consider the following problem:
Is it possible to construct a spacecraft from *components available today* (either space-qualified or space-proven), that is able *to orbit Neptune* after a *flight time of not more than 18 years* and *without any gravity assisted flyby* ?
The last demand is necessary for an uninterrupted launch window. Without loss of generality we want to start in the year of 2001 (ref. to S. Kubrick). A typical scientific payload of 50 kg should be transferred to Neptune; the total initial mass of the spacecraft must be compatible with present launching systems.

This problem is solved by a tight coupling of trajectory optimization and the optimization of spacecraft design. The usual approach – to construct a spacecraft, then optimize the trajectory, modify the spacecraft slightly according to the results, and iterate this process – does not yield the desired result.

2. Nomenclature and Coordinate System

Independent variable
t time

State variables
r radius
φ angle of azimuth
ϑ polar angle
m mass of the space probe
v_r radial component of $\vec{v}$
v_φ azimuthal component of $\vec{v}$
v_ϑ polar component of $\vec{v}$
$\vec{v}$ $= (v_r, v_\varphi, v_\vartheta)$: velocity
x $= (r, \varphi, \vartheta, m, v_r, v_\varphi, v_\vartheta)^T$

Adjoint variables
$\lambda = (\lambda_r, \lambda_\varphi, \lambda_\vartheta, \lambda_m, \lambda_{vr}, \lambda_{v\varphi}, \lambda_{v\vartheta})^T$

Control variables
$\Psi(t)$ thrust angle
$\Xi(t)$ thrust angle
$\beta(t)$ "related" thrust magnitude

Subscripts
max maximum value
$0,\ f$ initial, final value

Other important quantities
v_A effective exhaust velocity of the thruster
ce normalized exhaust velocity of the thruster (dimensionless)
F thrust
P electric power
$\vec{V}_{HL}$ hyperbolic launch velocity
T_f $= t_f - t_0$: total flight time
T_S launch date
M $= m/m(t_0)$: relative mass
ξ solar pressure

u $= (\Psi, \Xi, \beta)^T$

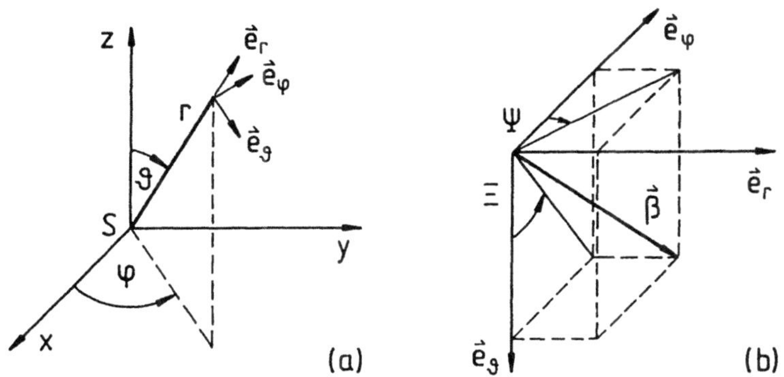

Fig. 1. Definition of the coordinate system (a) and the angles between the thrust direction and the coordinate axes (b); S denotes the Sun.

Numerical calculations are performed in a heliocentric ecliptic coordinate system.

3. The Mass Model

$$m_{system} = m_{upper\,stage} + m_{spacecraft}$$

$$m_{upper\,stage} = m_{fix} + m_{thruster} - m_{fuel}(1 + \delta_{fuel})(1 + \Delta_{residual\,fuel})$$

$$m_{spacecraft} = m_{PAYLOAD} + m_{GUIDANCE} + m_{STRUCTURE} + m_{POWER}(1 + \delta_{POWER})$$
$$+\ m_{THRUST}(1 + \delta_{THRUST}) + m_{FUEL}(1 + \delta_{FUEL})(1 + \Delta_{RESID.\ FUEL})$$

$$m_{POWER} = a_1 \cdot (\alpha_{SOL} \cdot P_{0,SOL} + m_{0,SOL}) + a_2 \cdot (\alpha_{NUC} \cdot P_{0,NUC} + m_{0,NUC})$$

$$P_{SOL}(t) = P_{0,SOL}\, \frac{1}{r^{1.7}}\, e^{-x_S t}$$

$$P_{NUC}(t) = P_{0,NUC} \cdot (1 - x_N \cdot (t/[years])) \cdot e^{-\lambda t}$$

$$m_{THRUST} = m_{0,THRUSTER} + \beta_{THRUSTER} F_{max}(t_0)$$

$$F_{max}(t) = \frac{\epsilon_{THRUSTER}}{1 + \frac{C^2}{v_A^2}} \cdot P_{max}(r,t) \cdot (1 - \beta_{lifetime}(t))$$

The total *system* consists of the *upper stage* and the *spacecraft*. The *upper stage* (a conventional rocket) injects the *spacecraft* into the interplanetary trajectory with the hyperbolic launch velocity $\vec{V}_{HL}$ and is then jettisoned.

For the installation of an additional component with a mass of m_{COMP} onboard the satellite (e.g. the installation of a power production unit with a mass of m_{POWER}), additional support structure with a mass of $m_{COMP,struc}$ is needed. $m_{COMP,struc} = m_{COMP,struc}(m_{COMP})$ depends specifically on the component and nonlinearly on its mass m_{COMP}; by definition $\delta_{COMP} := m_{COMP,struc}/m_{COMP}$. Experimental data for

δ_{COMP} vs. m_{COMP} are collected from literature; a smoothened cubic spline function is fitted to these experimental data and used as $\delta_{COMP}(m_{COMP})$ in the mass model.

Electric power production is either by solar cells (SOL) or by nuclear batteries (RTGs, subscript NUC). The a_i are free parameters, that determine the contribution to the total power production. x_S and x_N are fixed numerical values (engineering constants) describing system degradation; λ denotes the radioactive decay constant.

Analytical functions are fitted to experimental data which describe the mass of power production systems (m_{POWER}) vs. their electric power output ($P_{0,\cdot}$) at begin of mission (e.g. [12]). In the neighborhood of the solution these analytical functions are linearized with the gradient $\alpha_\cdot$ and the offset $m_{0,\cdot}$. If the solution changes (e.g. in the course of a homotopy), the linearization is iteratively corrected.

$m_{0,THRUSTER}$ and $\beta_{THRUSTER}$ are determined in a similar way and characteristic of each type of thruster. The last relationship is valid only for ion-thrusters. v_A is a free parameter (within certain boundaries), $\beta_{lifetime}(t)$ an analytical fit-function and $\epsilon_{THRUSTER}$ and C measured values characterizing the respective thruster type.

$\Delta_{RESID.\ FUEL}$ denotes that part of fuel that remains in the tank and cannot be used (about $1 - 2$ %). The other subscripts should be self-explaining. In addition, various lifetime criteria have been considered; moreover "outer" parameters (V_{HL}, T_S, T_f, $\dots$) interact with boundary conditions and model parameters.

4. The Mathematical Approach

4.1 The Generalized Problem. Let us find a state function $x : [t_0, t_f] \longrightarrow \mathrm{R}^n$ and a control function $u : [t_0, t_f] \longrightarrow \mathrm{U} \subset \mathrm{R}^m$, which minimizes the functional

$$\mathrm{I}(u) := m(t_0)$$

subject to the conditions

$$\dot{x} = f(x, u, t) = f_1(x, u_1, t) + f_2(x, u_1, t) \cdot u_2, \ u = (u_1, u_2)$$

$$x(t_0) = x_0, \quad r(t_0, t_f, x(t_f)) = 0 \in \mathrm{R}^k, \ k < n$$

In our case, $\Psi(t)$ and $\Xi(t)$ form the nonlinear controls, whereas $\beta(t)$ is the linear control.

4.2 The Related Boundary Value Problem. The above defined problem of optimal control theory is transformed in a well-known manner (see e.g. [2],[3]) into a multi-point boundary value problem. There, the following system of coupled nonlinear differential equations results ($H := \lambda^T f$)

$$\begin{aligned} \dot{x} &= f(x, u, t) \\ \dot{\lambda} &= -H_x(x, \lambda, u, t). \end{aligned}$$

The controls can be derived from $H_u(x, \lambda, u, t) = 0$ and have to satisfy the Legendre-Clebsch condition: $H_{uu}(x, \lambda, u, t)$ *pos.(semi)definite*. If the switching function S is defined by $S := \lambda^T f_2$, then in case of a linear control one gets

$$u_2 = \begin{cases} 0 & \text{if } S > 0 \\ u_{2,max} & \text{if } S < 0 \end{cases}$$

The boundary conditions are either prescribed a priori or can be obtained from the first variation of the augmented functional. The solution of this boundary value problem satisfies the necessary conditions for an optimal solution.

4.3 The Differential Equations for the State Variables. The basic set of equations of motion of the spacecraft is

$$\dot{r} = v_r$$
$$\dot{\varphi} = \frac{v_\varphi}{2\pi r \sin \vartheta}$$
$$\dot{\vartheta} = \frac{v_\vartheta}{r}$$
$$\dot{m} = -\frac{\beta}{r^\kappa}$$
$$\dot{v}_r = \frac{ce \cdot \beta}{m \cdot r^\kappa} \cdot \sin \Psi \sin \Xi + \frac{v_\varphi^2 + v_\vartheta^2}{r} - \frac{1}{r^2} + \frac{\xi}{mr^2} + G_{vr}(r, \varphi, \vartheta, t, T_s)$$
$$\dot{v}_\varphi = \frac{ce \cdot \beta}{m \cdot r^\kappa} \cdot \cos \Psi \sin \Xi - \frac{v_\varphi v_\vartheta}{r} - \frac{v_\varphi v_\vartheta}{r} \cdot \cot \vartheta + G_{v\varphi}(r, \varphi, \vartheta, t, T_s)$$
$$\dot{v}_\vartheta = \frac{ce \cdot \beta}{m \cdot r^\kappa} \cdot \cos \Xi - \frac{v_r v_\vartheta}{r} + \frac{v_\varphi^2}{r} \cdot \cot \vartheta + G_{v\vartheta}(r, \varphi, \vartheta, t, T_s)$$

with $\kappa \in \mathbb{R}$ given and fixed and

$$G_{vr}(r, \varphi, \vartheta, t, T_s) := -\sum_{j=1}^{n} \frac{m_j}{s_j^3}[r - r_j \cos \vartheta \cos \vartheta_j - r_j \sin \vartheta \sin \vartheta_j \cos(2\pi(\varphi - \varphi_j))]$$

$$G_{v\varphi}(r, \varphi, \vartheta, t, T_s) := -\sum_{j=1}^{n} \frac{m_j}{s_j^3}[r_j \sin \vartheta_j \sin(2\pi(\varphi - \varphi_j))]$$

$$G_{v\vartheta}(r, \varphi, \vartheta, t, T_s) := -\sum_{j=1}^{n} \frac{m_j}{s_j^3}[r_j \sin \vartheta \cos \vartheta_j - r_j \cos \vartheta \sin \vartheta_j \cos(2\pi(\varphi - \varphi_j))]$$

$$s_j := \sqrt{r^2 + r_j^2 - 2r_j r(\cos \vartheta \cos \vartheta_j + \sin \vartheta \sin \vartheta_j \cos(2\pi(\varphi - \varphi_j)))}$$

The functions $G_{vr}, G_{v\varphi}, G_{v\vartheta}$ contain the contributions of the gravitational forces of other celestial bodies like planets, moons or planetoids to the differential equations for the state variables $v_r, v_\varphi, v_\vartheta$. m_j denotes the relative mass of the j-th celestial body relative to the Sun, $\vec{r}_j(t, T_S) = (r_j, \varphi_j, \vartheta_j)^T(t, T_S)$ its current position and $s_j(t, T_S)$ its distance from the space probe (the real date $T_{real} := t + T_S$). Lengths are scaled to 1 AU, velocities are scaled to the mean velocity of the Earth on its way around the Sun.

4.4 The Determination of the Thrust and the Thrust Angles. The minimum principle yields the following conditions for the control variables

$$\beta = \begin{cases} 0 & \text{if } S > 0 \\ \beta_{max} & \text{if } S < 0 \end{cases}, \quad S := -\frac{ce}{m}\sqrt{\lambda_{vr}^2 + \lambda_{v\varphi}^2 + \lambda_{v\vartheta}^2} - \lambda_m$$

$$\sin(\Psi) = -\frac{\lambda_{vr}}{\sqrt{\lambda_{vr}^2 + \lambda_{v\varphi}^2}}, \quad \sin(\Xi) = \frac{\sqrt{\lambda_{vr}^2 + \lambda_{v\varphi}^2}}{\sqrt{\lambda_{vr}^2 + \lambda_{v\varphi}^2 + \lambda_{v\vartheta}^2}}$$

Despite of extensive numerical tests, S was never found to satisfy $S = 0$ (so-called *singular control*) on any interval.

4.5 The Numerical Techniques. The number of free parameters of the mass model is minimized, partly by the definition of auxiliary parameters. For the remaining free parameters *trivial* differential equations ($\dot{y} = 0$) are added to the basic set of equations of motion. In addition, model functions are directly substituted into the equations of motion. Boundary conditions are set up similar to those in [4].

The numerical treatment of boundary value problems for systems of nonlinear differential equations is by the well-known multiple shooting method [5]–[7]. For the calculations a modified form of the variant BOUNDSCO [8] is utilized. A high precision method is needed for the integration of complicated systems of ordinary differential equations; therefore extrapolation methods have been chosen to solve the arising initial value problems [9]. For these complex problems arising from astronautics additional numerical techniques [10] have been developed and applied.

Often only discrete values of model parameters are allowed (for instance, only a few types of thrusters – each with fixed maximum thrust – exist). In this case, the discrete values next to the optimal value of the respective parameter are tested separately; the best one is selected as the design parameter.

5. The Results

In 2001 there exists an uninterrupted launch window. Maximum permitted flight time is 18 years, maximum V_{HL} is 3.5 km/s: On all optimal trajectories these maximum values are taken on. The upper stage is a conventional rocket with bipropellant hydrazine thrusters. Initial mass of the ion driven spacecraft varies between 2250 kg and 1680 kg according to the launch date; thus m_{System} is well below the limits of Ariane V.

The optimal launch date in 2001 is December 9th. In this case the interplanetary spacecraft has an initial mass of 1680 kg including 760 kg of fuel (the noble gas Xe) and

50 kg of pure scientific payload. Calculated fuel consumption for the flight to Neptune is 728.6 kg.

The optimal spacecraft is equipped with one ion thruster RIT-30 [11] with 3 additional ionisator chambers (due to lifetime considerations). Maximum thrust is 135 mN (bom), $v_{A,opt.}$ is 48 km/s. In agreement with published data, thruster lifetime is assumed to be $\leq$ 25000 h and power consumption is $(4417 - 0.0042 \cdot t_{op}/[\text{h}])$ W. 4.5 kW of electric power are necessary; power production is purely nuclear by a couple of Pu-RTGs ("isotope batteries") of Galileo-type, despite their high specific mass of 155 kg/kW with light-weight shielding [12]. The optimal thrust level is almost by a factor 10 lower than that given by classical engineering rules of thumb.

For a one-stage vehicle the power production by solar cells is not competitive, even in combination with RTGs. Nuclear reactors fail lifetime demands. Chemical thrusters are not useful – if no driven swing-bys are considered – due to their low specific impulse.

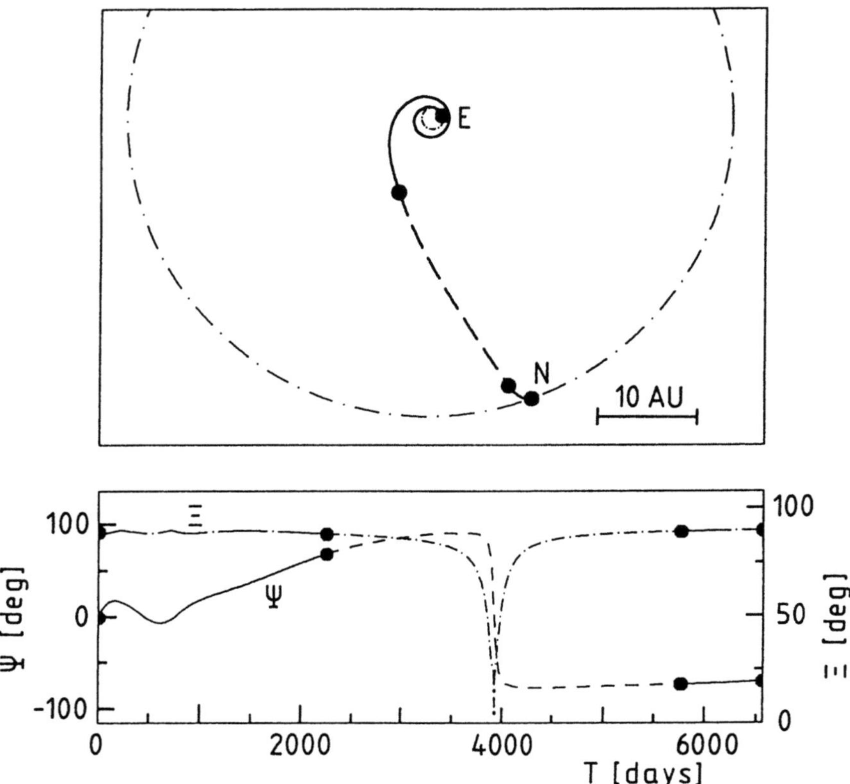

Fig. 2. Projection of the three-dimensional flight trajectory from Earth to Neptune ($M_f = 0.566$) and thrust angles vs. flight time for a launch on Dec. 9th, 2001. Solid lines – thrust on, broken lines – thrust off; N denotes Neptune, E the Earth.

Acknowledgements. The author would like to thank Prof. R. Bulirsch who always encouraged and supported this work. This research was supported by the DFG.

References

1. Brown, R.H., *Triton, Voyager's Finale*, Planetary Rep. $\underline{2}$ (1992) 17.
2. Hestenes, M.R., *Calculus of variations and optimal control theory*, John Wiley, New York, 1966.
3. Bryson, A.E., Ho, Y.-C., *Applied Optimal Control*, Revised Printing, Hemisphere Publishing Corp., Washington D.C., 1975.
4. Bulirsch, R., Callies, R., *Optimal Trajectories for a Multiple Rendezvous Mission with Asteroids*, IAF Paper IAF-91-342, 1991.
5. Osborne, M.R., *On Shooting Methods for Boundary-Value Problems*, J. Math. Anal. Appl. $\underline{27}$ (1969).
6. Bulirsch, R., *Die Mehrzielmethode zur numerischen Lösung von nichtlinearen Randwertproblemen und Aufgaben der optimalen Steuerung*, Report of the Carl-Cranz-Gesellschaft e.V., Oberpfaffenhofen, 1971.
7. Deuflhard, P., *Ein Newton-Verfahren bei fast singulärer Funktionalmatrix zur Lösung von nichtlinearen Randwertaufgaben mit der Mehrzielmethode*, Thesis, Cologne, 1972.
8. Oberle, H.J., *Numerical Computation of Minimum-Fuel Space-Travel Problems by Multiple Shooting*, Report TUM-MATH-7635, Dept. of Mathematics, Munich Univ. of Technology, Germany, 1976.
9. Bulirsch, R., Stoer, J., *Numerical Treatment of Ordinary Differential Equations by Extrapolation Methods*, Num. Math. $\underline{8}$ (1966) 1-13.
10. Callies, R., *Optimale Flugbahnen einer Raumsonde mit Ionentriebwerken*, Thesis, Munich, 1990.
11. Loeb, H.W., *New Interplanetary Mission Plans with the 200 mN Ion Thruster RIT-35*, IEPC Paper 84-45, 1984.
12. Bennet, G.L., et al., *On the Development of Power Sources for the Ulysses and Galileo Missions*, ESA SP-294, p. 117-121, 1989.

Author's address

Rainer Callies
Department of Mathematics
Munich University of Technology
P.O. Box 20 24 20
8000 München 2
Germany

Systems and Control:
Foundations and Applications

A series of monographs and advanced graduate texts

Edited by
Christopher Byrnes, Washington University, St. Louis, MO, USA

Systems and Control is designed for the publication of research level monographs and advanced graduate textbooks in all areas of systems and control theory and its applications to a wide variety of scientific disciplines.

J.-P. Aubin: Viability Theory (ISBN 3-7643-3571-8)

J.-P. Aubin /H. Frankowska: Set–Valued Analysis / SC 2 (ISBN 3-7643-3478-9)

H.T. Banks / K. Kunisch: Estimation Techniques for Distributed Parameter Systems / SC 1 (ISBN 3-7643-3433-9)

T. Basar / P. Bernhard: H∞–Optimal Control and Related Minimax Design Problems (ISBN 3-7643-3554-8)

A. Bensoussan / G. Da Prato / M.C. Delfour / S.K. Mitter:
Representation and Control of Infinite Dimensional Systems
Volume I (ISBN 3-7643-3641-2)
Volume II (ISBN 3-7643-3642-0)

H.F. Chen / L. Guo: Identification and Stochastic Adaptive Control (ISBN 3-7643-3597-1)

P. Falb: Methods of Algebraic Geometry in Control Theory I. Scalar Linear Systems
and Affine Algebraic Geometry / SC 4 (ISBN 3-7643-3454-1)

B. van Keulen: H∞-Infinity Control for Distributed Parameter Systems:
A State-Space-Approach (ISBN 3-7643-3709-5)

H.-J. Kushner: Weak Convergence Methods and Singularly Perturbed Stochastic Control
and Filtering Problems / SC 3 (ISBN 3-7643-3437-1)

J. Zabczyk: Mathematical Control Theory (ISBN 3-7643-3645-5)

If you have any concerns about our products,
you can contact us on
ProductSafety@springernature.com

In case Publisher is established outside the EU,
the EU authorized representative is:
**Springer Nature Customer Service Center GmbH
Europaplatz 3, 69115 Heidelberg, Germany**

Printed by Libri Plureos GmbH
in Hamburg, Germany